NEUROMETHODS

Series Editor
Wolfgang Walz
University of Saskatchewan
Saskatoon, SK, Canada

For further volumes:
http://www.springer.com/series/7657

Experimental Neurosurgery in Animal Models

Edited by

Miroslaw Janowski

The Russell H. Morgan Department of Radiology and Radiological Science, Division of MR Research Institute for Cell Engineering, Johns Hopkins University School of Medicine, Baltimore, MD, USA; NeuroRepair Department, Mossakowski Medical Research Centre, Warsaw, Poland; Department of Neurosurgery, Mossakowski Medical Research Centre, Warsaw, Poland

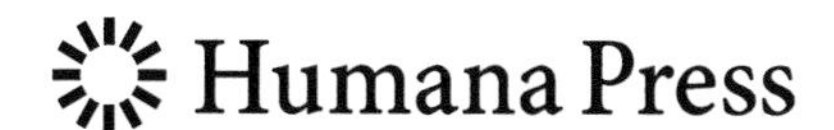

Editor
Miroslaw Janowski
The Russell H. Morgan Department of Radiology
and Radiological Science
Division of MR Research Institute for Cell Engineering
Johns Hopkins University School of Medicine
Baltimore, MD, USA

NeuroRepair Department
Mossakowski Medical Research Centre
Warsaw, Poland

Department of Neurosurgery
Mossakowski Medical Research Centre
Warsaw, Poland

ISSN 0893-2336 ISSN 1940-6045 (electronic)
Neuromethods
ISBN 978-1-4939-8122-9 ISBN 978-1-4939-3730-1 (eBook)
DOI 10.1007/978-1-4939-3730-1

Printed on acid-free paper

This Humana Press imprint is published by Springer Nature
The registered company is Springer Science+Business Media LLC New York

Preface to the Series

Experimental life sciences have two basic foundations: concepts and tools. The *Neuromethods* series focuses on the tools and techniques unique to the investigation of the nervous system and excitable cells. It will not, however, shortchange the concept side of things as care has been taken to integrate these tools within the context of the concepts and questions under investigation. In this way, the series is unique in that it not only collects protocols but also includes theoretical background information and critiques which led to the methods and their development. Thus it gives the reader a better understanding of the origin of the techniques and their potential future development. The *Neuromethods* publishing program strikes a balance between recent and exciting developments like those concerning new animal models of disease, imaging, in vivo methods, and more established techniques, including, for example, immunocytochemistry and electrophysiological technologies. New trainees in neurosciences still need a sound footing in these older methods in order to apply a critical approach to their results.

Under the guidance of its founders, Alan Boulton and Glen Baker, the *Neuromethods* series has been a success since its first volume published through Humana Press in 1985. The series continues to flourish through many changes over the years. It is now published under the umbrella of Springer Protocols. While methods involving brain research have changed a lot since the series started, the publishing environment and technology have changed even more radically. Neuromethods has the distinct layout and style of the Springer Protocols program, designed specifically for readability and ease of reference in a laboratory setting.

The careful application of methods is potentially the most important step in the process of scientific inquiry. In the past, new methodologies led the way in developing new disciplines in the biological and medical sciences. For example, Physiology emerged out of Anatomy in the nineteenth century by harnessing new methods based on the newly discovered phenomenon of electricity. Nowadays, the relationships between disciplines and methods are more complex. Methods are now widely shared between disciplines and research areas. New developments in electronic publishing make it possible for scientists that encounter new methods to quickly find sources of information electronically. The design of individual volumes and chapters in this series takes this new access technology into account. Springer Protocols makes it possible to download single protocols separately. In addition, Springer makes its print-on-demand technology available globally. A print copy can therefore be acquired quickly and for a competitive price anywhere in the world.

Saskatoon, Canada *Wolfgang Walz*

Preface

In the field of neuroscience, animal surgeries are often performed both to develop animal models and to test the application of various therapies. Surgery has always been considered an art and individual variability is high. However, proper surgical preparation is directly related to the achieved results. While there has been progress in the automation of tissue sampling and image analysis, surgery is still a manual procedure. Thus, animal surgery is somewhat of a bottleneck because the great variation in individual skill cannot ensure that the experimental results will always be of the highest quality. Despite the importance of the surgical procedure, the technical description in journal articles is usually very brief, which may also cause difficulties when someone attempts to reproduce the experiment. Moreover, the technical obstacles that a researcher might encounter, as well as tips and tricks about how to overcome them, are usually not mentioned in the literature. In addition, surgical techniques are not sold as kits with instructions included. Thus, a reference book, in which procedures in the form of a corpus of instructions, is highly desired. Considering that surgical technique is meticulous and deserves a full explanation of the technical details to perform procedures properly, the book *Experimental Neurosurgery in Animal Models* has been prepared to address these challenges.

For many years, small-animal models were favored mostly due to low cost, ease of care, and the possibilities for high throughput. While they are still valuable for answering some basic research questions, the translation of therapeutic approaches from bench to bed is usually unsuccessful. Thus, there is a growing awareness that therapies should be tested in large-animal models prior to clinical application. Although, currently, very few laboratories perform neurosurgical procedures on large animals, there is a growing interest in using these animals. Therefore, it would be of great value to have access to the operative expertise of leaders in the field of large-animal surgery. This book answers that need and also highlights the experienced laboratories that could serve as a reference for newcomers. Thus, the part of the book devoted to large animals is especially compelling and sets the standard for state-of-the-art translational research. While the initial chapters of the book present the standard small-animal models now used in neuroscience, these are later followed by a description of procedures in large-animal models.

While manual precision is equally important in all models, the complexity of surgical dissection varies. The first six chapters focus primarily on the brain, while the next six chapters concern the spinal cord in rodents. The last four chapters provide a description of operative procedures in large animals. The book begins with three chapters that describe rapid procedures that do not always require the use of a scalpel or in which the use of a scalpel is very limited, but all these chapters are related to the major neurosurgical disciplines, such as neurotrauma, radiosurgery, and stereotaxy. The next two chapters describe the very complicated craniotomies in small animals. The sixth chapter deals with the advances in the use of robotics, which is expected to have growing role in animal models. The next two chapters are devoted to the presentation of models of spine injury, and the following chapter describes microsurgical access to the spinal cord. Chapters 10 and 11 present various methods of injection to the spine and CSF through the cisterna magna. Chapter 12 focuses on cranial and peripheral nerve dissection using a very advanced water-jet dissection method.

The large-animal section begins by detailing the performance of a craniotomy in swine. This procedure is universal and can be used for wide brain access for various purposes, as well as for surgical training. Thus, this would be of interest not only to researchers but also to neurosurgical residents and neurosurgeons. The next chapter presents stereotaxy, which is far more complex in a large-animal setting, but, due to advances in sophisticated technology, is a very powerful method for the translation of animal research to the clinical scenario, particularly as monkeys are often used as the experimental species. The sheep model of stroke was long-awaited after many unsuccessful attempts to translate the positive small-animal data to the clinical setting. There is a good possibility that this model will have wide preclinical utility, particularly in the current climate that mandates that clinical tests be preceded by relevant animal studies. The last chapter is devoted to the extremely important neurosurgical disease, subarachnoid hemorrhage. The chapter introduces a reader to the complexity of pathological sequelae that can be directly related to the neurosurgical technique and provides both the successes and failures related to the use of various techniques, which allows researchers to build on the enormous experience of this group in studying this disease in a primate model.

This book is expected to gather the interest of various readerships. It will be very useful for basic researchers, who need to establish animal models in the field of neuroscience, especially in neurosurgery. The vast group of neurosurgical residents can treat it as a repository of research and training opportunities for the use of animal models. The book may also facilitate the selection of an appropriate animal model as well as serve as a basis for further technical improvements and refinements of these models for academic neurosurgeons running their own labs simultaneously with clinical practice. Thus, it is expected that the book will be warmly received and will serve frequently as a handbook during the planning and performance of surgical procedures on the central nervous system.

Baltimore, MD, USA ***Miroslaw Janowski***

Contents

Contributors

JOHN BACHER • *Division of Veterinary Resources, Office of Research Services, National Institutes of Health, Bethesda, MD, USA*
KRYSTOF S. BANKIEWICZ • *Department of Neurosurgery, University of California San Francisco, San Francisco, CA, USA*
HENRYK BARTHEL • *Department of Nuclear Medicine, University of Leipzig, Leipzig, Germany*
JOHANNES BOLTZE • *Department of Medical Cell Technology, Fraunhofer Research Institution for Marine Biotechnology, Lubeck, Germany; Neuroscience Center, Massachusetts General Hospital and Harvard Medical School, Charlestown, MA, USA*
DOROTHÉE CANTINIEAUX • *GIGA-Neuroscience, University of Liège, Liège, Belgium*
FREDRIK CLAUSEN • *Section of Neurosurgery, Department of Neuroscience, Uppsala University, Uppsala, Sweden*
MÁTÉ DÖBRÖSSY • *Department of Stereotactic and Functional Neurosurgery, University of Freiburg Medical Center – Neurocentre, Freiburg, Germany*
ANTJE Y. DREYER • *Department of Cell Therapy, Fraunhofer Institute for Cell Therapy and Immunology, Leipzig, Germany*
CHARLA C. ENGELS • *Division of MR Research, Russell H. Morgan Department of Radiology and Radiological Science, Cellular Imaging Section, Institute for Cell Engineering, The Johns Hopkins University School of Medicine, Baltimore, MD, USA; Department of Radiology, Faculty of Medical Sciences, University of Warmia and Mazury, Olsztyn, Poland*
MICHAEL G. FEHLINGS • *Spine Program, University Health Network, Toronto Western Hospital, University of Toronto, Toronto, ON, Canada*
WENDY FELLOWS-MAYLE • *The Department of Neurological Surgery, The University of Pittsburgh, Pittsburgh, PA, USA; The Center for Image-Guided Neurosurgery, University of Pittsburgh Medical Center, Pittsburgh, PA, USA*
MASSIMO S. FIANDACA • *Department of Neurosurgery, University of California San Francisco, San Francisco, CA, USA*
JOHN FORSAYETH • *Department of Neurosurgery, University of California San Francisco, San Francisco, CA, USA*
RACHELLE FRANZEN • *GIGA-Neuroscience, University of Liège, Liège, Belgium*
MORTIMER GIERTHMUEHLEN • *Department of Neurosurgery, University of Freiburg, Freiburg, Germany*
VICTORIA HOFFMANN • *Division of Veterinary Resources, Office of Research Services, National Institutes of Health, Bethesda, MD, USA*
HIDEAKI IMAI • *Department of Neurosurgery, Faculty of Medicine, The University of Tokyo, Tokyo, Japan*

MIROSLAW JANOWSKI • *The Russell H. Morgan Department of Radiology and Radiological Science, Division of MR Research Institute for Cell Engineering, Johns Hopkins University School of Medicine, Baltimore, MD, USA; NeuroRepair Department, Mossakowski Medical Research Centre, Warsaw, Poland; Department of Neurosurgery, Mossakowski Medical Research Centre, Warsaw, Poland*

ULF D. KAHLERT • *Department of Neurosurgery, Heinrich-Heine University Duesseldorf, Duesseldorf, Germany*

JAN KAMINSKY • *Department of Neurosurgery, University of Freiburg, Freiburg, Germany*

DOERTHER KEINER • *Neurochirurgische Klinik, Universitaetsklinikum des Saarlandes Homburg/Saar, Homburg/Saar, Germany*

DOUGLAS KONDZIOLKA • *The Department of Neurological Surgery, The University of Pittsburgh, Pittsburgh, PA, USA; The Center for Image-Guided Neurosurgery, University of Pittsburgh Medical Center, Pittsburgh, PA, USA*

DONALD LOBSIEN • *Department of Neuroradiology, University of Leipzig, Leipzig, Germany*

L. DADE LUNSFORD • *Department of Neurological Surgery, The University of Pittsburgh, Pittsburgh, PA, USA; The Center for Image-Guided Neurosurgery, University of Pittsburgh Medical Center, Pittsburgh, PA, USA*

JAROSLAW MACIACZYK • *Department of Neurosurgery, Heinrich-Heine University Duesseldorf, Duesseldorf, Germany*

I. MHAIRI MACRAE • *Institute of Neuroscience and Psychology, College of Medical Veterinary and Life Sciences, University of Glasgow, Glasgow, Scotland, UK*

JASON W. MOTKOSKI • *Division of Neurosurgery, Seaman Family MR Research Centre, Foothills Medical Centre, Calgary, AB, Canada*

GUIDO NIKKHAH • *Department of Stereotactic and Functional Neurosurgery, University of Freiburg Medical Center – Neurocentre, Freiburg, Germany*

AJAY NIRANJAN • *The Department of Neurological Surgery, The University of Pittsburgh, Pittsburgh, PA, USA; The Center for Image-Guided Neurosurgery, University of Pittsburgh Medical Center, Pittsburgh, PA, USA*

BJÖRN NITZSCHE • *Department of Cell Therapy, Fraunhofer Institute for Cell Therapy and Immunology, Leipzig, Germany; Department of Nuclear Medicine, University of Leipzig, Leipzig, Germany*

JOACHIM OERTEL • *Neurochirurgische Klinik, Universitaetsklinikum des Saarlandes Homburg/Saar, Homburg/Saar, Germany*

DENNIS T.T. PLACHTA • *Laboratory for Biomedical Microtechnology, Department of Microsystems Engineering, University of Freiburg – IMTEK, Freiburg, Germany*

RYSZARD M. PLUTA • *Surgical Neurology Branch, National Institute of Neurological Disorders and Stroke, National Institutes of Health, Bethesda, MD, USA; Fishbein Fellow, JAMA, Chicago, IL, USA*

JAN REGELSBERGER • *Department of Neurosurgery, University Medical Center Hamburg Eppendorf, Hamburg, Germany*

NOBUHITO SAITO • *Department of Neurosurgery, Faculty of Medicine, The University of Tokyo, Tokyo, Japan*

JEAN SCHOENEN • *GIGA-Neuroscience, University of Liège, Liège, Belgium*

BORIS SKOPETS • *Division of Veterinary Resources, Office of Research Services, National Institutes of Health, Bethesda, MD, USA*

GARNETTE R. SUTHERLAND • *Division of Neurosurgery, Seaman Family MR Research Centre, Foothills Medical Centre, Calgary, AB, Canada; Department of Clinical Neurosciences, University of Calgary, AB, Canada*
CHRISTOPH A. TSCHAN • *Klinik für Neurochirurgie, Ludmillenstift Meppenr, Germany*
PIOTR WALCZAK • *Division of MR Research, Russell H. Morgan Department of Radiology and Radiological Science, Cellular Imaging Section, Institute for Cell Engineering, The Johns Hopkins University School of Medicine, Baltimore, MD, USA*
JARED T. WILCOX • *Spine Program, University Health Network, Toronto Western Hospital, University of Toronto, Toronto, ON, Canada*
DALI YIN • *Department of Neurosurgery, University of California San Francisco, San Francisco, CA, USA*
VILIA ZEISIG • *Department of Nuclear Medicine, University of Leipzig, Leipzig, Germany*

Chapter 1

Animal Models of Traumatic Brain Injury

Fredrik Clausen

Abstract

Animal models of traumatic brain injury (TBI) have been the core of the research on the molecular, cellular, and functional effects of the disease. To be able to simulate the heterogeneous aspects of TBI several models have been designed. This chapter aims to describe the three most commonly used experimental models of TBI in rodents.

Key words Traumatic brain injury, Mice, Rats, Controlled cortical impact, Fluid percussion injury, Weight drop injury, Stereotaxy

1 Introduction

1.1 The Need for Animal Models in Basic Research on Traumatic Brain Injury

Basic traumatic brain injury (TBI) research is reliant on animal models to study the multitude of effects the event has on the brain. The brain is simply too complex to simulate in vitro or in silico. Add to this, the interaction with the blood stream and immune system and the complexity grows another magnitude.

The first animal models of TBI were setup in larger species such as dogs and cats, though presently most experiments are done in rodents. Porcine models of TBI have become more common and they are more clinically relevant as the pig brain is closer to the human in regard to size and anatomy, but the differences in cost and effort between using pigs compared to rodents are substantial.

Most studies are done in young male adult rats and mice, which actually is quite relevant as young males are over represented in TBI, as they are more prone to experience vehicle accidents, sports injuries, and violence. Interestingly, also the elderly suffer a higher risk of TBI due to falls, though very few studies in aged rodents have been performed. One large difference between human TBI and most animal models is that while human TBI is heterogenous, experimental TBI is most often homogenous. In human patients there are often secondary complications such as multitrauma, intoxication, and preexisting disease.

Miroslaw Janowski (ed.), *Experimental Neurosurgery in Animal Models*, Neuromethods, vol. 116,
DOI 10.1007/978-1-4939-3730-1_1, © Springer Science+Business Media New York 2016

Animal welfare is a big issue in several countries and it is of course recommended that the experimenters abide to the current laws and regulation regarding animal experiments. It should be noted that even though experimental TBI may sound very brutal, even in rodent models at a severe setting most animals recuperate quickly. The loss of motor function and cognitive acuity is hard to spot in a standard animal cage and requires special tests to assess properly.

1.2 Most Commonly Used Models of TBI

1.2.1 Controlled Cortical Impact

Controlled cortical impact (CCI) was developed by Dixon et al. [1] to use as a rat model of TBI and was later adapted to be used in mice [2]. It is one of the most widely used TBI models. The model relies on a pneumatically driven piston to compress the exposed brain, and the severity of the injury is determined by the depth of compression and speed of the piston (Fig. 1). This results in a predominantly focal contusion at the site of impact, but have effects throughout the brain and gives rise to contralateral changes in the hippocampus and thalamus.

There are several manufacturers of CCI devices, but the original model made at Virginia Commonwealth University (Richmond, Virginia) is still in production by Amscien Inc [3].

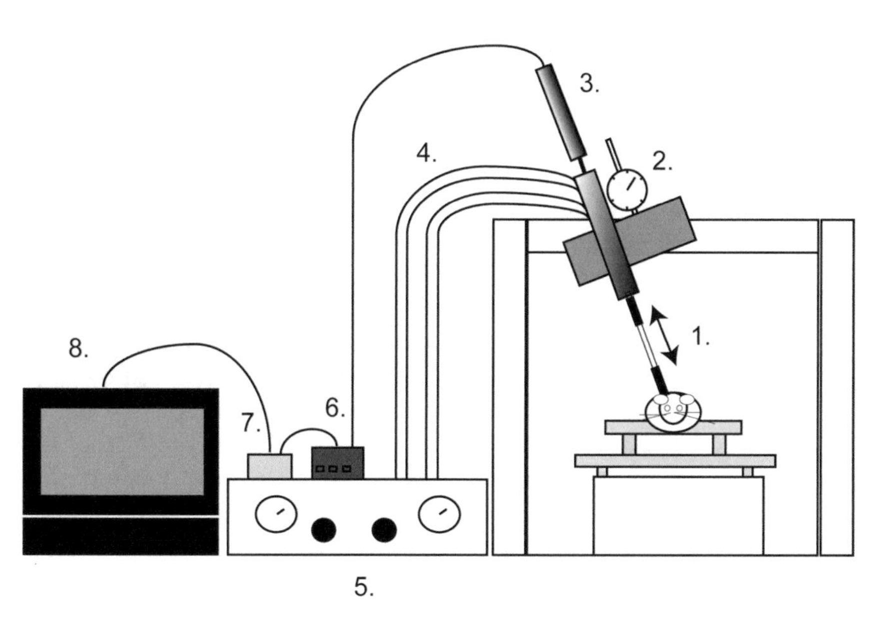

Fig. 1 Schematic drawing of the controlled cortical injury device. Piston (1), micrometer gauge (2), oscilloscope (3), pressure hoses (4), control unit (5), signal conditioner (6), automatic average rod speed measurement unit (7), and computer (8). The stereotactic frame is placed in the CCI device. The piston (1) is extended to its maximum and placed perpendicularly against the exposed dura mater. The piston is then retracted and lowered to the desired compression depth, measured by the micrometer gauge (2). The piston is firmly locked in position and the trauma induced. The acceleration of the piston is registered by the oscilloscope (3), amplified by the signal conditioner (6), measured by the rod speed measurement unit (7), and the recorded by the computer (8)

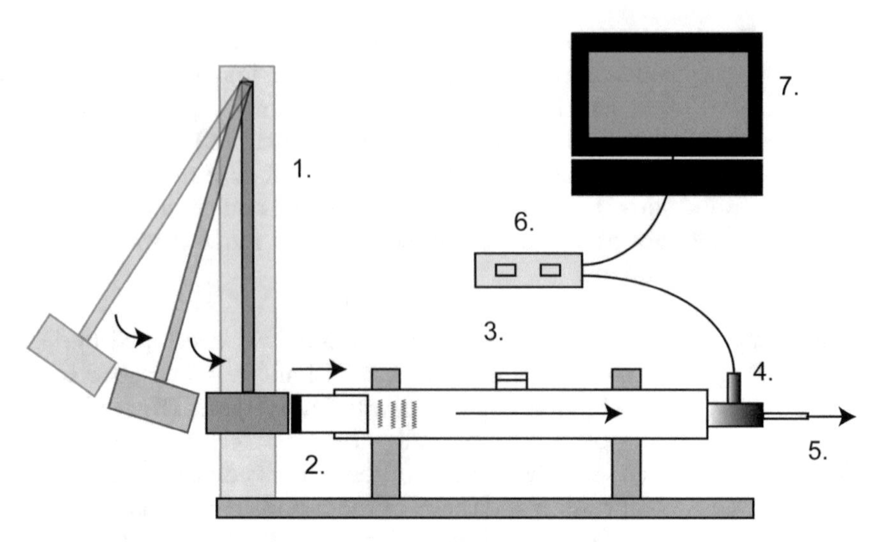

Fig. 2 Schematic drawing of the fluid percussion injury device. Pendulum with hammer head (1), piston (2), fluid-filled cylinder (3), pressure transducer (4), nozzle (5), pressure monitor (6), and computer (7). The injury is induced by attaching the animal to the nozzle (5) and releasing the pendulum (1) from the desired height. The hammer on the pendulum will hit the piston (2) that causes a rise in pressure in the cylinder (3) which sends a pressure pulse into the skull of the animal. The pressure is registered by the pressure transducer (4), measured by the pressure monitor (6), and recorded on the computer (7)

1.2.2 Fluid Percussion Injury

Fluid percussion injury (FPI) was originally developed in cats [4] and was later adapted to be used in rat [5, 6] and mice [7]. The model is based on the rapid injection of a fluid pulse on the exposed dura mater of the brain with a subsequent displacement of the brain. The fluid pulse is generated as a hammer head mounted on a pendulum pushes a piston into a fluid-filled cylinder and is directed into a nozzle coupled either directly or via a hose to the skull cavity (Fig. 2).

The injury severity depends on the pressure of the fluid pulse, which can be changed by adjusting the height of the fall of the hammer head. The pressure is recorded close to the end of the nozzle spout and has been found to correlate well to the pressure inside the skull in rats [8]. The position of the craniotomy and coupling to the FPI device majorly affects the outcome [9]. To that end two variants of FPI are referred to. In lateral FPI (LFPI), the injury is centered over the parietal cortex, resulting in a more focal injury than central FPI (CFPI) that is centered over the midline.

FPI results in more global brain effects than CCI and if set at moderate or severe injury it elicits apnea and can shock the brain stem fatally. In the most severe injury settings of LFPI, it is expected that around 25 % of the animals do not survive the trauma. CFPI has an even greater global effect and is in general used with lower pressure than LFPI as it easily shocks the brain stem causing extended apnea or death. CFPI is used when a more diffuse axonal injury (DAI) is wanted without a defined contusion.

There are currently two commercially available FPI devices that differ in the design. The first model is produced by Amscien [3] and has a continuous water pillar through the device into the brain allowing the pressure wave to travel in a straight line into the skull cavity. Dragonfly R&D Inc. [10] manufactures a model with a flexible hose attached to the nozzle, making it possible to deliver the pressure pulse into the skull cavity with the rodent in the stereotaxic frame. Frey et al. have developed a novel design where the pendulum and piston are replaced by a picospritzer [11], though this model is not yet commercially available.

1.2.3 Weight Drop Injury

Weight drop injury (WDI), also called closed head injury (CHI), differs from CCI and FPI as it is an acceleration/deceleration model of TBI. Out of the three TBI models described in this chapter it is the simplest equipment wise. The basic concept with a weight falling through a guide tube, hitting the head of the rodent, and accelerating it into a foam bead underneath the animal (Fig. 3) was first described by Shapira et al. [12]. The model can be used with or without protecting the skull bone. Foda and Marmarou developed the model by attaching a steel disk to the skull of the rodent to reduce fracturing of the skull [13, 14].

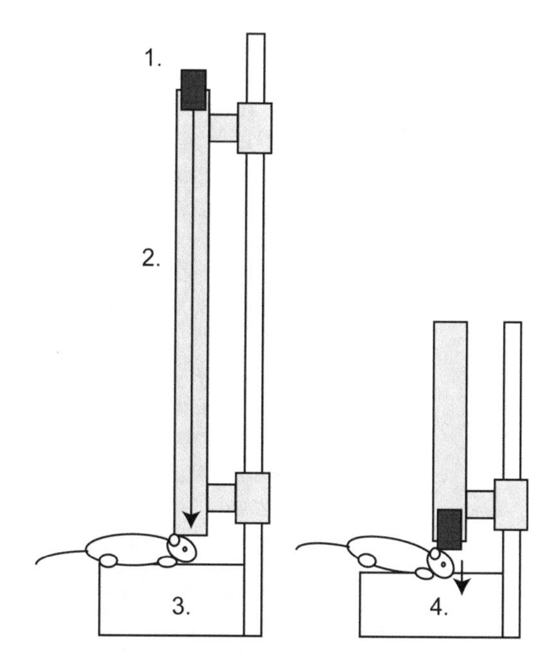

Fig. 3 Schematic drawing of the weight drop injury device. Weight (1), guide tube (2), foam bed (3), and impact (4). The weight (1) is released into the guide tube (2) and hits the head of the animal. The impact accelerates the head of the animal into the foam bed (3)

This is the most heterogeneous model of the three discussed in this chapter, especially if there are skull fractures involved. However, this is not necessarily a bad thing, as human TBI is very heterogeneous. The severity of the injury is determined by the size of the weight, the distance it falls, and the flexibility of the material underneath the animal. WDI causes a predominantly diffuse injury, but many studies also show visible lesions and cavities.

1.3 General Considerations

1.3.1 Stereotaxy

To be able to perform the craniotomy and other procedures on the rodent skull a stereotaxic frame is needed to secure the head of the animal properly. There are several manufacturers of stereotaxic frames for rodents, but it is advisable to consider the needs for the laboratory when it comes to species as some frames are more easily converted from use for rats to mice. A need for micromanipulators attached to the frame can also influence the choice of manufacturer.

1.3.2 Craniotomy

CCI and FPI both require that a craniotomy is made to deliver the force that causes the trauma. In CCI, the craniotomy is typically made slightly larger than the tip of the piston used. For FPI, the craniotomy is preferably performed with a trephine of a size that offers the best possible fit for the coupling to the device.

The placement of the craniotomy, and subsequently the place of the injury, is of great importance. Studies have shown that moving the craniotomy for FPI results in different outcome in regards to lesion size and functional deficits [9]. It is also one of the largest factors when it comes to inter-operator and inter-laboratory differences.

After the injury is made, the bone piece removed during the craniotomy is replaced. To more accurately model a CHI, the bone piece can be fastened using tissue glue and/or bone cement. This will result in a larger injury as the contused brain won't swell out of the cavity and a higher intracranial pressure (i.c.p.) will be higher, causing a reduction in blood flow to the injured area [15].

1.3.3 Anesthesia and Temperature

There are many different ways to keep the rodent sedated and anesthetized during surgery, and the rules and regulation varies between different countries as to what constitutes satisfactory anesthesia in small animals. All forms of anesthesia have strengths and weaknesses.

Gas anesthesia with isoflurane with nitrous oxide and oxygen has been shown to be neuroprotective in itself [16], which can mask smaller treatment effects. Though isoflurane so far is deemed harmless to humans, proper ventilation of the operating table should be uses to protect the experimenter. Halothane should not be used as it is carcinogenous to humans.

Pentobarbital and chloral hydrate offer adequate sedation, but does not offer pain relief making them illegal to use in some countries due to animal welfare concerns. Interestingly, pentobarbital decreases the metabolism of the brain, whereas chloral hydrate increases it [17].

Hypnorm/dormicum offers both sedation and pain relief but needs frequent injections to keep the animal sedated.

Local anesthesia can be injected into the scalp before it is opened to reduce the discomfort and stress in the animal. It is recommended to use local anesthesia in any other surgical wound made during the operation to diminish the postoperative pain as much as possible.

In most experimental setups, normothermia is desired and since several forms of anesthesia decreases metabolism and body temperature a way to keep the animal heated is necessary. This can be achieved by a heating pad and/or a heating lamp. For longer experiments a combination of the two is recommended. As cerebral hypothermia lowers the metabolism of the brain and is considered neuroprotective, it is useful to monitor the brain temperature as well as the core body temperature and keep it above 36.5 °C. There are thin probes available that can be placed between the brain and the skull bone or alternatively between the temporal muscle and the skull bone if there is no room for the probe in the craniotomy.

As the animal is unconscious from the anesthesia and trauma, after the surgery is finished it is essential to setup a heated recovery cage where the animal can regain consciousness in a warm and calm environment to reduce stress. It is necessary to heat the cage with a heating lamp to avoid postoperative hypothermia that could influence the outcome.

1.3.4 Physiological Monitoring

In shorter experiments (less than 30 min anesthesia), physiological monitoring could be restricted to core temperature or blood pressure if there is noninvasive measuring available.

In longer experiments (i.e., microdialysis, imaging, and complicated treatments), it is advisable to monitor blood pressure, blood gases, and brain temperature to make sure that the animal is within normal physiological parameters (or nonphysiological if that is part of the design of the experiment).

1.3.5 Selecting a TBI Model

Naturally the primary concern when choosing TBI model is which outcome measures that is being studied. If DAI is to be studied, the choice is between WDI and MFPI. WDI is easier and less costly to setup and the experimental procedure is quicker than MFPI. However, the inherent heterogeneity of the model makes it necessary to do larger groups of animals. If a more focal injury is of interest then the choice is between CCI and LFPI. Once again, FPI is the more labor intensive and costly method, but if the laboratory also is interested in DAI the same FPI device can be used for both applications.

2 Materials

2.1 Controlled Cortical Impact

2.1.1 Materials Needed for CCI

Surgical tools: Dumont forceps, delicate scissors, flat small forceps (8–9 cm), hemostatic forceps (10–12 cm), scalpel, dental drill or trephine, stereotaxic frame (some models of CCI come with a basic stereotaxic setup, but it is recommended to use a free-standing stereotaxic frame, especially if you do other surgical procedures on the rat, i.e., micro dialysis or stereotaxic injections), sutures, and needle. Tissue adhesive to reattach the bone piece if so desired.

2.2 Fluid Percussion Injury

2.2.1 Materials Needed for FPI

Surgical tools: Dumont forceps, flat small forceps (8–9 cm), hemostatic forceps (10–12 cm), scalpel, dental drill or trephine (recommended), stereotaxic frame, sutures, and needle.

Trauma coupling: Luer lok injection needle adapted to fit the craniotomy (note that the needle part is entirely removed), tissue adhesive, bone/dental cement, 2 mm screw and appropriate screwdriver, and 2 mm drill.

2.3 Weight Drop Injury

The equipment is relatively easy to make in house as all that is needed is an appropriate weight, a guide tube that fits the weight and a flexible material to rest the animal on.

2.3.1 Materials

Surgery (if a protection is attached to the skull bone): scalpel, forceps, hemostatic forceps, protective disk, tissue adhesive, and stereotaxic frame.

3 Method

3.1 Controlled Cortical Impact

1. Sedate the animal and attach it to the stereotaxic frame.
2. Trim the fur on the head if wanted.
3. Inject local anesthesia under the scalp and open up the scalp along the midline using a scalpel or scissors. Retract the skin and expose the skull bone. Keep the scalp retracted using hemostatic forceps.
4. Use Dumont forceps to clear the skull bone from periost. If necessary retract the muscle lateral to the lateral ridges to achieve more space for the craniotomy.
5. Use the midline and bregma sutures on the skull bone to position the craniotomy (Fig. 4).
6. Use a dental drill or trephine to perform the craniotomy without causing a rift to the dura mater. Remove the bone piece and place it in sterile, isotonic saline if it is to be replaced later (Fig. 5).
7. Move the sterotaxic frame to the CCI device.

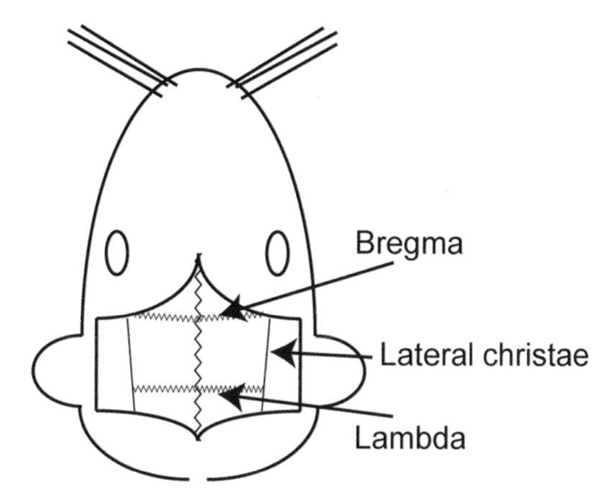

Fig. 4 Schematic drawing of the exposed rodent skull

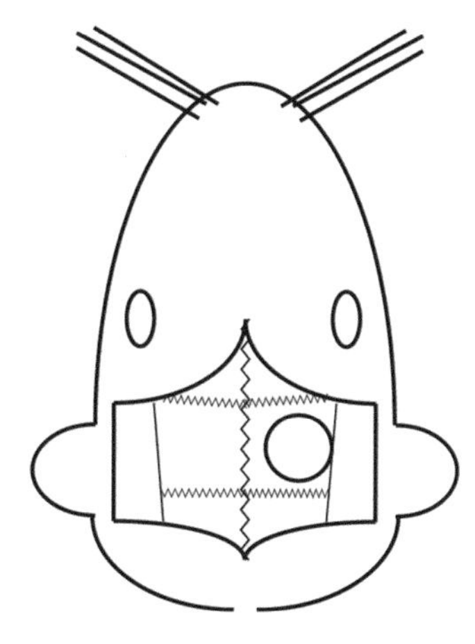

Fig. 5 Most common placement of the craniotomy for FPI or CCI

8. Find the null position and retract the piston. Lower the piston the desired distance. Perform the CCI and time the length of apnea (mostly applicable in mice).
9. Move the stereotaxic frame back to the operating table.
10. Replace the bone piece and use tissue adhesive to secure it in its former place.
11. Suture the scalp and move the animal to a recovery cage.

3.2 Fluid Percussion Injury

1. Sedate the animal and place it in the sterotaxic frame.
2. Trim the fur on the head and inject the scalp with local anesthesia. Open the scalp along the midline and retract the skin to expose the skull bone. Keep the scalp retracted using hemostatic forceps.
3. Use Dumont forceps to clean the skull from connective tissue and periosteum.

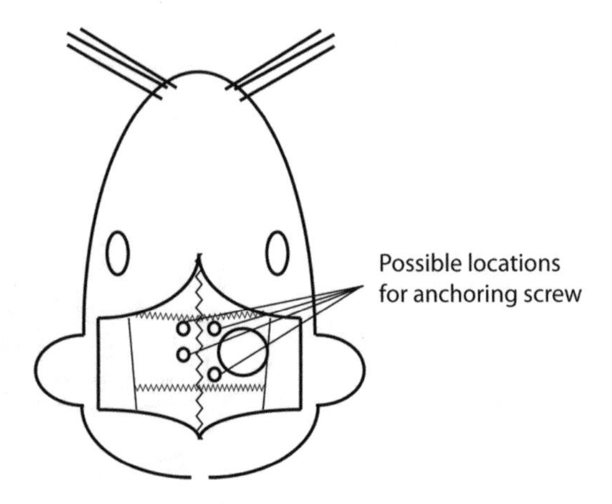

Fig. 6 Possible locations for the anchoring screws

4. Use the midline and bregma sutures as reference points to place the craniotomy in the desired place. Make the craniotomy using a trephine, be sure not to rupture the dura mater (Fig. 5).
5. Use a drill to make a hole for the anchor screw, without going all the way through the bone. Attach the screw to the skull without penetrating the skull bone (Fig. 6).
6. Place the trauma coupling in the craniotomy and secure it with tissue adhesive.
7. Pour dental cement around the trauma coupling. Make sure to include the anchor screw in the cast. Allow the dental cement to set properly.
8. Wean the animal off anesthesia enough for it to regain the pinch reflex in the paws.
9. Attach the animal to the FPI device and perform the injury.
10. Time the length of apnea if at a moderate or severe setting.
11. Return the animal to the heated pad and measure the time till it regains the pinch reflex in the contralateral paws.
12. Replace the bone piece and attach it with tissue adhesive and suture the scalp.
13. Move the animal to a recovery cage when the pinch reflex has returned.

3.3 Weight Drop Injury

1. Sedate the animal.
2. If a protective disk is used, place the animal in a stereotaxic frame.
3. Inject local anesthesia under the skin and expose the skull bone by opening the scalp along the midline, retract the skin to expose the skull bone. Keep the scalp retracted using hemostatic forceps.

4. Remove connective tissue and periost from the skull bone.
5. Make sure the bone is dry before applying the tissue adhesive to attach the disk in the preferred position, use the midline and bregma sutures as reference points.
6. Place the animal on the flexible bed and position the guide tube so the weight will hit the disk or the head of the animal according to the decided protocol.
7. Release the weight from the decided height.
8. Move the rat immediately after the weight hits to avoid a second impact as the weight rebounds.
9. Remove the disk and suture the scalp.
10. Move the animal to a recovery cage.

4 Notes

4.1 Controlled Cortical Impact

Firstly, it is important to remember that even though most experiments are made in age matched, inbred male subjects, these are biological experiments and there may be differences between individuals, both physiological and behavioral.

Aside from individual differences in the lab animals, there are two factors that can cause a great variability between operators and laboratories in CCI. The first is the placement of the trauma, a few mm difference in where the piston strikes can cause a different outcome. The second is the null position before the impact. This can be made depressing the cortex slightly or just touching it with the piston. The difference between those two positions is about 0.5 mm, which is the difference between a moderate and severe injury both in mice and rats.

Another possible source of variability is the angle of the piston to the exposed brain. It should be perpendicular and the angle is easily adjusted on the CCI device if needed.

If a severe injury is desired, a rift on the dura mater is expected after impact, whereas on moderate or mild setting ruptures should be avoided.

4.2 Fluid Percussion Injury

The FPI device itself can cause variability if there are air bubbles present in the cylinder or nozzle. Before starting the experiment a few test hits should be made and the pressure curve checked for irregularities. If the pressure curve is full of spikes it is a sign of air bubbles in the system and they should be removed before starting the experiment. Make sure that the pressure peak is at the desired value.

When applying tissue adhesive to the trauma coupling after attaching it to the craniotomy it is possible for the tissue adhesive to leak onto the brain if the trauma coupling does not fit well. The trauma coupling then has to be removed and cleaned or discarded, since the tissue adhesive will set on the brain surface and decrease the pressure of the trauma. When doing FPI in mice, the tissue adhesive can attach hard enough to the dura mater to cause a rift when the trauma coupling is removed. In mice, it may be advisable to use a very fast acting adhesive on just one or two points of the trauma coupling to avoid this problem.

4.3 Weight Drop Injury

To make the experiments as reproducible as possible it is important that the weight hits the animals as similarly as possible. This can to some extent be achieved by forming the flexible bed so that the animals rests in the same position. Keeping the guide tube properly lubricated to allow the weight to accelerate as similarly as possible from experiment to experiment is also of importance. If a protective disk is used it should be mounted as similar as possible in all animals. To avoid a second impact from the rebound of the weight the animal can be moved or a catch for the weight built into the device.

References

1. Dixon CE, Clifton GL, Lighthall JW, Yaghmai AA, Hayes RL (1991) A controlled cortical impact model of traumatic brain injury in the rat. J Neurosci Methods 39:253–262
2. Smith DH, Soares HD, Pierce JS, Perlman KG, Saatman KE, Meaney DF et al (1995) A model of parasagittal controlled cortical impact in the mouse: cognitive and histopathologic effects. J Neurotrauma 12:169–178
3. www.amscien.com (2004) AmScien Instruments, Richmond, VA
4. Stalhammar D, Galinat BJ, Allen AM, Becker DP, Stonnington HH, Hayes RL (1987) A new model of concussive brain injury in the cat produced by extradural fluid volume loading: I. Biomechanical properties. Brain Inj 1:73–91
5. Dixon CE, Lyeth BG, Povlishock JT, Findling RL, Hamm RJ, Marmarou A et al (1987) A fluid percussion model of experimental brain injury in the rat. J Neurosurg 67:110–119
6. McIntosh TK, Vink R, Noble L, Yamakami I, Fernyak S, Soares H et al (1989) Traumatic brain injury in the rat: characterization of a lateral fluid-percussion model. Neuroscience 28:233–244
7. Carbonell WS, Grady MS (1999) Regional and temporal characterization of neuronal, glial, and axonal response after traumatic brain injury in the mouse. Acta Neuropathol 98:396–406
8. Clausen F, Hillered L (2005) Intracranial pressure changes during fluid percussion, controlled cortical impact and weight drop injury in rats. Acta Neurochir (Wien) 147:775–780
9. Floyd CL, Golden KM, Black RT, Hamm RJ, Lyeth BG (2002) Craniectomy position affects morris water maze performance and hippocampal cell loss after parasagittal fluid percussion. J Neurotrauma 19:303–316
10. www.dragonflyinc.com (2010) Dragonfly Research & Development Incorporated, Ridgeley, WV
11. Frey LC, Hellier J, Unkart C, Lepkin A, Howard A, Hasebroock K et al (2009) A novel apparatus for lateral fluid percussion injury in the rat. J Neurosci Methods 177:267–272
12. Shapira Y, Shohami E, Sidi A, Soffer D, Freeman S, Cotev S (1988) Experimental closed head injury in rats: mechanical, pathophysiologic, and neurologic properties. Crit Care Med 16:258–265
13. Foda MA, Marmarou A (1994) A new model of diffuse brain injury in rats. Part II: Morphological characterization. J Neurosurg 80:301–313
14. Marmarou A, Foda MA, van den Brink W, Campbell J, Kita H, Demetriadou K (1994)

A new model of diffuse brain injury in rats. Part I: Pathophysiology and biomechanics. J Neurosurg 80:291–300

15. Zweckberger K, Eros C, Zimmermann R, Kim SW, Engel D, Plesnila N (2006) Effect of early and delayed decompressive craniectomy on secondary brain damage after controlled cortical impact in mice. J Neurotrauma 23: 1083–1093
16. Goren S, Kahveci N, Alkan T, Goren B, Korfali E (2001) The effects of sevoflurane and isoflurane on intracranial pressure and cerebral perfusion pressure after diffuse brain injury in rats. J Neurosurg Anesthesiol 13:113–119
17. Uematsu M, Takasawa M, Hosoi R, Inoue O (2009) Uncoupling of flow and metabolism by chloral hydrate: a rat in-vivo autoradiographic study. Neuroreport 20:219–222

Chapter 2

Experimental Radiosurgery in Animal Models

Ajay Niranjan, Wendy Fellows-Mayle, Douglas Kondziolka, and L. Dade Lunsford

Abstract

Lars Leksell described stereotactic radiosurgery as a method to destroy intracranial targets using a single, high dose of focused, ionizing radiation administered using stereotactic guidance. Radiosurgery is an impressive blend of minimally invasive technologies guided by a multidisciplinary team of surgeons, oncologists, medical physicists, and engineers. The long-term results of radiosurgery are now available and have established it as an effective noninvasive management modality for intracranial vascular malformations and many tumors. A variety of experimental models have been used to study the effect of radiosurgery in brain. The results of experimental radiosurgery have enhanced our understanding of the biological impact of radiosurgery on different tissues. Additional applications of radiosurgery in the management of malignant tumors and functional disorders are being assessed.

Key words Experimental, Animal models, Epilepsy, Radiosurgery, Functional disorders, Tumors, Vascular malformations

1 Introduction

Radiosurgery is a surgical technique which is designed to produce a specific radiobiological effect within a sharply defined target volume using a single, high dose of focused, ionizing radiation administered using stereotactic guidance. Normal tissue effects are limited by the highly focused nature of the radiosurgical beams. Whereas conventional fractionated radiotherapy is generally most effective in killing rapidly dividing cells, radiosurgery induces biological responses irrespective of the mitotic activity, oxygenation, and inherent radiosensitivity of target cells. The field of stereotactic radiosurgery represents one of the fundamental shifts in the neurological surgery over the last two decades. Compared to conventional neurosurgery techniques, stereotactic radiosurgery is minimally invasive and relies on biological response of tissues in order to eradicate or inactivate them. Considering the unique biological response of tissues to radiosurgery, it is important to

Miroslaw Janowski (ed.), *Experimental Neurosurgery in Animal Models*, Neuromethods, vol. 116,
DOI 10.1007/978-1-4939-3730-1_2, © Springer Science+Business Media New York 2016

study the biological effects of radiosurgery in both normal and pathological nervous system tissues in animal models. Information gained from radiosurgical research studies is useful in devising future strategies to prevent damage to normal tissue without compromising treatment efficacy.

1.1 History of Experimental Radiosurgery Using Animal Models

Initial radiosurgery experiments were performed using the rabbit and goat central nervous system (CNS) models. These experiments were designed to investigate the use of focused radiation as neurosurgical tool. The early histological results (third to eighth day) using proton beam radiation on rabbit spinal cord tissue showed complete transection of the spinal cord with 400 Gy using 1.5 mm beam diameter and with 200 Gy using 10 mm beam diameter [1]. The goat brain model was also used to document sharply defined lesions in deep parts of the brain, 4–7 weeks after 200 Gy of stereotactic multiple port proton beam radiation. Rexed et al. preformed proton beam radiosurgery on rabbit brain with 200 Gy using a 1.5 mm collimator [2]. These investigators documented a well-demarcated lesion at the target. In a similar model, Leksell et al. used cross-fired irradiation with a narrow beam of high-energy to create well-circumscribed intracerebral lesions of appropriate size and shape [2]. Andersson et al. in a study of long-term effect of proton radiosurgery on goat brain documented that there were no adverse effects in or around the lesion after 1.5–4 years after 200 Gy radiosurgery [3]. Nilsson et al. irradiated (100–300 Gy) the basilar artery of cats by stereotactic technique using 179-source cobalt-60 prototype gamma unit [4]. Histology demonstrated endothelial wall injury, and hyalinization and necrosis of the muscular layers. These investigations demonstrated that radiosurgery could be potentially used to create sharply defined lesions in deep parts of the brain.

1.2 Radiosurgery as a Lesioning Technique

Lunsford et al. [5] and Kondziolka et al. [6] studied the radiobiological effects of stereotactic radiosurgery using a baboon model. A central dose of 150 Gy using an 8 mm collimator was delivered to the caudate, thalamus, or pons regions using the gamma knife. No imaging changes were noted at 4 weeks after irradiation. MR imaging documented a circumscribed contrast enhanced lesion by 6–8 weeks and frank necrosis at the irradiated target by 24 weeks.

Kondziolka et al. [7] irradiated the frontal lobe of rats with maximal doses of 30–200 Gy using a 4 mm collimator and studied histologic changes 90 days after radiosurgery. While detectable histologic alterations were noted with doses of more than 70 Gy, necrosis was seen only in tissues irradiated with more than 100 Gy. Blatt et al. [8] evaluated serial tissue changes after 125 Gy using linear accelerator (LINAC) radiosurgery of internal capsule of cats. Serial imaging and histopathological evaluations showed tissue necrosis accompanied by vascular proliferation and edema by

3.5 weeks. Kamiryo et al. studied the radiosensitivity of brain by evaluating the effects of radiosurgery dose and time rat brain. The rat brain was irradiated using the maximum doses of 50, 75, or 120 Gy and analyzed for histologic changes and blood–brain barrier integrity up to 12 months [9]. Whereas tissue irradiated with higher doses (120 Gy) showed alterations in astrocytic morphology by 3 days, such changes were not observed until 3 months with lower doses (50 Gy). Blood–brain barrier breakdown was noted within 3 weeks of 120 Gy irradiation but was not seen even up to 12 months after 50 Gy. These findings suggested that the latent period between irradiation and detection of pathologic alterations was dependent on both the dose and the biological end point used. The impact of dose and biological end points on latency was also reported by Karger et al. [10], who evaluated the rat brain using MRI at 15, 17, or 20 months after treatment with 26–50 Gy of LINAC-based radiosurgery. No radiation-induced effect on MRI was noted at any time point for doses less than 30 Gy. After 40 Gy radiosurgery, the latency of detectable MRI changes was 19–20 weeks, whereas the latency after 50 Gy was 15–16 weeks. In a similar study focusing on vascular changes after a maximal dose of 75 Gy delivered to the rat brain using a gamma knife, it was noted that vascular changes preceded necrosis [11]. This finding suggests that the vascular response is also an important component in the biologic effect of radiosurgery (Table 1).

1.3 Radiation Protection Studies in Animal Models

A few strategies for radioprotection of normal tissue have already been explored. The initial strategies included use of cerebral protective agents while delivering a high dose to tumor cells. Oldfield et al. [12] documented protection from radiation-induced brain injury using pentobarbital. Buatti et al. [13] found that 21-aminosteroids (21-AS) protected the cat brain from injury due to radiosurgery and was significantly more effective than corticosteroids [13]. Kondziolka et al. showed that 15 mg/kg but not 5 mg/kg of U-74389G (a 21-AS) was effective at reducing brain injury in the rat when administered 1 h prior to radiosurgery. U74389G ameliorated vasculopathy and regional edema and delayed the onset of necrosis, while gliosis remained unaffected [14]. Preliminary evidence suggests that this agent may be acting through reduction of the cytokines induced by brain irradiation.

1.4 Enhancing the Effect of Radiosurgery in Animal Models

Although benign tumor radiosurgery is associated with high tumor control rates malignant glial tumors often recur. Additional strategies to improve cell kill of malignant brain tumors are needed. Niranjan et al. studied the synergistic effect of tumor necrosis factor alpha (TNF-α) on enhancing the tumor response to radiosurgery. TNF-α can act as a tumoricidal agent with direct cytotoxicity mediated through binding to its cognate cell-surface receptors and a variety of activities triggering a multifaceted immune attack on tumors

Table 1
Central nervous system response to radiosurgery

Author	Year of study	Animal model	Radiosurgery target	Central dose (Gy)	Collimator size (mm)	Radiosurgery technique	Results
Lunsford	1990	Baboon	Caudate, thalamus, pons	150	8	Gamma knife	MR imaging and histology documented lesion 45–60 days posttreatment
Kondziolka	1992	Rat	R. Frontal lobe	30–200	4	Gamma knife	Histology at 90 day showed tissue changes at lower doses (60 Gy) and necrosis at higher doses (100 Gy)
Blatt	1994	Cat	Internal capsule	149	10	LINAC	MRI and serial histopathology indicated mass effect and neurologic deficits at 3.5–4.5 weeks, some necrosis 12–29 weeks, and late resorption of necrosis
Kamiryo	2001	Rat	Parietal cortex	75	4	Gamma knife	Electron microscopy at 3.5 months showed decreased vascularity and increased capillary diameter in irradiated regions; basement membrane changes precede vascular damage
Karger	2002	Rat	Parietal cortex	26–50	3	LINAC	MR imaging documented contrast enhancement at 15 weeks after 50 Gy and 19 weeks after 40 Gy radiosurgery

[15–20]. In addition, locally produced TNF-α has been reported to enhance the sensitivity of tumors to radiation in nude mice [15]. We employed a replication defective herpes simplex virus (HSV), as a vector to deliver thymidine kinase (TK) and/or tumor necrosis factor (TNF-α) genes to U-87 MG tumors in nude mice. Radiosurgery was performed 48 h after gene transfer using 15 Gy to the tumor margin (21.4 Gy to the center). Daily ganciclovir therapy (GCV) was started after gene transfer and continued for 10 days. The combination of radiosurgery with TNF-α or with HSV-TK-GCV (suicide gene therapy) and TNF-α significantly improved median survival of animals [21]. In additional experiments, the connexin-43 gene was added to enhance the formation of gap junctions between tumor cells, which should facilitate the intercellular dissemination of TK-activated GCV from virus-infected cells to noninfected surrounding cells. This creates a bystander effect that can improve tumor cell killing [22]. Addition of connexin-43 gene to this paradigm (TK-GCV + TNF-α + radiosurgery) further improved survival (90% survival in tumor-bearing mice). We also studied this strategy in a 9 L rat glioma model and found that addition of radiosurgery to suicide gene therapy (SGT) significantly improved animal survival compared to SGT alone. The combination of HSV-based SGT (TK-GCV), TNF-α gene transfer, and radiosurgery was more effective than SGT or radiosurgery alone. The combination of SGT with radiosurgery was also more effective than SGT or radiosurgery alone. Although, the exact mechanism of this effect is unclear and remains the subject of future investigations, these experiments indicate that gene therapy could be an effective strategy for enhancing the radiobiological impact of radiosurgery. In other studies, tumor sensitization to radiation was apparently mediated by extracellular TNF-α promoting the destruction of tumor vessels, whereas HSV vector mediated TNF-α enhanced killing of malignant glioma cell cultures is presumably a consequence of an intracellular TNF-α activity [20, 23] (Table 2).

1.5 Functional Radiosurgery in Animal Models

Radiosurgery is rapidly expanding beyond its use as a treatment of brain tumors and AV malformations. It has been found effective for other neurologic disorders, such as epilepsy, movement disorders, and trigeminal neuralgia. The promise of "functional" radiosurgery has led to a need to investigate its efficacy, limitations, and potential drawbacks.

The potential efficacy of radiosurgery for the treatment of epilepsy has been evaluated using rat models. Kainic acid reproducibly induces epilepsy in rats when injected into the hippocampus. Mori et al. [24] treated kainic acid-induced epilepsy in rats with doses of 20–100 Gy radiosurgery using gamma knife. The efficacy of the treatment on epilepsy was evaluated by direct observation and scalp EEG for 42 days. Even 20 Gy significantly reduced the number of seizures, and the efficacy improved with increasing dose. Only doses higher than 60 Gy induced histologic changes. Maesawa et al. [25] irradiated epileptic rats with a single dose of 30 or 60 Gy. Both doses significantly reduced EEG-defined seizures. The latency to this effect was less after the higher dose (5–9 weeks for 60 Gy versus 7–9 Gy for 30 Gy). While kainic acid injection alone reduced performance of rats on the water maze task, the performance of rats that had radiosurgery after kainic acid administration was not

Table 2
Experimental radiosurgery for malignant brain tumors

First author	Year of study	Animal model	Maximum dose (Gy)	Tumor model	Collimator size (mm)	Experimental treatment	Results
Kondziolka	1992	Rat	30–100	C6 Glioma	4	Radiosurgery	Treated animals survived 39 days (control 29 days). Treated tumors had hypocellular appearance with cellular edema
Niranjan	2000	Nude mouse	21.4	U 87 MG	4	Radiosurgery + HSV-based gene therapy	The combination treatment enhanced median survival (75 days) with 89% animal surviving
Nakahara	2002	Rat	32	MADB 106 cells	4	Radiosurgery + cytokine transduced tumor cell vaccine	The combination treatment significantly prolonged animal survival and protected animals from a subsequent challenge by parental tumor cells placed in the CNS
Niranjan	2003	Rat	21.4	9 L Glioma	4	Radiosurgery + HSV-based gene therapy	The combination of radiosurgery and multigene therapy enhanced median animal survival (150 days) with 75% animal surviving

different from controls. Liscak et al. [26] evaluated the effects of radiosurgery on normal hippocampus in an effort to identify potential normal tissue complications and determine dose limits for hippocampal radiosurgery. This study employed four separate 4 mm isocenters to irradiate the entire hippocampus with 25–100 Gy. Doses <50 Gy did not cause any perceptible changes based on histology, MRI, and Morris water maze testing. In contrast, the performance on the Morris water maze was significantly worse for animals who were treated with >50 Gy. These investigations support the concept that radiosurgery may be an effective method for treating epilepsy, but they also suggest that doses to the hippocampus should be limited to reduce potential effects on learning and memory.

The effect of radiosurgery on potential targets for the treatment of movement disorders has been evaluated. De Salles and colleagues [27] used a LINAC and 3 mm collimator to deliver a maximal dose of 150 Gy to the subthalamic nucleus of one vervet monkey and to the substantia nigra of another. Follow-up MRI detected a 3 mm lesion that did not increase in size throughout the course of the study. Kondziolka et al. [28] examined the effects of thalamic radiosurgery in a baboon model, and reported that a dose of 100 Gy (central dose using 4-mm collimator) was sufficient to induce contrast enhancement of MR images and coagulative necrosis as evaluated by histology.

Radiosurgery has significant potential as an effective, noninvasive method for treatment of trigeminal neuralgia. Kondziolka et al. investigated the effect of gamma knife irradiation on the trigeminal nerve in the baboon [29]. A central dose of 80 or 100 Gy using a 4 mm collimator was delivered to the normal proximal trigeminal nerve. Follow-up MRI at 6 months showed a 4 mm region of contrast enhancement on the nerve. Histology showed that both large and small fibers were affected with axonal degeneration occurring after 80 Gy and necrosis after 100 Gy. Neither dose was effective at selectively damaging fibers responsible for transmission of pain while maintaining those responsible for other sensations, which would be optimal for effective treatment of trigeminal neuralgia. This study demonstrated that it was possible to noninvasively and precisely affect specific nerves using the gamma knife. In a recent study Zhao et al. investigated the effect of dose and single versus two isocenters on trigeminal nerve radiosurgery in rhesus monkeys [30]. These authors delivered a central radiation dose of 60, 70, 80, or 100 Gy using 4-mm collimator at trigeminal nerve root. One side of the nerve was exposed to single-target-point irradiation, and the contralateral side was exposed to double-target-point irradiation. Histological examination at 6 months revealed that the target doses of 80 Gy resulted in partial degeneration and loss of axons and demyelination. The extent of histological changes was identical with the single-target-point and the double-target-point irradiation (Table 3).

Table 3
Experimental functional radiosurgery

First author	Year of study	Animal model	Maximum dose (Gy)	Region(s) irradiated	Irradiation technique	Collimator size (mm)	Results
Ishikawa	1999	Rat	200	Medial temporal lobe	Gamma knife	4	Sequential MRI and histopathology showed consistent necrosis at 2 weeks after 200 Gy radiosurgery.
Mori	2000	Rat	20–100	Hippocampus	Gamma knife	4	Reduction in seizure frequency after ≥20 Gy radiosurgery.
Maesawa	2000	Rat	30–60	Hippocampus	Gamma knife	4	Reduction in seizure frequency after 30–60 Gy radiosurgery, shorter latency after higher dose, learning and memory unaffected.
Kondziolka	2000	Baboon	80–100	Trigeminal nerve	Gamma knife	4	Axonal degeneration on electron microscopy 6 months after radiosurgery at all doses
Chen	2001	Rat	20–40	Hippocampus	Gamma knife	4	Substantially reduction in seizure frequency and duration by subnecrotic (20–40 Gy) radiosurgery
De Salles	2001	Monkey	150	Subthalamic nucleus, substantia nigra	LINAC	3	MRI and histology showed that necrotic lesion remained at <3 mm size after LINAC radiosurgery.

(continued)

Table 3 (continued)

First author	Year of study	Animal model	Maximum dose (Gy)	Region(s) irradiated	Irradiation technique	Collimator size (mm)	Results
Kondziolka	2002	Baboon	100	Thalamus	Gamma knife	4	MRI, histology showed necrosis at 6 months
Liscak	2002	Rat	25–150	Hippocampus	Gamma knife	4	Altered memory performance after >50 Gy radiosurgery
Zerris	2002	Rat	140	Caudate–putamen complex	Gamma knife	4	Radiosurgery significantly reduced 6-OHDA-induced hemiparkinsonian behavior. Areas surrounding necrotic lesions were highly positive for GDNF
Brisman	2003	Rat	5–130 CGE	Hippocampus	Proton beam	NA	Proton radiosurgery with doses 90 CGE or higher resulted in adverse behavioral effects and necrosis in 3 months. 30 or 60 CGE radiosurgery led to marked increase in HSP-72 staining but no necrosis
Zhao	2011	Monkey	60–100	Trigeminal nerve	Gamma knife	4	Irradiation at 80 Gy can cause partial degeneration and loss of axons and demyelination. A 100-Gy dose can cause some necrosis of neurons. No additional effect of double-target-point irradiation was seen.

Vincent et al. performed hypothalamic radiosurgery on genetically obese Zucker rats and studied the effect of subnecrotic hypothalamic radiosurgery on body weight set point [31]. These investigators performed radiosurgery using a total dose of 40 Gy delivered to two nearby targets in the medial hypothalamus. These investigators noted significant and sustained reductions in weight set point for animals that received radiosurgery compared to sham-treated animals after a latency of 7 weeks. No gross behavioral abnormalities were noted. Histopathological analysis showed no abnormalities except a small area of necrosis in one animal. At the University of Pittsburgh, the authors are investigated the feasibility of hypothalamic radiosurgery using a primate model. The results showed that hypothalamic radiosurgery is feasible and safe.

2 Materials and Methods

2.1 Small Animal Radiosurgery Models (Mice Model/Rat Model)

2.1.1 Animal Preparation for Radiosurgery

1. The rats/mice are anesthetized with Ketamine and Acepromazine administered intramuscularly.
2. Anesthetized animals are placed in a stereotactic head frame (David Kopf Instruments, Tujunga, CA).
3. A small craniotomy is drilled 2 mm to the right of midline and 1 mm anterior to the coronal suture. Dura was not opened.
4. A predetermined number of cells (1×10^5 U-87MG glioblastoma cells in a 3-μl volume) is implanted stereotactically in the right frontal lobe region 3 mm below the dura mater. This area corresponds to the lateral portion of the right striatum of the mouse.
5. A drug or viral vector can also be injected using the above technique.
6. The injection needle is removed.
7. A 2 mm section of a 25-gauge needle is placed in the craniotomy site over the dura for later stereotactic targeting.
8. The craniotomy is then sealed with bone wax and the scalp is closed with a 3-0 silk suture.

2.1.2 Radiosurgery Technique for Small Animals

1. Animals are anesthetized and placed on a small animal specially modified platform which is attached to stereotactic frame.
2. Animals are secured in place using transparent adhesive tape.
3. Angiography fiducial box is attached to the stereotactic frame.
4. Lateral and posteroanterior plain X-rays are taken. It is important to make sure that all nine fiducial markers as well as a metal marker place on animal skull are visible on X-ray films (Fig. 1).
5. X-ray films are scanned into a Gamma Knife planning computer.

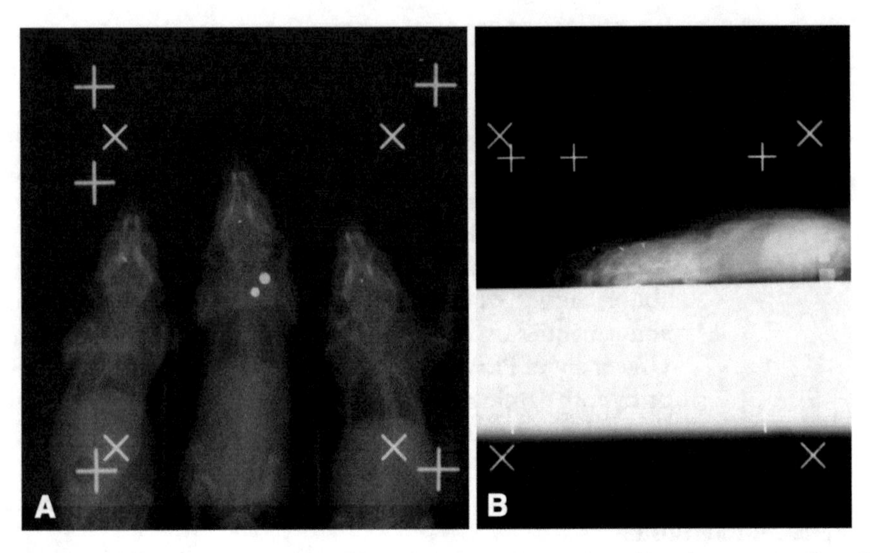

Fig. 1 Figure showing anteroposterior (**a**) and lateral (**b**) views of plain X-ray films for stereotactic radiosurgery of small animals. Three anesthetized rats are placed on a small animal specially modified platform which is attached to stereotactic frame. Angiography fiducial box is attached to the stereotactic frame. Lateral and anteroposterior plain X-rays are taken. Note that the fiducial markers from Angiography fiducial box as well as a metal marker placed on animal skull are visible on X-ray films

6. Target is defined based on its predetermined distance from the metal skull marker. We routinely used a point 3 mm perpendicular to an extradural metal marker in the right frontal brain region, which corresponded to the center of the tumor cell injection.
7. Radiosurgery planning is performed using Leksell Gamma Plan®.
8. For small animals, a plan is achieved using one 4-mm radiation isocenter.
9. Depending upon the goal of radiosurgery a central or a margin dose is selected. For our experiments involving tumor radiosurgery, we selected a margin dose of 15 Gy (center dose, 21.4 Gy) and delivered it to the 70 % isodose line using a 4-mm collimator (Fig. 2).
10. The final plan is printed and exported to Leksell Gamma Unit. The print out shows the *x*, *y*, and *z* coordinates of the planned isocenter as well as time required at that position in order to deliver the desired dose.
11. Radiosurgery is performed by positioning the small animal frame at the *x*, *y*, and *z* coordinates of the isocenter in the Leksell gamma knife unit (Elekta Instruments, Atlanta, GA) and setting the time obtained from dose plan.
12. At the end of the treatment, animal platform with stereotactic frame is removed from the Gamma Unit. Animals are removed from the frame and put in their respective cages.
13. Animals are observed till the regain consciousness and are full awake.

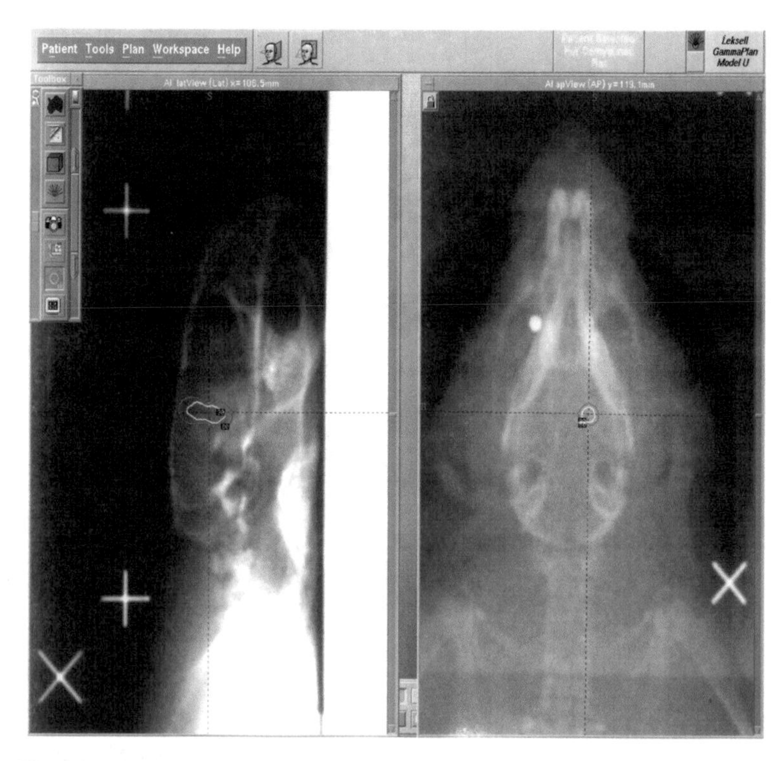

Fig. 2 Figure showing radiosurgery dose plan for rat model of demyelination. For radiosurgery planning, stereotactic X-ray films are scanned into a Gamma Knife planning computer. Target is defined based on its predetermined distance from the metal skull marker or skull suture. Radiosurgery planning is performed using Leksell Gamma Plan® using 4-mm radiation isocenter

2.1.3 Animal Observation Protocol

1. All animals are observed twice daily to monitor external appearance, feeding behavior, and locomotion (ability to walk to a distance of 50 cm in 10 s).
2. The contralateral limbs are observed daily for the development of paresis both passively and actively.
3. Animals are sacrificed at the first sign of an adverse event (paresis, inability to feed) and brains are removed for histological examination.
4. Animals surviving through the 75-day observation period are euthanized and the brains removed for histological examination.

2.2 Large Animal Radiosurgery Models (Baboon Model, Monkey Model)

2.2.1 Animal Preparation and Radiosurgery

1. Large animals (monkeys) are individually housed in stainless steel cages in air-conditioned and temperature and light-cycle-controlled rooms.
2. Animals are anesthetized using intravenous Propofol (2,6-diisopropylphenol) infusion and were intubated and maintained on inhalation anesthetic agents.
3. The animal is brought in the laboratory adjacent to MR unit. The standard Model-G Leksell Head frame is used for large

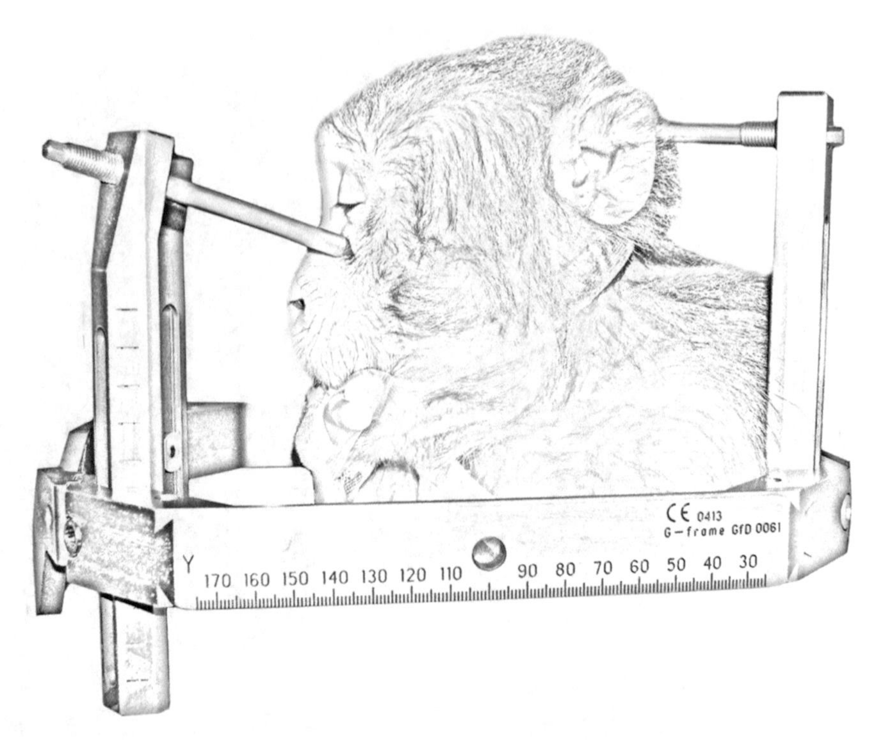

Fig. 3 Diagrammatic representation showing monkey with standard Model-G Leksell head frame anchored to his head. Large animals are anesthetized using intravenous Propofol (2,6-diisopropylphenol) infusion and the Leksell head frame is anchored to their head using two pins on the forehead and two pins on the back of head. Two front posts are attached to cheek (maxilla) using 60–70 mm long pins. The two back posts are fixed on the occipital ridge using long 80–90 mm pins. The MR fiducial box is then attached on top of the head ring and stereotactic MRI is performed. These stereotactic MR images are imported into the dose planning computer for radiosurgery dose planning

animals. This frame is widely used clinically for stereotactic applications of brain.

4. In humans, the frame is anchored to head using two pins on the forehead and two pins on the back of head. Because monkeys do not have a convex forehead and their head size is much smaller compared to human, special technique was used to anchor Leksell head frame. Two front posts were attached to cheek (maxilla) using 60–70 mm long pins. The two back posts were fixed on the occipital ridge using long 80–90 mm pins (Fig. 3).
5. After securing the head ring, the fiducial box is attached on top of the head ring.
6. The animal is taken to MR unit for the second set of images (stereotactic images).
7. Head frame is stabilized using a modified head holder that fits MR table and head coil.
8. The stereotactic MR images using an FOV (field of view) of 250×250 mm are obtained using standard head coil.

9. The images are transferred to Gammaplan computer via ethernet system.
10. Stereoatctic images are defined using GammaPlan software.
11. The target is selected based on the goals of the experiment. The target could be thalamus, trigeminal nerve, caudate nucleus, hypothalamus, etc.
12. An optimum margin and central dose is selected.
13. The plan is printed and exported to Leksell Gamma unit.
14. The animal is observed and maintained under anesthesia throughout the planning period.
15. Once the dose plan is ready, the animal is secured to Gamma unit by attaching his head frame to the unit.
16. x, y, and z coordinates obtained from the dose planning are set on the Gamma Unit.
17. The time obtained for desired dose delivery is set on the unit and the treatment is initiated.
18. At the completion of treatment animal is removed from the unit.
19. Head frame is removed and head is dressed with antiseptic and antibiotic crèmes.
20. Anesthesia reversal is given and endotracheal tube is removed once animal starts breathing on her own.
21. Animal is placed in the cage and monitored till she is fully awake and active.

2.2.2 Animal Observation Protocol

1. All animals are observed daily to monitor external appearance, feeding behavior, and locomotion (ability to walk to a distance of 50 cm in 10 s).
2. The contralateral limbs are observed daily for the development of paresis both passively and actively.
3. Depending upon the goals of research follow-up MR imaging is performed.
4. Animals are sacrificed at the first sign of an adverse event (paresis, inability to feed) and brains are removed for histological examination.
5. Depending upon the protocol Animals are euthanized and the brains removed for histological examination.

3 Notes

1. Anesthesia is a critical component of small animal radiosurgery. Anesthesia needs to be titrated judiciously in order to keep mice or rats immobile during the whole experiment which can

last up to 3 h. While higher doses of anesthetic agents can kill the animals, the lower doses can make experiment ineffective if animal moves during the radiosurgery treatment. Every animal is weighed prior to experiment and doses of anesthetics are titrated accordingly. Rats were usually anesthetized with an intramuscular or Intraperitoneal injection of Ketamine and Acepromazine (9:1) at a dose of 44 mg/kg. Animals are given additional doses at the slightest sign of movement of any body parts or wakefulness.

2. While placing the animals on stereotactic frame care is taken to make sure that when angiography fiducial box is attached to the frame, the animal is located in the middle of the frame. Special precaution is taken to immobilize the animal by gently taping there head to the frame platform.
3. One plain X-rays are taken, we always ensure that all nine fiducial markers (pluses and crosses) are visible on both anteroposterior and lateral view. Visualization of these markers is critical to scanning these films into gamma plan computer. Once these films are digitized these can be used for dose planning.
4. Dose planning is performed using the gamma plan software. For small animal radiosurgery, a 4 mm beam diameter collimator is selected for radiation delivery. If a larger lesion is desired (such as for demyelination experiment), two shot dose plan can be designed. The margin dose is prescribed to 50 % isodose line. For most tumor radiosurgery, however, target is much smaller than 4 mm. In such cases, the volume covered by delivered radiation can be decreased by prescribing the margin dose to a higher isodose line (such as 70 %).
5. Proper technique for target selection is important in order to treat normal or tumor brain. If the goal is to treat the normal brain (to study the effect of radiation) then a target can be selected based on distances from the coronal suture, which can be visualized easily in plain X-rays. However, if the goals is to treat a previously implanted tumor, then the exact coordinates of the tumor center are needed. To identify the location of the implanted tumor we use a metal marker on the dura mater of the animals. We usually implant the tumor cells 3 mm below the dura. On the day of tumor implantation the anesthetized animal is placed in a stereotactic head frame (David Kopf Instruments, Tujunga, CA) and a small craniotomy is drilled, 2 mm to the right of midline and 1 mm anterior to the coronal suture. The dura is not opened. Tumor cells in a 3-μl volume are implanted stereotactically in right frontal lobe region 3 mm below the dura mater. This area corresponds to the lateral portion of the right corpus striatum of the mouse. After removal of the injection needle, a 2 mm section of a 25-gauge needle is

placed in the craniotomy site over the dura for later stereotactic targeting. The craniotomy was then sealed with bone wax and the scalp is closed with 3-0 silk suture. On the day of radiosurgery this marker is seen on the X-ray films. Because we know that the implanted tumor is 3 mm below the marker we can center the radiosurgery target 3 mm below the metal marker.

4 Future Experimental Radiosurgery

Rapid developments in computer hardware and software, imaging technologies, and stereotactic techniques have contributed to improvement in radiosurgery technology. As a result, the role of radiosurgery has expanded beyond its initial application for functional neurosurgery, pain management, arteriovenous malformations, and selected skull base tumors. The clinical spectrum of radiosurgery now includes a wide variety of primary and secondary brain tumors, vascular malformations, and functional brain disorders. Although radiosurgery provides survival benefits in diffuse malignant brain tumors, cure is still not possible. Additional strategies are needed to specifically target tumor cells while sparing normal CNS tissue. Further animal model-based research is needed to develop new treatment strategies that would maximize the effectiveness of radiosurgery on target tissue and minimize injury to other areas.

References

1. Larsson B, Leksell L, Rexex B, Sourander P, Mair W, Anderson B (1958) The high-energy proton beam as neurosurgical tool. Nature 182(6):1222–1223
2. Leksell L, Larsson B, Anderson B, Rexed B, Sourander P, Mair W (1960) Lesions in the depth of the brain produced by a beam of high-energy protons. Acta Radiol Ther Phys Biol 53:251–264
3. Andersson B, Larsson B, Leksell L et al (1970) Histopathology of late local radiolesions in the goat brain. Acta Radiol Ther Phys Biol 9(5): 385–394
4. Nilsson A, Wennerstrand J, Leksell D, Backlund EO (1978) Stereotactic gamma irradiation of basilar artery in cat. Preliminary experiences. Acta Radiol Oncol Radiat Phys Biol 17(2):150–160
5. Lunsford LD, Altschuler EM, Flickinger JC, Wu A, Martinez AJ (1990) In vivo biological effects of stereotactic radiosurgery: a primate model. Neurosurgery 27(3):373–382
6. Kondziolka D, Lunsford LD, Altschuler EM, Martinez AJ, Wu A, Flickinger JC (1992) Biological effects of stereotactic radiosurgery in the normal primate brainstem. In: Lunsford LD (ed) Stereotactic radiosurgery update. Elsevier, New York, pp 291–294
7. Kondziolka D, Lunsford LD, Claassen D, Maitz AH, Flickinger JC (1992) Radiobiology of radiosurgery: part I. The normal rat brain model. Neurosurgery 31(2):271–279
8. Blatt DR, Friedman WA, Bova FJ, Theele DP, Mickle JP (1994) Temporal characteristics of radiosurgical lesions in an animal model. J Neurosurg 80(6):1046–1055
9. Kamiryo T, Kassell NF, Thai QA, Lopes MB, Lee KS, Steiner L (1996) Histological changes in the normal rat brain after gamma irradiation. Acta Neurochir (Wien) 138(4):451–459
10. Karger CP, Munter MW, Heiland S, Peschke P, Debus J, Hartmann GH (2002) Dose-response curves and tolerance doses for late functional changes in the normal rat brain after stereotactic radiosurgery evaluated by magnetic resonance imaging: influence of end points and follow-up time. Radiat Res 157(6): 617–625

11. Kamiryo T, Lopes MB, Kassell NF, Steiner L, Lee KS (2001) Radiosurgery-induced microvascular alterations precede necrosis of the brain neuropil. Neurosurgery 49(2):409–414, discussion 14–15
12. Oldfield EH, Friedman R, Kinsella T et al (1990) Reduction in radiation-induced brain injury by use of pentobarbital or lidocaine protection. J Neurosurg 72(5):737–744
13. Buatti JM, Friedman WA, Theele DP, Bova FJ, Mendenhall WM (1996) The lazaroid U74389G protects normal brain from stereotactic radiosurgery-induced radiation injury. Int J Radiat Oncol Biol Phys 34(3): 591–597
14. Kondziolka D, Somaza S, Martinez AJ et al (1997) Radioprotective effects of the 21-aminosteroid U-74389G for stereotactic radiosurgery. Neurosurgery 41(1):203–208
15. Staba MJ, Mauceri HJ, Kufe DW, Hallahan DE, Weichselbaum RR (1998) Adenoviral TNF-alpha gene therapy and radiation damage tumor vasculature in a human malignant glioma xenograft. Gene Ther 5(3):293–300
16. Cao G, Kuriyama S, Du P et al (1997) Complete regression of established murine hepatocellular carcinoma by in vivo tumor necrosis factor alpha gene transfer [comment]. Gastroenterology 112(2):501–510
17. Han SK, Brody SL, Crystal RG (1994) Suppression of in vivo tumorigenicity of human lung cancer cells by retrovirus-mediated transfer of the human tumor necrosis factor-alpha cDNA. Am J Respir Cell Mol Biol 11(3):270–278
18. Ostensen ME, Thiele DL, Lipsky PE (1989) Enhancement of human natural killer cell function by the combined effects of tumor necrosis factor alpha or interleukin-1 and interferon-alpha or interleukin-2. J Biol Response Mod 8(1):53–61
19. Owen-Schaub LB, Gutterman JU, Grimm EA (1988) Synergy of tumor necrosis factor and interleukin 2 in the activation of human cytotoxic lymphocytes: effect of tumor necrosis factor alpha and interleukin 2 in the generation of human lymphokine-activated killer cell cytotoxicity. Cancer Res 48(4):788–792
20. Gridley DS, Archambeau JO, Andres MA, Mao XW, Wright K, Slater JM (1997) Tumor necrosis factor-alpha enhances antitumor effects of radiation against glioma xenografts. Oncol Res 9(5):217–227
21. Niranjan A, Moriuchi S, Lunsford LD et al (2000) Effective treatment of experimental glioblastoma by HSV vector-mediated TNF alpha and HSV-tk gene transfer in combination with radiosurgery and ganciclovir administration. Mol Ther 2(2):114–120
22. Marconi P, Tamura M, Moriuchi S et al (2000) Connexin 43-enhanced suicide gene therapy using herpesviral vectors. Mol Ther 1(1):71–81
23. Moriuchi S, Oligino T, Krisky D et al (1998) Enhanced tumor cell killing in the presence of ganciclovir by herpes simplex virus type 1 vector-directed coexpression of human tumor necrosis factor-alpha and herpes simplex virus thymidine kinase. Cancer Res 58(24):5731–5737
24. Mori Y, Kondziolka D, Balzer J et al (2000) Effects of stereotactic radiosurgery on an animal model of hippocampal epilepsy. Neurosurgery 46(1):157–165, discussion 65–68
25. Maesawa S, Kondziolka D, Dixon CE, Balzer J, Fellows W, Lunsford LD (2000) Subnecrotic stereotactic radiosurgery controlling epilepsy produced by kainic acid injection in rats. J Neurosurg 93(6):1033–1040
26. Liscak R, Vladyka V, Novotny J Jr et al (2002) Leksell gamma knife lesioning of the rat hippocampus: the relationship between radiation dose and functional and structural damage. J Neurosurg 97(5 Suppl):666–673
27. De Salles AA, Melega WP, Lacan G, Steele LJ, Solberg TD (2001) Radiosurgery performed with the aid of a 3-mm collimator in the subthalamic nucleus and substantia nigra of the vervet monkey. J Neurosurg 95(6):990–997
28. Kondziolka D, Conce M, Niranjan A, Maesawa S, Fellows W (2002) Histology of the 100 Gy thalomotomy in the baboon. Radiosurgery 4(4):279–284
29. Kondziolka D, Lacomis D, Niranjan A et al (2000) Histological effects of trigeminal nerve radiosurgery in a primate model: implications for trigeminal neuralgia radiosurgery. Neurosurgery 46(4):971–976, discussion 6–7
30. Zhao ZF, Yang LZ, Jiang CL, Zheng YR, Zhang JW (2010) Gamma Knife irradiation-induced histopathological changes in the trigeminal nerves of rhesus monkeys. J Neurosurg 113(1):39–44
31. Vincent DA, Alden TD, Kamiryo T et al (2005) The baromodulatory effect of gamma knife irradiation of the hypothalamus in the obese Zucker rat. Stereotact Funct Neurosurg 83(1):6–11

Chapter 3

Stereotactic Surgery in Rats

Jaroslaw Maciaczyk, Ulf D. Kahlert, Máté Döbrössy, and Guido Nikkhah

Abstract

Animal models represent the final step to complete preclinical investigations. Here, we describe in detail the principles and procedures for the surgical, toxin-induced animal models for Parkinson's disease (PD), and Huntington's disease (HD). Using highly precise stereotactic intracerebral injections of toxins into the nigrostriatal pathway and basal ganglia, we are able to target specific neural circuits in different regions of the dopaminergic and GABAergic system. In addition, validated protocols for adult and neonatal cell transplantation to reconstruct the destructed neuronal circuits as models for neural repair are described.

Key words Stereotactic neurosurgery, Rat, Cell transplantation, Parkinson's disease model, Huntington's disease model, Neural stem cells, Neurorepair

1 Introduction

Stereotactic surgery has been an invaluable tool applied since early 20th century in both experimental and clinical neuroscience to precisely target deep located brain regions for lesioning, injection of anatomical tracers, implantation of electrodes, neurotransplantation, sampling of various brain pathologies, gene delivery, etc. Its history begins with the development of the first stereotactic apparatus by Horsley and Clarke in 1908 [1], who also coined the term "stereotaxis" deriving from Greek words "stereos" meaning "three dimensional" and "taxis" meaning "to touch or to move." Despite further technical development and refinement of stereotactic equipment the principles of the surgical procedure enabling minimal invasive targeting of the deep brain structures based on the three-dimensional Cartesian coordinate system remained constant over the last hundred years. Whereas in humans the accuracy of this minimal invasive method relies on the preoperative imaging, in rodents the precision of the stereotactic targeting is based on the relation between bony landmarks of the cranium with deep located structures, which can be easily determined from stereotactic atlases such as *The Mouse Brain in Stereotaxic Coordinates* [2] and *The*

Miroslaw Janowski (ed.), *Experimental Neurosurgery in Animal Models*, Neuromethods, vol. 116,
DOI 10.1007/978-1-4939-3730-1_3,

Rat Brain in Stereotaxic Coordinates [3]. The coordinates describe the distance of the target in three dimensions (*x*, *y*, and *z*) from bregma—an intersection of the coronal and sagittal sutures on the surface of the skull, where the *x* plane represents a mediolateral (ML), the *y* plane rostrocaudal distance from the reference point, and *z* plane the dorsoventral (DV) distance from dura level that can be easily calculated upon the immobilization of the head of experimental animal in the stereotactic apparatus.

In this chapter, we are going to focus on the stereotactic surgical procedures in both adult and neonatal rats routinely performed in our laboratory to create rodent models of neurodegenerative disorders, i.e., Parkinson's disease (PD), Huntington's disease (HD), and for experimental neurotransplantation approaches.

2 PD Rat Models

Human idiopathic PD is a progressive neurodegenerative disorder, primarily characterized by degeneration of the dopaminergic neurons of the *substantia nigra* (SN) and subsequent loss of dopaminergic innervation of the striatum, leading to deregulation of the complex neurotransmitter system of the basal ganglia. Therefore, modeling of PD in rodents requires the depletion of the dopamine (DA) releasing terminals in their projection areas. In rodents, a standardized DA depletion of the basal ganglia can be reached by axotomy of nigral afferents (traditional model), by systemic application of 1-methyl-4-phenyl-1,2,5,6-tetrahydropyridine (MPTP) in mice, or by unilateral intracerebral stereotactic injection of 6-hydroxydopamine [4]—being a most common rodent model of PD. The neurotoxicity of 6-OHDA is based on its potent inhibitory effect on the mitochondrial respiratory enzymes (chain complexes I and IV, 2) that consequently leads to loss of dopaminergic neurons [5, 6]. Based on different injection sites of the 6-OHDA within the nigrostriatal pathway various grades of severity of lesion-induced PD can be achieved (Fig. 1):

1. The caudate–putamen unit (CPu)—referred as to "partial" or "terminal lesion," which also leads to relatively mild, local DA depletion resembling an early and intermediate stage of PD.
2. The medial forebrain bundle (MFB) which leads to extensive DA depletion recapitulating a later PD status, referred as to "complete lesion."
3. The substantia nigra pars compacta (SNc), which leads to more specific and moderate DA depletions for the representation of earlier PD status.

Independently from the model chosen the lesion coordinates are set according to bregma as a reference point for the anteroposterior

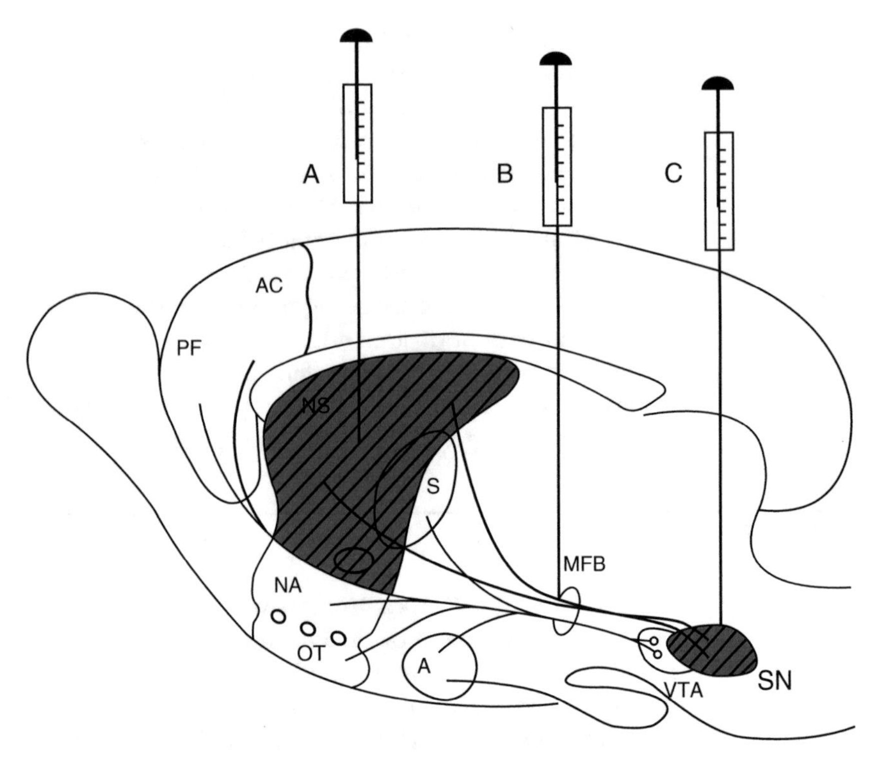

Fig. 1 Sagittal schematic diagram of the rodent brain indicating the three main targets for the 6-OHDA induced dopamine (DA) depletion. Partial DA loss can be achieved by infusion of the toxin in the striatum at the presynaptic level of the nigrostriatal projections (A), whereas complete DA depletion is produced if the toxin is injected into the medium forebrain bundle (B). Infusion of toxin into the substantia nigra has also been used (C) but this option gives less consistent striatal DA pathology. *SN* substantia nigra, *VTA* ventral tegmental area, *A* amygdala, *OT* olfactory tubercle, *NA* nucleus accumbens, *MFB* medium forebrain bundle, *S* septum, *NS* neostriatum (striatum), *PF* prefrontal cortex, *AC* anterior cingulated cortex

(AP) and ML coordinates and the dura as reference for the DV coordinate, using the rat brain atlas Paxinos and Watson [7]. The 6-OHDA is used in a concentration of 3.6 μg/μl in saline containing 0.2% (w/v) ascorbic acid.

3 Injection of 6-OHDA into the CPu

6-OHDA injection into the CPu, referred to as "partial" or "terminal lesion" results in more selective damage of the nigrostriatal dopaminergic system (Fig. 2c, d). Following the injection of the neurotoxin, due to its retrograde transport to SN *pars compacta* (SNc), exclusively nigral dopaminergic neurons projecting to the injection area undergo degeneration and cell death. In this animal model, usually ventrolateral and dorsomedial striatum have been targeted. In rodents, the ventrolateral portion of the CPu receives input from motor and sensorimotor areas of the neocortex whereas its DA innervations exclusively projects from the SNc. In contrast,

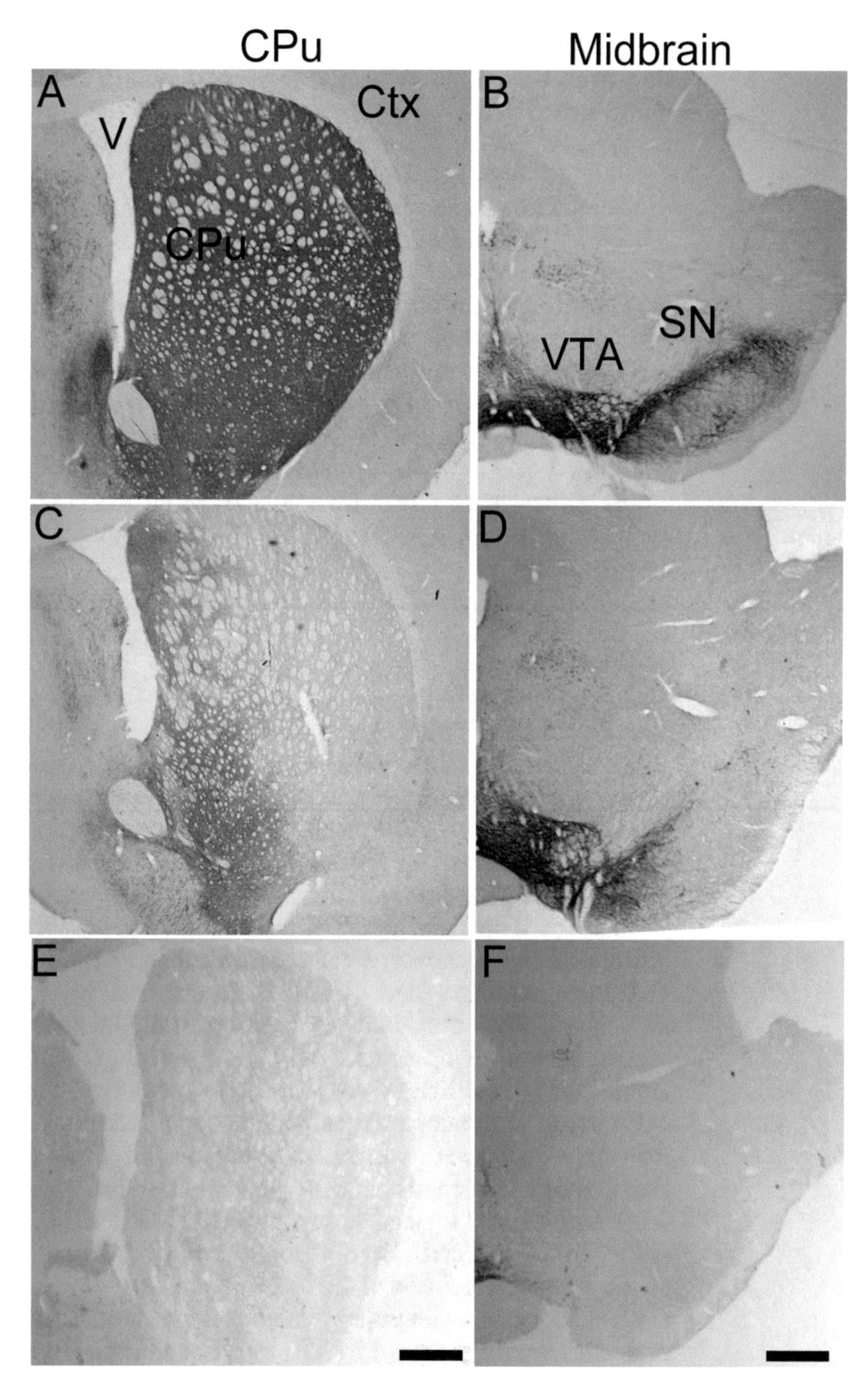

Fig. 2 (**a**–**f**) Following 6-OHDA lesion, sections are stained for DA afferents and cell body using tyrosine hydroxylase as the marker. Control animals with an intact nigrostriatal projections show complete DA innervation throughout the caudate putamen unit (CPU, **a**), and full complement of cell body staining in the midbrain both in the substantia nigra and the ventral tegmental area (**b**). Following partial/terminal 6-OHDA lesion, striatal DA depletion will occur selectively in the areas targeted but will remain intact in other striatal regions (**c**); and the partial lesion will be reflected at the midbrain level as well (**d**). Medial forebrain bundle lesions result in the complete loss of DA in the striatum (**e**) and at the midbrain level affecting both the substantia nigra and the ventral tegmental area (**f**). *V* ventricle, *Ctx* cortex, *Str* striatum, *VTA* ventral tegmental area, *SN* substantia nigra. Scale bar = 1 mm (**a**, **c**, **e**), and 500 μm (**b**, **d**, **f**)

the dorsal part of the CPu is characterized by a mixed DA innervation from both SN and ventral tegmental area (VTA), receiving inputs from frontal cortical areas as well as from the limbic system being an equivalent of the nucleus caudatus in humans.

Interestingly, more pronounced effects on locomotion and drug-induced rotation behavior can be achieved by lesioning the dorsomedial part of the CPu, whereas injection of the neurotoxin into the ventrolateral parts of the CPu provokes predominantly difficulties with movement initiation, sensorimotor orientation, and skilled motor behavior. Therefore, lesions in rodent ventrolateral CPu resemble most closely the depletion of DAnergic innervation in putamen of patients with PD [8]. However, to achieve long-lasting contralateral behavior deficits the neurotoxin has to be distributed over multiple injection sites along the rostrocaudal axis of the ventrolateral CPu with a bilateral CPu lesion as a closest parallel to the human disease [9, 10].

4 Injection of 6-OHDA in the MFB

Unilateral 6-OHDA injection into the MFB leads to an almost total destruction of the DAnergic neurons of the SNc projecting to striatum as well as of the VTA projecting to the nucleus accumbens [52] (Fig. 2e, f), causing a postsynaptic denervation supersensitivity of DA receptors. As a result of the lesion-induced imbalance between the nigrostriatal systems of both hemispheres, the animals show unilateral sensorimotor deficits enabling the evaluation of the lesion by behavioral analysis. The most robust one is a spontaneous postural motor asymmetry, causing the animals to rotate toward their impaired hemisphere. This can be enhanced by stress and in particular during drug-induced rotation using D1/D2 receptor agonist apomorphine and/or DA reuptake inhibitor d-amphetamine (for further details, see [11–13]). In contrast, the bilateral MFB lesion of the dopaminergic nigrostriatal system in adult animals leads to a severe sensorimotor impairment, rapid deterioration of the general condition with aphagia and adipsia, necessitating intensive care, and total parenteral nutrition [14, 15]. However, an advantage of bilateral injection is the avoidance of sprouting of axons from an intact side of the brain accounting for reported partial compensation of neurological deficits after the unilateral lesion. Nevertheless, the standard 6-OHDA rat model, caused by unilateral injection of 6-OHDA into the MFB is not only more pragmatic, but also permits a direct comparison of lesion effects and therapeutic results within one animal by the comparison of both hemispheres.

5 Injection of 6-OHDA into the SNc

Injection of the neurotoxin into SNc results in a more selective depletion of DAnergic neurons and less dramatic damage in the DA system. Animals receive unilaterally either medial and/or a single lateral injection into the SNc. The loss of dopaminergic neurons within the SNc, as measured with tyrosine hydroxylase (TH) immunohistochemistry reaches approximately 90%. Interestingly, also VTA seems to be affected by the injection of neurotoxin showing only about 70% of surviving DAnergic neurons as compared to the unlesioned side. The single lateral injection of the 6-OHDA spares the dopaminergic cells in the medial SNc and reflects a neuropathological finding of PD patients with the DAnergic cell loss present mainly within the lateral aspect of the SN [16]. In consequence, at the lesion side the density of DAnergic fibers within the lateral CPu diminishes dramatically as compared to medial parts of the caudate–putamen unit [17]. Moreover, the remaining DAnergic CPu innervation corresponds clearly to the extent of DAnergic cell depletion in SNc and correlates with the number of rotations in response to apomorphine, but not to d-amphetamine. Some authors reported bilateral SNc neurotoxin injections better resembling a clinical picture of PD with both hemispheres affected by the pathological process [18]. One of the major difficulties of this model is the small size of the target structure making the injection of the toxin into the SNc without lesioning adjacent structures, i.e., VTA [17] a very challenging task and restricting its application to very rare experimental designs.

6 Neonatal 6-OHDA Injections

To produce a bilateral degeneration of the DAnergic nigrostriatal pathway in neonatal rats, 6-OHDA solution is injected transcutaneously into both lateral ventricles on postnatal day 1 (P1) [19–21]. Anatomical landmark to define the correct coordinates is the bregma, in this developmental stage still visible through the skin. Other than in the adult MFB-lesion model, this bilateral lesion surgery does not result in severe akinesia and severe sensorimotor deficits. In most cases, the lesioned neonatal rats can be raised by their mother without additional support; however, it has also been observed that the mother can reject lesioned pups in which case the affected pups need to be hand feed until they gain autonomy.

7 Rodent Model of HD

HD is a hereditary progressive neurodegenerative disorder characterized by the development of emotional, behavioral, and psychiatric abnormalities; loss of previously acquired intellectual or cognitive functions; and motor disturbances. The gene involved in HD, called huntingtin (Htt) or Interesting Transcript 15 (IT15) gene (historically known as HD gene), is located on the short arm of chromosome 4 (4p16.3) [22] The sequence of three DNA bases, cytosine–adenine–guanine (CAG) coding for amino acid glutamate [23] located at the 5′end of the HD gene show increased number of repeats corresponding to the onset of the clinical HD symptoms [24]. For CAG repeats equal to or greater than 35, the HD gene is thought to have 100% penetrance [25]. The neuropathological hallmark of HD is the loss of DARPP-32 positive, GABA-ergic striatal medium spiny projection neurons leading to the degeneration of the basal ganglia, and development of the clinical symptoms of the disease [26–28]. There are numerous rodent—mainly mice—transgenic models of HD but these models are best to study molecular pathways and mechanisms that are affected by the mutant huntingtin as they do not reliable reproduce the striatal pathology. Therefore, experimental modeling of the HD requires a selective depletion of this striatal cell subpopulation achieved by stereotactic, uni-, or bilateral injection of a neurotoxin into the CPu. The most widely used is an excitotoxic quinolinic acid (QA), an agonist of the *N*-methyl-d-aspartate (NMDA) receptor and an endogenous metabolite of tryptophan, selectively affecting striatal medium spiny neurons containing GABA, dynorphin, and enkephalin [29] causing striatal degeneration in rats [30, 31] similar to the one seen in human HD. For a variety of reasons detailed below, QA became the preferred excitotoxin for use in HD studies. Several experimental data favor an excitotoxic hypothesis as a possible pathogenesis of HD. It has been reported that in brains of HD patients, the activity of 3-hydroxyanthranilate oxygenase, which is the biosynthetic enzyme for QA, is increased [32] in parallel with reduced levels of kynurenic acid, an antagonist of NMDA receptors that may modulate QA-induced neurotoxicity [33]. Moreover, there are obvious similarities between pathology observed in the HD brain and in the QA model. In QA-induced striatal lesions, the levels of GABA and substance P, both of them localized in the medium spiny neurons, are reduced and the content of somatostatin present in the medium aspiny neurons is preserved [29]. In accordance with these neurochemical findings, some morphological studies in rats have shown that intrastriatal injections of QA selectively depletes medium spiny neurons with relative sparing of medium aspiny neurons and large neurons [29, 32]. Another reason for the widespread use of QA in HD research is that the cell death caused by the excitotoxin may closely resemble oxidative damage-related mechanism of neuronal death seen in HD brains [34–37]. QA injections in rodents produce the most reliable and morphologically reproducible lesions as well as motor, cognitive, and motivational impairments manifested by HD patients in early (but not later) stages of HD. Motor-related behavioral deficits in the rat model include: spontaneous and apomorphine-induced ipsilateral rotation behavior upon unilateral QA injection [38] due to the loss of the ipsilateral striatal projection neurons and the modulatory effect that the dopamine signaling has on the output neurones; animals also show hyperlocomotion, impairments in skilled paw use, and lateralized sensory motor deficits [39–42]. Cognitive functions, such as discriminative capability and response selection [43–45] as well as the water maze and the T-maze based tests [46–48], indicating impaired visuospatial skills and memory recall, are also impaired in rats with bilateral striatal lesion. At the morphological level, QA-induced neuronal loss leads to a dose-dependant progressive atrophy of the striatum, starting 4 weeks after the toxin injection. Depending on the lesion parameters, by 3–4 months post-lesioning, the volume of the entire striatum is reduced by around 30–70 % with concomitant enlargements of the ipsilateral ventricle [40].

Taken together, the QA-induced lesions of the striatum in rodents resemble some crucial neuropathological features of HD (Fig. 3a–f). The model is therefore suitable for exploring the feasibility of intrastriatal grafts to replace for lost striatal medium spiny neurons.

8 Other Stereotactic Procedures in Rats

Beside the stereotactic injections of the neurotoxic substances to selectively lesion certain brain structures, as described above; the stereotactic procedure may also be applied for intracerebral cell implantation. This technique is an important part of an methodological armamentarium of experimental neurooncology aiming at the implantation of malignant cells, derived either directly from the postoperative tumor specimen or from in vitro culture of the tumor-derived cell line, and regenerative neurosurgery, enabling the precise intracranial implantation of stem cells of different origin or predifferentiated precursors in order to reconstitute the pathologically changed cytoarchitecture of the recipients' brain or to deliver cells ectopically producing neurotransmitters and growth factors. With the development of new experimental fields the emergence of novel applications of the stereotactic procedures could be observed. This includes viral and short-hairpin RNA implantations for in vivo gene manipulation, microdialysis for studies of brain metabolism with stereotactically implanted probes, interstitial radiation, or implantation of electrodes for both registering of the electric activity of specific brain regions and their stimulation.

The general principles of the stereotactic procedure do not significantly differ in all of the above mentioned applications; therefore further description of the technique will be based on stereotactic lesioning and neurotransplantation in adult and neonatal animals.

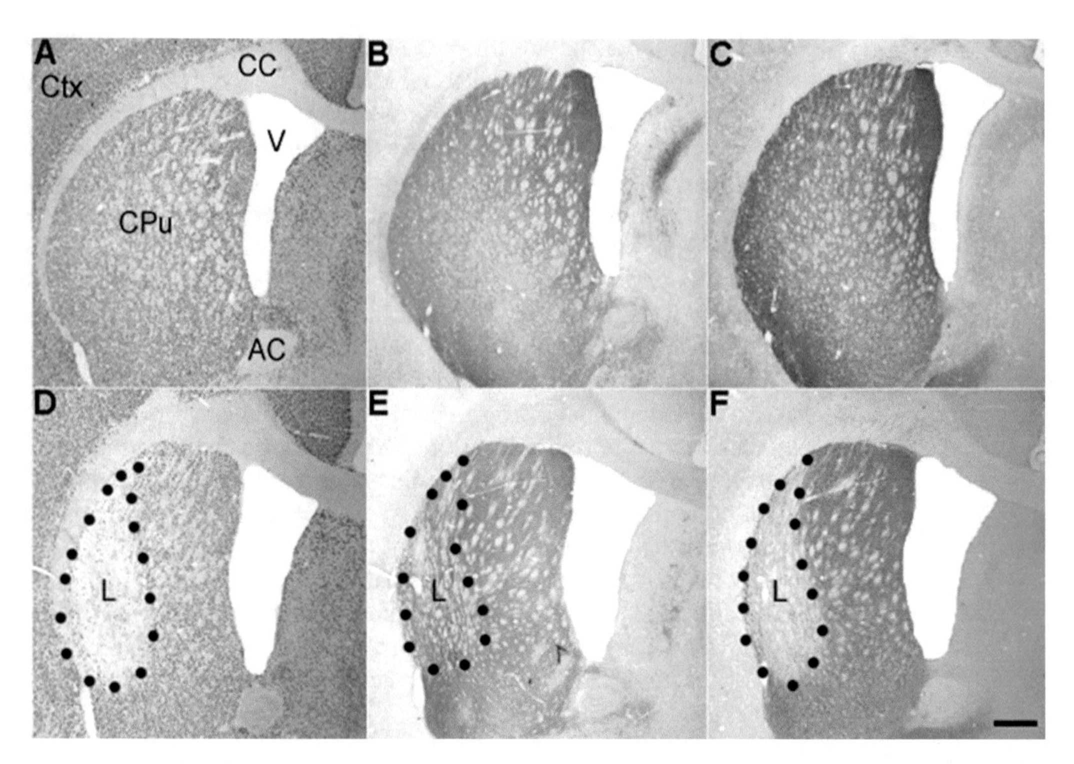

Fig. 3 (**a**–**f**) Following quinolinic acid (QA) lesion, sections from the control (**a**–**c**) and the lesion (**d**–**f**) groups are typically stained with a selection of markers specific for neurones (NeuN; **a**, **d**), dopaminergic afferents (TH, tyrosine-hydroxylase; **b**, **e**), and medium spiny neurones (DARPP-32; **c**, **f**). The control sections show regular staining, and no anatomical deformation. However, lesioned sections have enlarged ventricles, collapsed axons of passage, shrunk striatal tissue, a necrotic core, and exhibit a general reduction of staining in the striatum. *Ctx* cortex, *CC* corpus collasum, *V* ventricle, *AC* anterior commisure. Scale bar = 1 mm

9 General Stereotactic Surgical Procedure in Adult and Neonatal Rats

In the following paragraphs, the general stereotactic surgical procedure in adult and neonatal rats are explained.

10 Materials

The choice of experimental animal, its gender and age, will depend entirely on the experimental objective of the investigator. Typically, within a given investigation, surgical procedures are performed on young adult rats (200–250 g at the start of the study) housed under 12-h light/12-h dark conditions, temperature controlled facilities (22–24 °C), with food and water ad libitum, and up to five rats per cage. We advise, following the fixation of the experimental animal in the stereotactic apparatus, in order to optimize the precision of procedures to perform all steps under an operating microscope such as the SMED-Studer Medical, Engineering-AG, Switzerland Yasargil System, VM-900 as used in our laboratory.

All experiments must be carried out according to guidelines and regulations of the relevant local and national authorities, and it is the direct responsibility of the PI to be aware of these responsibilities.

10.1 Equipment

Disposable scalpel No. 10 (Feather Safety Razor, Japan), Wullstein retractors (No. 17018-11), adson forceps (No. 91106-12), MORIA forceps (Straight, No. 11370-40) MORIA forceps (Curved, No. 11370-42), Micro-Mosquito (Straight, serrated, No. 13010-12), Hartman Hemostatic forceps (No. 13002-10), Michel suture clips (No. 12040-02), applying forceps for Michel suture clips (No. 12018-12), Ear punch for animal identification (No. 24210-02) from FST (Fine Science Tools GmbH, Germany) catalog, a bone scraper and "surgical hooks" made from clipped and bent needles. small animal stereotactic apparatus, i.e., Stoelting stereotactic frame no. 51600 and Cunningham neonatal rat adaptor no 51625 (Stoelting, USA) [49], high speed microdrill: Proxxon Micromot 40 with small dental drill bits (Proxxon, Germany), 5 μl-calibrated borosilicate glass capillaries (i.e., BF100-50-7.5, Sutter, USA), micropipette puller (Sutter P-97, Sutter, USA), Hamilton microliter syringes 2 and 10 μl (Hamilton Europe, Switzerland), operating microscope (SMED-Studer Medical, Engineering-AG, Switzerland Yasargil System, VM-900). Isoflurane-vaporizer for neonatal surgeries, micro pump (World Precision Instruments Inc., UK), cotton swabs, a microdrill holder attached to the arm of the frame.

10.2 Reagents

Ethanol 70% vol/vol, anesthetics and analgesics (isofluran, ketamine, xylazine, lidocaine, buprenorphine), lubricant eye ointment (i.e., Dexpanthenol), sterile PBS, neuotoxins (6-OHDA, QA) for lesioning or cell suspension for neurotransplantation approaches, Borgal anibiotic solution 24% (Sulfadoxin + Trimethoprim) (Intervet, Germany). When choosing the anesthetic, the investigator needs to be aware that some products might interfere with the action of the neurotoxin. For example, ketamine, a noncompetitive NMDA antagonist, has shown to mitigate the lesion induced by QA, a glutamate analog acting on the NMDA receptor, as the two products have the same principle target [50].

10.3 Surgical Procedure

Prior to procedure the surgical area has to be cleaned and disinfected with 70% ethanol and the tools used either for lesioning or stereotactic intracerebral cell implantation should be sterilized by autoclaving or immersion in the ethanol solution (and then air dried).

10.3.1 Anesthesia

Using a vaporizer (Fig. 4a), the isoflurane solution is converted into its gaseous form and delivered to the animal by O_2 into an induction box (Fig. 4b) with isoflurane at a gas flow rate of approximately 5.0 vol% with 1.5 l O_2/min. It is essential that once the animals are induced—but still waiting for the surgery—the percentage of isoflurane is reduced from 5 vol% to around 3 vol% as keeping the animals too long on the higher percentage will result in possible severe respiratory impairment and significantly increase a perioperative risk of. In case of adult animals, standard intraperitoneal injection of 10 mg/kg i.p. Ketamine hydrochloride (Ketamine® 10% Essex Pharma GmbH, Germany) and 5 mg/kg i.p. Xylazine hydrochloride (Rompun®, Bayer AG, Germany) have been chosen. In our experience, it has never been necessary to neither intubate the rats nor to control the blood gases and body temperatures. For performing the QA intrastriatal injections, it is important to use inhalation anesthesia, i.e., with isofluran, usually induced at 5.0 vol% with 1.5 l O_2/min, and maintained at approximately 2.0 vol% with 0.8–1.0 l O_2/min, due to significant reduction of QA excitotoxic effect on the rat striatum when using ketamine, as described earlier [50]. During the surgeries performed on neonatal rodents, hypothermia has been the method of choice for reliably anesthetizing animals up to the eighth day of age. For this purpose, neonatal rats have been covered with crushed ice for approximately 5–7 min depending on size of the animal (usually 1 min/g body weight suffices). Following the induction the hypothermic anesthesia can be safely maintained for up to 30 min by adding 10 g of dry ice every 10 min to a 50% ethanol bath into the reservoir of the Cunningham neonatal rat adaptor no. 51625 (Stoelting, USA) [49]. This will maintain the temperature of the instrument at approximately 5 °C. For longer surgical procedures, the animal should be removed from the stereotactic apparatus after 20–30 min and warmed up to the point of slight responsiveness to nociceptive stimuli (i.e., a pinch of the tail or paw). Afterwards, the hypothermia can be induced again as described above and the animal repositioned in the apparatus for the next phased of the surgery. For surgery in older neonates (beyond P10), the animals have been anesthetized with isoflurane/oxygen-ventilation: 2.5 vol% of isoflurane given with 3 l O_2/min. Some authors advocate for administration of 0.02 mg/kg body weight of atropine given subcutaneously prior the anesthesia to reduce bronchial secretion and improve breathing.

The effectiveness of the induction and depth of anesthesia can be monitored by the responsiveness of the animals to nociceptive stimuli, as described above.

Fixation of the Experimental Animal in the Stereotactic Apparatus

In anesthetized animals, the fur on the skull has to be shaved and the skin disinfect with 70% ethanol solution. Afterwards, one ear bar of the apparatus should be fixed in the stereotactic frame (Stoelting stereotactic frame no. 51600) and the animal's head should be gently positioned so, that the ear canal is lead onto the fixed ear bar. Keeping the head of the animal without changing position the second ear bar should be introduced into the ear canal to complete the fixation (Fig. 4c, d). It is important to apply only moderate pressure and nonrupture ear bars with wide angle-tip in

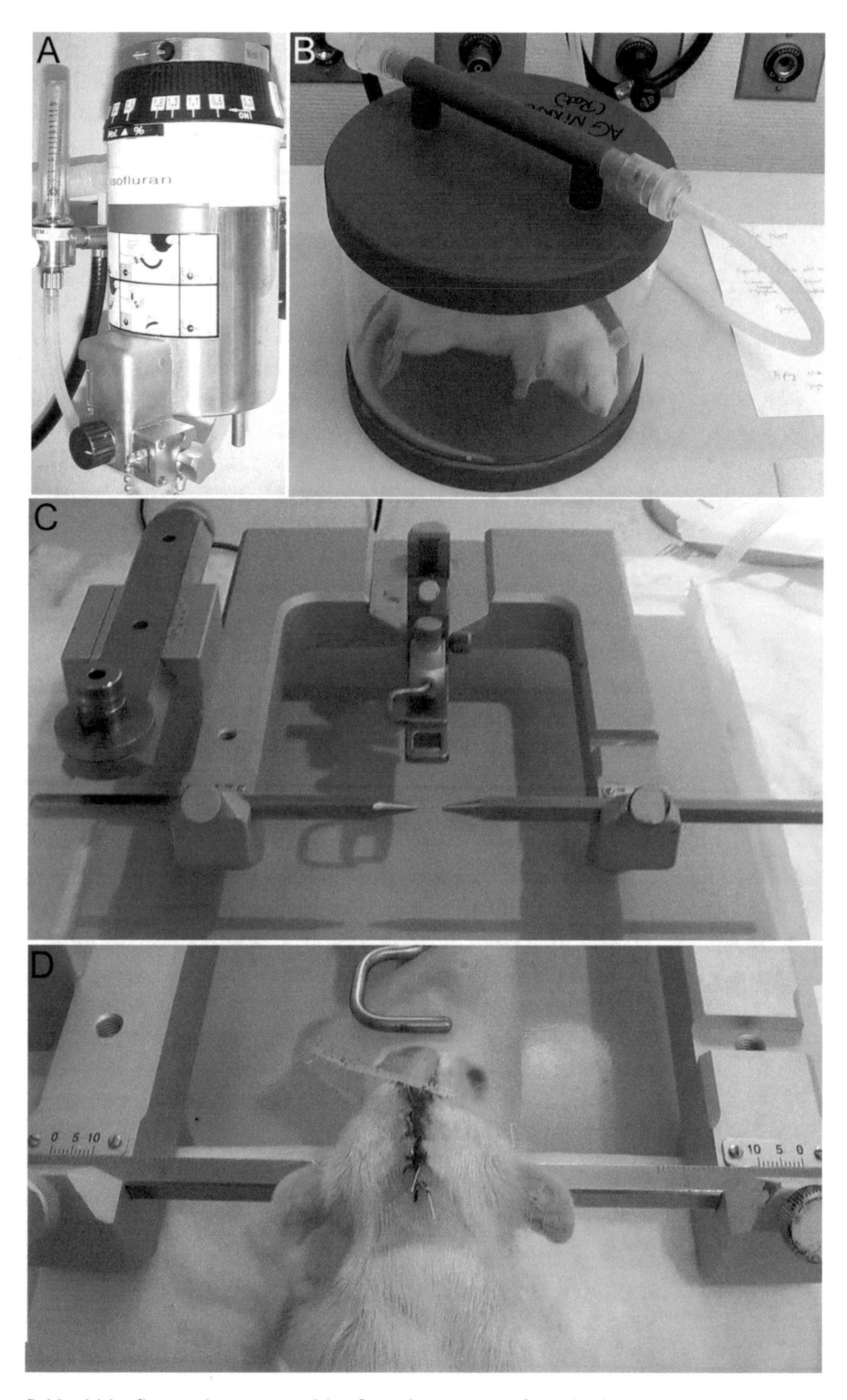

Fig. 4 (**a–d**) Liquid Isoflurane is converted by O_2 to its gaseous form in the vaporizer (**a**) and carried to the induction box (**b**) by plastic tubing. The principal components of a stereotactic frame are two ear bars that move laterally and a tooth bar that is moveable backwards and forwards; the adjustable arm to which the drill, the lesioning, or the transplantation equipment can be attached is not depicted here (**c**). The head of the anesthetized animal is fixed in the frame using the ear bars and the tooth bar with the nose clamp gently fastened (**d**). More detailed description can be found in the text

order to avoid injuring of the tympanic membrane. In case of continuation of the gas anesthesia, as soon as the head is fixed in the ear bars, the gas inlet needs to be secured close to the animal's nose to ensure that the required level of anesthesia is maintained. Correct fixation positions the head of the animal horizontal and symmetrical to the ear bars as shown in Fig. 4d and enables its free vertical movement precluding at the same time movements lateral to the ear bar axis. Further, the lower jaw of the animals should be gently pull down with small forceps to allow the incisor adapter (tooth bar) be introduced into the animal's mouth deep enough to place the incisors in the opening of the adapter. After that, we suggest to pull the adapter slightly backward to exert traction on the animal's head, which significantly improves the stability of fixation. For targeting striatum either for neurotoxin injections (QA HD model or terminal 6-OHDA lesion) or for cell implantation tooth bar is usually set at 0.0 mm. In case of MFB lesion, a "flat skull position" with tooth bar set at +3.4/−2.3 mm for the first and second trajectory, respectively, is necessary. The final step of mounting the animal into the stereotactic apparatus is the fixation with the nose clamp applied with a very low pressure on the animal's nostrils. In some cases, especially when working with very young and small animals the nose clamp could be completely omitted. Finally, in order to prevent the obstruction of the upper respiratory tract of the animal and secure unproblematic breathing throughout the procedure the tongue should be pulled out and aside.

10.3.2 Craniectomy, Coordinates, and Stereotactic Injection

After proper fixation of the animal in the stereotactic apparatus, we routinely apply the operating microscope for further steps of the procedure to maximize precision. Using scalpel a midline incision of 1–2 cm exposing the bregma and lambda as anatomical landmarks should be made (Fig. 5a). The subcutaneous tissue should be carefully removed using a small bone scraper and margins of the wound should be retracted leaving the skull exposed. It is important to keep the skull moist with sterile PBS throughout the surgery, as mentioned above. The horizontal position of the skull depends on the position of the tooth bar and differs according to performed procedure. The most critical step for calculating the coordinates is the proper measuring of the x and y coordinates of the bregma. For this purpose, the tip of the Hamilton syringe or the tip of the drill bit, mounted to the holder arm of the stereotactic frame has to be lowered to the level of the skull pointing the intersection of the coronal and sagital sutures (Fig. 5b and inset). The coordinates of this point can be read from the x (anterior–posterior, AP) and y (mediolateral, ML) arms the frame. To calculate the coordinates of the cannula entry point for further craniotomy, the coordinates of skull entry point, as determined from a stereotactic brain atlas, has to be added to the coordinates of the bregma. The standard coordinates for stereotactic targets used in our laboratory are listed in Sect. 11. In the next step, the skull over the target area is going to be thinned using the high speed drill, usually leaving a thin bonny lamella through which blood vessels and the dura are visible. It is important not to drill through the bone, as it would probably cause an injury to the surface of the brain. To remove the last part of the skull, we use self-made "surgical hooks" from clipped and bent 27G needles enabling the elevation of the carefully perforated edges of the craniotomy and their removal with fine forceps. Afterwards, applying the same "surgical hooks" very careful perforation of the dura is going to be performed. An alternative to holding the drill in the hand is to have it attached with an adaptor to a stereotactic arm using the tip of the drill bit to locate the bregma, measure out the appropriate

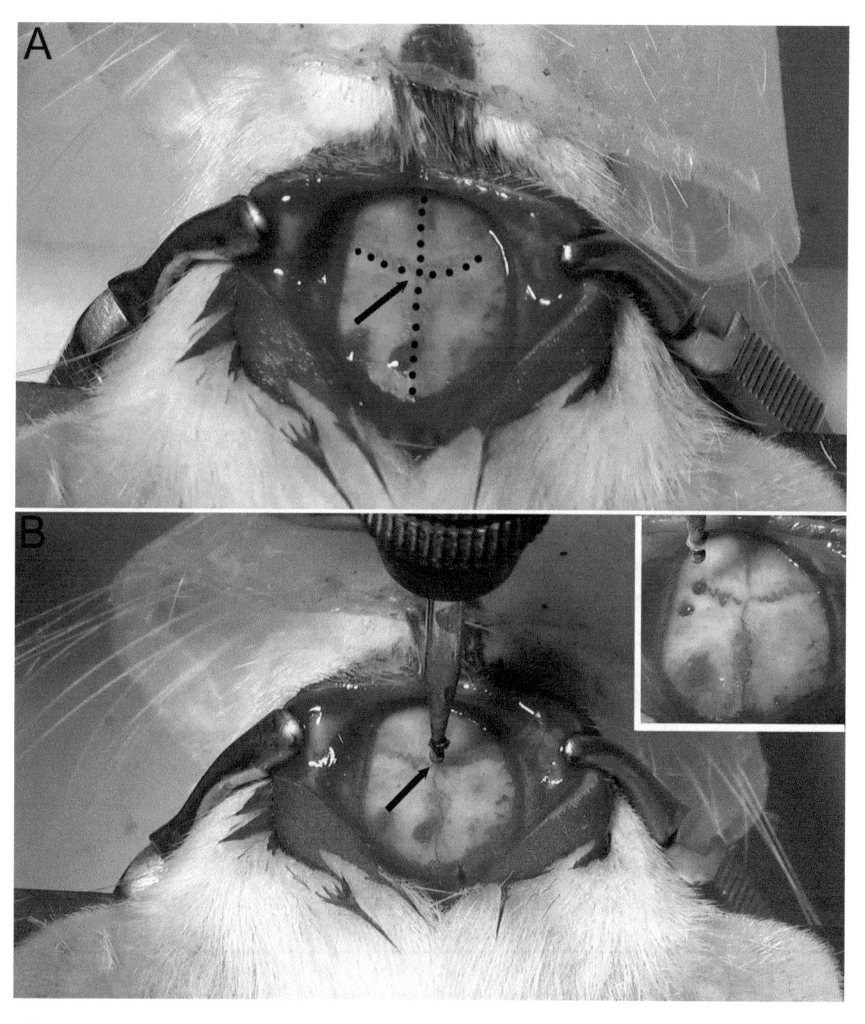

Fig. 5 (**a**, **b**) The exposed skull reveals the skull plates that join up at the bregma (at the intersection of the *dots*, **a**). The bregma is used as the point of reference for the anterior–posterior and the medial–lateral coordinates. If using a fixed drill with a fine drill bit, the burr holes at the required coordinates can be measured out with the drill (**b**, and *inset*)

coordinates, and make the burr holes (Fig. 6a). Further steps of the stereotactic procedure depend on the type of surgery. In case of the injection of neurotoxins for either PD or HD model, a 10-μl Hamilton syringe with a 30-gauge steel cannula, mounted to the holder of the stereotactic apparatus is going to be applied. After filling the Hamilton syringe with the neurotoxin solution, the tip of the attached needle is lowered to the level of the dura, which is a reference for the *z*-axis, i.e., the DV coordinate of the target. Afterwards, the needle is slowly introduced into the brain parenchyma and a deposit of the neurotoxin is injected with an injection rate of approximately 2 μl/min, although this is a parameter that depends on the discretion of the investigator. Using a minipump system for the toxin injection is an option that can improve consistency (Fig. 6b–f). The cannula should be then held in place for 3 min before retraction to prevent a retrograde flux of the neurotoxin along the trajectory canal. Preparation of 6-OHDA is described in detail in Sect. 11 of the chapter. To prevent oxidation, 6-OHDA solution needs to be kept in dark on ice being made up fresh from powder after every 3 h of surgery. The cannula needs to be reloaded with fresh toxin after each animal. Similar to the 6-OHDA, details concerning the preparation and handling of QA is described in Sect. 11. QA is more stable then the 6-OHDA solution, nevertheless similar precautions are taken such as protecting it from light and keeping it on ice. QA can be made up and aliquoted in units of 50 μl up to 12 months in advance if stored at −20 °C. If kept on ice, a single aliquot can be used for an entire lesioning session but then must be disposed of. To ensure consistent toxin quality throughout the day, the lesion cannula needs to be reloaded between each animal.

10.3.3 Transplantation

Implantation of the cell suspension differs in some steps significantly from described above, standard lesioning procedure. One of the most critical phases is the preparation of the tissue for grafting. Depending on the experimental paradigm graft can be composed of pieces of the tissue of interest or be prepared as a cell suspension [51]. The latter requires usually enzymatic and mechanical dissociation of the tissue/cell culture. The types of enzymes, length of incubation, and subsequent mechanical separation of cells depend strictly on the cell type and usually have to be determined empirically prior to implantation, and the reader needs to refer to key publications (for example, [15]). Due to the relatively low rate of cell survival following the stereotactic implantation, especially in case of dopaminergic precursors it is important to monitor the viability of the single cell suspension, i.e., according to standard Trypan blue exclusion method or using automatic cell counters. This parameter seems to be critical for the survival of grafted cells, so that the viability of the sample amenable for transplantation in our laboratory must not be lower then 90–95 %. After the counting, the cells are resuspended in a desired volume of the transplantation

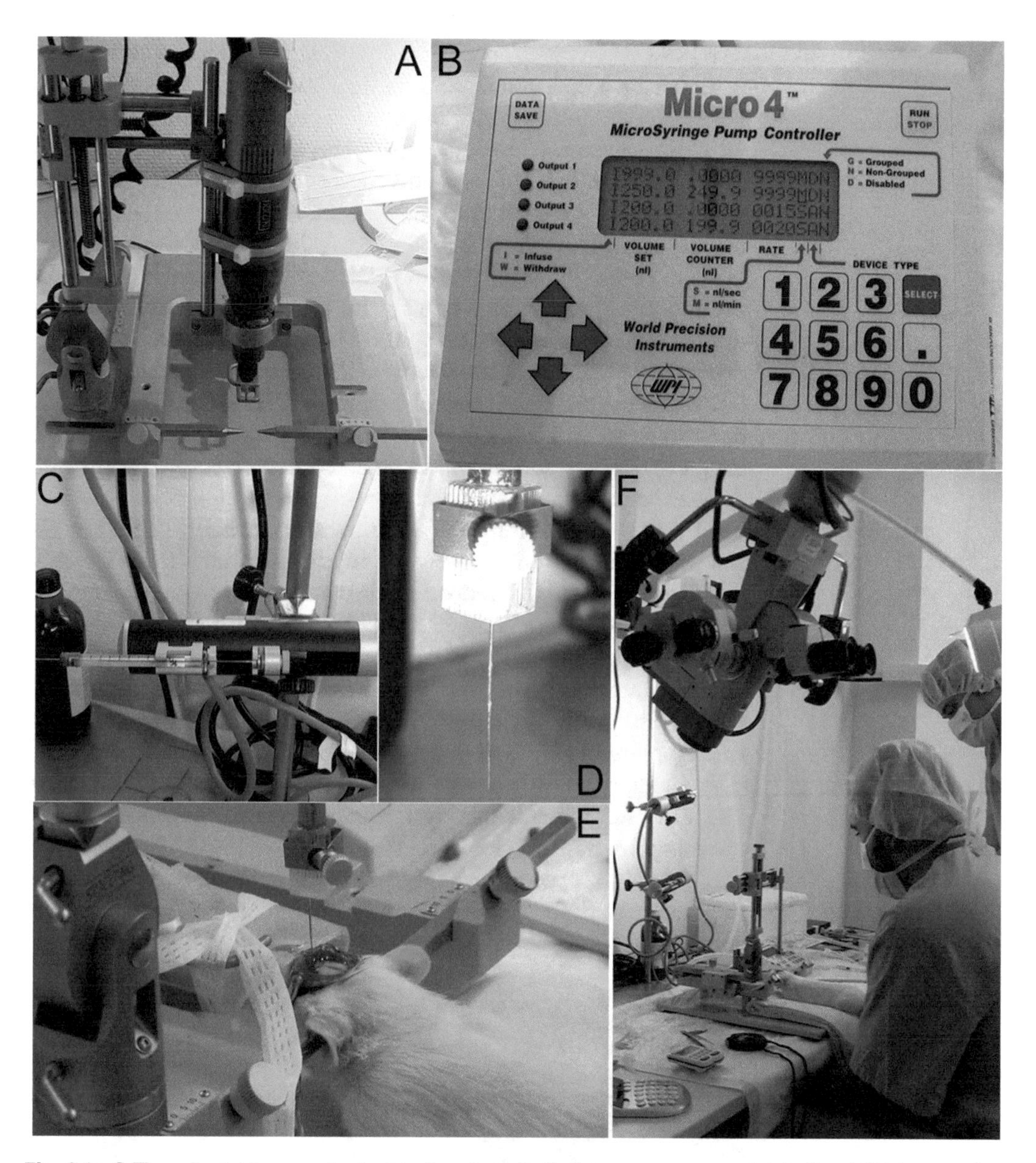

Fig. 6 (**a–f**) The adjustable arm attached to the stereotactic frame can accommodate the drill (**a**), as well as other instrument. In the case of QA lesions, a micropump (**b**) is used to exert precise pressure onto the plunger of a 10 μl Hamilton syringe (**c**) which has a 280 μm thick (internal diameter) polythene tube filled with saline. The 30 gauge lesioning cannula that penetrates the brain is attached to the adjustable arm (**d**, **e**). To ensure precision during the all surgical procedure, the use of a microscope is recommended (**f**)

solution (usually cell culture medium in case of serum-free growth conditions or Hank's balanced salt solution—HBSS—for cells cultivated in serum supplemented medium) to the final density. In order to prevent a reaggregation of the cells that may cause a plugging of the syringe precluding the reproducible implantation of cell deposits 0.05% DNase is routinely added to the transplantation solution in our laboratory. Standard cell implantation procedure requires a 2 μl Hamilton microsyringe with 26-gauge steel cannula. After performing a craniectomy as described above, the cell suspension is slowly drawn up to fill the syringe with the desired volume plus 10% of the total syringe capacity. Following that the syringe is lowered to the level of dura and the DV coordinates of the target are calculated. After perforating the dura with the bevel of the small needle syringe, cannula is slowly introduced into the recipient's brain to reach the required vertical coordinate and left at this position about 1 min before starting injection. In the next step, the desired volume of the cell suspension is injected at a rate of approximately 0.5 μl/min and the needle is left thereafter in place for additional 2 min prior the careful withdrawal. If more than one cell deposits is to be placed at different depths along the same trajectory, the deepest one should be injected first, followed by the next deepest, etc. [52].

In order to minimize the brain trauma inevitably caused by the cell implantation and to allow the injection of small graft deposits ranging 50–500 nl in a reproducible manner a microtransplantation approach can be applied [53, 54]. The most crucial modification represents the introduction of the glass capillary connected to the end of blunt-end of the steel Hamilton syringe cannula using a cuff made of the polyethylene tubing (Fig. 7a, b [52, 55]). The glass micropipette of a desired diameter is prepared from the borosilicate glass capillary using a pipet puller (Sutter P-97). The temperature and time settings are usually empirically determined to obtain a micropipette having a long (8–10 mm) slowly tapering shank with final tip diameter of 50–75 μm. The tip of the pipette must be broken square at the level of the desired inner diameter, which is done

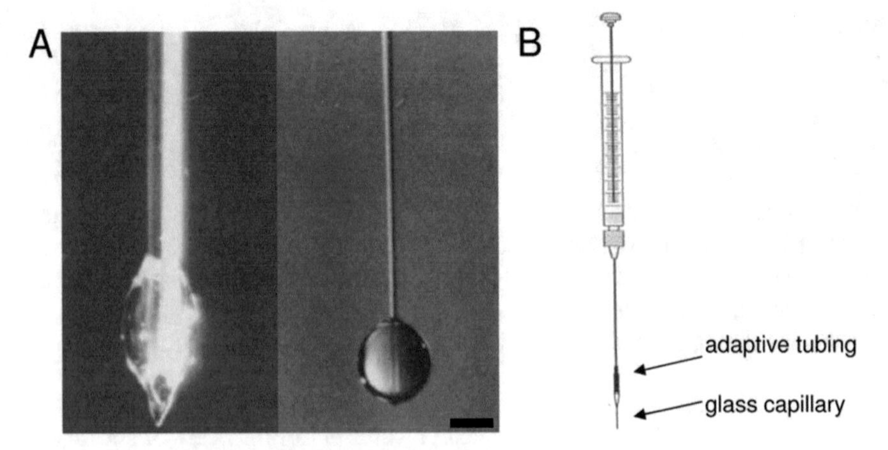

Fig. 7 (**a**, **b**) Cells can be introduced into the brain using either regular or microtransplantation method. The tips of the regular metal Hamilton cannula (*left side*, 500 μm outer diameter) and the glass capillary (*right side*, 50–70 μm) are depicted with 1.0 μl of medium being extruded (**a**). The microtransplantation instrument consists of a 2 μl Hamilton microsyringe fitted with the glass capillary using a cuff of polyethylene tubing as an adapter (**b**). Scale bar = 500 μm

easily under dissecting microscope. Therefore, the tip of the capillary can be tailored to accommodate various types of cell suspensions. After assembling of the whole system, it has to be completely filled with fluid (e.g., sterile saline or implantation medium) and devoid of air collections by removing the plunger and backfilling the syringe to eliminate dead space. A special care must be taken no to damage a very fragile tip of the glass capillary and to keep its lumen open. For the latter, any contact of the tip with the blood should be avoided. Following retraction from the brain the capillary should be immediately rinsed with the implantation medium to counteract the aggregation of the cell in the tip. This microtransplantation approach has significantly improved the grafting procedure in the iso- (e.g., [55, 56]), allo- (e.g., [57]), and xenografting [58] paradigms in the rat model.

At the end of the session, the steel cannula or the glass capillary is gently removed, the exposed area carefully rinsed with the sterile saline solution. While still in the earbars, the scalp wound is closed either with the wound clips (adult animals) or 7-0 nylon suture (newborn rats). Antibiotic ointment on the wound can be added to prevent postoperative infection.

10.3.4 Neonatal Transplantation

Implantation of cells into a neonatal recipient is routinely performed using the microtransplantation technique, but requires some modifications of the procedure. As already mentioned, neonatal animals are to be operated under hypothermic anesthesia using a Cunningham neonatal rat adaptor (Stoelting, USA) (Fig. 8). Like as for adult rats, the midline skin incision should be made before fixation in the stereotactic apparatus and the skin, together with connective tissue should be pull downward at both sides of the skull to expose the premature external acoustic meatuses consisting of delicate tube-like membraneous part connecting the earbud to its cartilaginous part located vetral and anterior to the transverse and occipital sinuses easily seen through the skull. The tips of the ear bars should be gently inserted into the cartilaginous external acoustic meatus until the resistance is felt. Please note that applying an excessive force may distort the animal's head indicated by the disappearance of the blood, particularly form the transverse sinus, due to the sinus constriction. Next, the head is mounted in the apparatus by inserting the mouthpiece and tightening the nose clamp. Finally, the head positioning such as the bregma and lambda have the same vertical coordinate, and the points 3 mm on either side of the lambda are localized on the same horizontal plane, has to be accomplished. The calculation of the coordinates of the site of interest does not differ significantly from the one described for adult recipients. However, due to the fact that the cranial sutures used to determine the bregma and lambda point are less distinct in newborns than those in adults, the

Fig. 8 The Cunningham adapter shown with a neonatal rat. The adapter is used to allow the stereotactic intervention on neonatal rodents too small to operate on with the adult setup

consistence in choosing the reference point is absolutely essential and needs practice. Further steps of cell implantation are performed similarly to the procedure described for adult rats.

Key Recommendation Relating to Lesion and Grafting Surgery

Independently of the applied stereotactic procedure, i.e., either neurotoxic lesion or transplantation using the standard or microtransplantation approach in neonatal or adult animals, before performing an experimental intervention the investigator should carry out the following preparative steps:

1. Determine the lesion or transplantation coordinates of the target using an appropriate standard stereotaxic atlas.
2. Carry out pilot runs of the lesion, and or transplantation to ensure correct preparation of the toxin, the cells, and to validate the stereotactic coordinates used. This is particularly important when working with neonatal animals. Lesion coordinates can also be practiced by injecting a dye, ink, or Trypan blue. After sacrificing of the animal, the brain can be sectioned to visualize the lesion sites, the dyes, or the implanted cell deposits. Any misplacements of the lesion or graft can be then easily measured and corrected.

10.3.5 Postoperative Care

In general, the postoperative complications occur rarely after stereotatic procedures in adult experimental animals, particularly in rodents. After the surgery rats have to be kept warm. In our institution, we apply routinely a heat lamp during the recovery from the anesthesia. Furthermore, the breathing pattern of the animal should be carefully observed. By respiratory arrest in many cases, a successful resuscitation can be performed, though this problem appears to be more common in neonatal rats during the rewarming period rather than in adults. Another important issue of the neonatal surgery is maternal neglect that can be prevented to certain extent by proper preoperative handling of the animals. Special attention has to be paid to adequate analgesic treatment after the surgery. We use routinely 0.05 mg/kg body weight of buprenorphine (Temgesic) applied subcutaneously with the first injection prior to regaining consciousness. Additionally, to avoid postoperative dehydration 30 ml/kg body weight of sterile saline should be injected as a subcutaneous deposit. The sufficient food intake during the first phase of the recovery should be facilitated using moist food pellets put on the dishes inside the animal's cage for easy food access. The clinical status of the operated animals should be closely monitored with special attention to any signs of distress.

11 Notes

1. Typical parameters for the 6-OHDA lesion: Preparation, doses, coordinates

   The lesion coordinates are set according to bregma as a reference point for the AP and ML coordinates and the dura as reference for the DV coordinate, using the rat brain atlas Paxinos and Watson 1998.

   Powder form of 6-OHDA is weighed out and stored in 1.5 ml Eppendorfs in the fridge prior to use. When required for the lesion, appropriate amount of the toxin is freshly made up with 0.2 % ascorbic acid to a working dilution of 3.6 µg/µl. The solution is protected from light and kept on ice.

2. *Medial Forebrain Bundle*:

   Typical dose : 3.6 µg/µl 6-OHDA in saline containing 0.2 % (w/v) ascorbic acid

Track #	TB (mm)	AP (mm)	ML (mm)	DV (mm)	Volume (µl)
1	-2.3	-4.4	-1.2	-7.8	2.5
2	+3.4	-4.0	-0.8	-8.0	3

The injection rate should be 1.0 µl/min and the cannula is kept in place for an additional 4 min before it is slowly retracted.

3. *Terminal/partial DA lesions, injected into the striatum*:

   Typical dose: 3 × 7 µg/µl (3.6 µg/µl 6-OHDA in saline containing 0.2 % (w/v) ascorbic acid)

Track #	TB (mm)	AP (mm)	ML (mm)	DV (mm)	Volume (μl)
1	0.0	+1.0	–3.0	–5.0	2
2	0.0	–0.1	–3.7	–5.0	2
3	0.0	–1.2	–4.5	–50	2

The injection rate should be 1.0 μl/min and the cannula is kept in place for an additional 4 min before it is slowly retracted.

4. Typical parameters for the QA lesions: Preparation, doses, coordinates

   Preparing the toxin at the appropriate pH is essential, as this ensures the complete resolution of the toxin which is generally purchased in powder form. Measuring the pH is done using litmus paper, and if the agent is prepared in a too small volume, one can dip a needle tip into the toxin and spot the needle onto the litmus paper directly.

   Under typical circumstances a stock solution of 0.12 M QA (molecular weight = 167.12) is prepared. The aim is to prepare 6.25 ml of stock solution using 125 mg of research grade QA.

   Dissolve 125 mg of QA in 750 μl PBS (pH 7.4), add 50 μl of 10 M sodium hydroxide.

   Sonicate the above solution for 15 min.

   Add 3200 μl PBS. The total volume at this stage will be 4 ml, and this is will permit the use of a pH meter.

   Add 50 μl of 10 M sodium hydroxide to bring the solution to pH 7.4; if pH needs to be adjusted use sodium hydroxide or concentrated hydrochloric acid.

   Add 2200 μl of PBS to obtain the required concentration of 0.12 M QA.

   Check pH again, and if needed adjust to pH 7.4.

   Aliquot 50 μl into Eppendorfs, label and store in freezer at –20 °C.

   If the required concentration is 0.09 M QA, then add 16.7 μl of PBS to a 50 μl aliquot of 0.12 M QA. The stock can be stored at –20 °C safely for 12 months; beyond this time point a new batch should be made up.

   The amount of QA injected is typically expressed either as "*X*" number of deposits of "*Y*" μl each of "*Z*" M (molarity): for example, four deposits of 0.2 μl of 0.12 M QA; or as "*X*" nmol (molality): for example, 96 nmol QA. Each deposit is infused with the micropump over 90 s, with 1 min between different vertical deposits, and a 3 min wait prior to removal of the cannula from the brain to eliminate/ reduce lesion damage due to toxin reflux.

   Typical coordinates of QA striatal lesion:

Track #	TB (mm)	AP (mm)	ML (mm)	DV (mm)	Volume (μl)
1	0.0	+1.0	+2.9	−5.0/−4.0	0.2
2	0.0	−0.4	+3.3	−5.07−4.0	0.2

Under this two track and two deposits/track protocol a total of four deposits of 0.20 μl (total of 0.8 μl) of QA is released into the striatum.

References

1. Horsley V, Clarke RH (1908) The structure and functions of the cerebellum examined by a new method. Brain 31:45–124
2. Paxinos G, Franklin K (2012) The mouse brain in stereotaxic coordinates. Academic Press, San Diego
3. Paxinos G, Franklin K (2006) The rat brain in stereotaxic coordinates. Academic Press, San Diego
4. Cenci MA, Whishaw IQ, Schallert T (2002) Animal models of neurological deficits: how relevant is the rat? Nat Rev Neurosci 3:574–579
5. Olanow CW (1993) A radical hypothesis for neurodegeneration. Trends Neurosci 16:439–444
6. Blum D, Torch S, Lambeng N, Nissou M, Benabid AL, Sadoul R, Verna JM (2001) Molecular pathways involved in the neurotoxicity of 6-OHDA, dopamine and MPTP: contribution to the apoptotic theory in Parkinson's disease. Prog Neurobiol 65:135–172
7. Paxinos G, Watson C (2006) The rat brain in stereotaxic coordinates. Academic Press, San Diego
8. McGeer PL, Itagaki S, Akiyama H, McGeer EG (1988) Rate of cell death in parkinsonism indicates active neuropathological process. Ann Neurol 24:574–576
9. Amalric M, Moukhles H, Nieoullon A, Daszuta A (1995) Complex deficits on reaction time performance following bilateral intrastriatal 6-OHDA infusion in the rat. Eur J Neurosci 7:972–980
10. Kirik D, Rosenblad C, Björklund A (1998) Characterization of behavioral and neurodegenerative changes following partial lesions of the nigrostriatal dopamine system induced by intrastriatal 6-hydroxydopamine in the rat. Exp Neurol 152:259–277
11. Creese I, Burt DR, Snyder SH (1977) Dopamine receptor binding enhancement accompanies lesion-induced behavioral supersensitivity. Science 197:596–598
12. Marshall JF, Ungerstedt U (1977) Supersensitivity to apomorphine following destruction of the ascending dopamine neurons: quantification using the rotational model. Eur J Pharmacol 41:361–367
13. Dunnett SB, Robbins TW (1992) The functional role of mesotelencephalic dopamine systems. Biol Rev Camb Philos Soc 67:491–518
14. Dunnett SB, Björklund A, Schmidt RH, Stenevi U, Iversen SD (1983) Intracerebral grafting of neuronal cell suspensions. V. Behavioural recovery in rats with bilateral 6-OHDA lesions following implantation of nigral cell suspensions. Acta Physiol Scand Suppl 522:39–47
15. Dunnett SB (2010) Chapter 55: neural transplantation. Handb Clin Neurol 95:885–912
16. Goto S, Hirano A, Matsumoto S (1989) Subdivisional involvement of nigrostriatal loop in idiopathic Parkinson's disease and striatonigral degeneration. Ann Neurol 26:766–770
17. Carman LS, Gage FH, Shults CW (1991) Partial lesion of the substantia nigra: relation between extent of lesion and rotational behavior. Brain Res 553:275–283
18. van Oosten RV, Cools AR (2002) Differential effects of a small, unilateral, 6-hydroxydopamine-induced nigral lesion on behavior in high and low responders to novelty. Exp Neurol 173:245–255
19. Bentlage C, Nikkhah G, Cunningham MG, Björklund A (1999) Reformation of the nigrostriatal pathway by fetal dopaminergic micrografts into the substantia nigra is critically dependent on the age of the host. Exp Neurol 159:177–190
20. Nikkhah G, Cunningham MG, Cenci MA, McKay RD, Björklund A (1995) Dopaminergic microtransplants into the substantia nigra of neonatal rats with bilateral 6-OHDA lesions. I. Evidence for anatomical reconstruction of the nigrostriatal pathway. J Neurosci 15:3548–3561

21. Nikkhah G, Eberhard J, Olsson M, Björklund A (1995) Preservation of fetal ventral mesencephalic cells by cool storage: in-vitro viability and TH-positive neuron survival after microtransplantation to the striatum. Brain Res 687:22–34
22. Gusella JF, Wexler NS, Conneally PM, Naylor SL, Anderson MA, Tanzi RE, Watkins PC, Ottina K, Wallace MR, Sakaguchi AY (1983) A polymorphic DNA marker genetically linked to Huntington's disease. Nature 306:234–238
23. (1993) A novel gene containing a trinucleotide repeat that is expanded and unstable on Huntington's disease chromosomes. The Huntington's Disease Collaborative Research Group. Cell 72:971–983.
24. Kieburtz K, MacDonald M, Shih C, Feigin A, Steinberg K, Bordwell K, Zimmerman C, Srinidhi J, Sotack J, Gusella J (1994) Trinucleotide repeat length and progression of illness in Huntington's disease. J Med Genet 31:872–874
25. Brinkman RR, Mezei MM, Theilmann J, Almqvist E, Hayden MR (1997) The likelihood of being affected with Huntington disease by a particular age, for a specific CAG size. Am J Hum Genet 60:1202–1210
26. Borrell-Pagès M, Zala D, Humbert S, Saudou F (2006) Huntington's disease: from huntingtin function and dysfunction to therapeutic strategies. Cell Mol Life Sci 63:2642–2660
27. Deng YP, Albin RL, Penney JB, Young AB, Anderson KD, Reiner A (2004) Differential loss of striatal projection systems in Huntington's disease: a quantitative immunohistochemical study. J Chem Neuroanat 27:143–164
28. Nakamura K, Aminoff MJ (2007) Huntington's disease: clinical characteristics, pathogenesis and therapies. Drugs Today (Barc) 43:97–116
29. Beal MF, Kowall NW, Ellison DW, Mazurek MF, Swartz KJ, Martin JB (1986) Replication of the neurochemical characteristics of Huntington's disease by quinolinic acid. Nature 321:168–171
30. Bordelon YM, Chesselet MF, Nelson D, Welsh F, Erecińska M (1997) Energetic dysfunction in quinolinic acid-lesioned rat striatum. J Neurochem 69:1629–1639
31. Ribeiro CAJ, Grando V, Dutra Filho CS, Wannmacher CMD, Wajner M (2006) Evidence that quinolinic acid severely impairs energy metabolism through activation of NMDA receptors in striatum from developing rats. J Neurochem 99:1531–1542
32. Schwarcz R, Okuno E, White RJ, Bird ED, Whetsell WO (1988) 3-Hydroxyanthranilate oxygenase activity is increased in the brains of Huntington disease victims. Proc Natl Acad Sci U S A 85:4079–4081
33. Beal MF, Matson WR, Swartz KJ, Gamache PH, Bird ED (1990) Kynurenine pathway measurements in Huntington's disease striatum: evidence for reduced formation of kynurenic acid. J Neurochem 55:1327–1339
34. Beal MF, Ferrante RJ, Swartz KJ, Kowall NW (1991) Chronic quinolinic acid lesions in rats closely resemble Huntington's disease. J Neurosci 11:1649–1659
35. Beal MF, Matson WR, Storey E, Milbury P, Ryan EA, Ogawa T, Bird ED (1992) Kynurenic acid concentrations are reduced in Huntington's disease cerebral cortex. J Neurol Sci 108:80–87
36. Ferrante RJ, Kowall NW, Cipolloni PB, Storey E, Beal MF (1993) Excitotoxin lesions in primates as a model for Huntington's disease: histopathologic and neurochemical characterization. Exp Neurol 119:46–71
37. Roberts RC, Ahn A, Swartz KJ, Beal MF, DiFiglia M (1993) Intrastriatal injections of quinolinic acid or kainic acid: differential patterns of cell survival and the effects of data analysis on outcome. Exp Neurol 124:274–282
38. Vazey EM, Chen K, Hughes SM, Connor B (2006) Transplanted adult neural progenitor cells survive, differentiate and reduce motor function impairment in a rodent model of Huntington's disease. Exp Neurol 199:384–396
39. Döbrössy MD, Dunnett SB (2003) Motor training effects on recovery of function after striatal lesions and striatal grafts. Exp Neurol 184:274–284
40. Döbrössy MD, Dunnett SB (2006) The effects of lateralized training on spontaneous forelimb preference, lesion deficits, and graft-mediated functional recovery after unilateral striatal lesions in rats. Exp Neurol 199:373–383
41. Döbrössy MD, Dunnett SB (2006) Morphological and cellular changes within embryonic striatal grafts associated with enriched environment and involuntary exercise. Eur J Neurosci 24:3223–3233
42. Döbrössy MD, Dunnett SB (2007) The corridor task: striatal lesion effects and graft-mediated recovery in a model of Huntington's disease. Behav Brain Res 179:326–330
43. Döbrössy MD, Svendsen CN, Dunnett SB (1996) Bilateral striatal lesions impair retention of an operant test of short-term memory. Brain Res Bull 41:159–165
44. Brasted PJ, Döbrössy MD, Robbins TW, Dunnett SB (1998) Striatal lesions produce distinctive impairments in reaction time performance in two different operant chambers. Brain Res Bull 46:487–493

45. Döbrössy MD, Svendsen CN, Dunnett SB (1995) The effects of bilateral striatal lesions on the acquisition of an operant test of short term memory. Neuroreport 6:2049–2053
46. Furtado JC, Mazurek MF (1996) Behavioral characterization of quinolinate-induced lesions of the medial striatum: relevance for Huntington's disease. Exp Neurol 138:158–168
47. Isacson O, Brundin P, Kelly PA, Gage FH, Björklund A (1984) Functional neuronal replacement by grafted striatal neurones in the ibotenic acid-lesioned rat striatum. Nature 311:458–460
48. Shear DA, Dong J, Gundy CD, Haik-Creguer KL, Dunbar GL (1998) Comparison of intrastriatal injections of quinolinic acid and 3-nitropropionic acid for use in animal models of Huntington's disease. Prog Neuropsychopharmacol Biol Psychiatry 22:1217–1240
49. Cunningham MG, McKay RD (1993) A hypothermic miniaturized stereotaxic instrument for surgery in newborn rats. J Neurosci Methods 47:105–114
50. Jiang W, Büchele F, Papazoglou A, Döbrössy M, Nikkhah G (2009) Ketamine anaesthesia interferes with the quinolinic acid-induced lesion in a rat model of Huntington's disease. J Neurosci Methods 179:219–223
51. Pruszak J, Just L, Isacson O, Nikkhah G (2009) Isolation and culture of ventral mesencephalic precursor cells and dopaminergic neurons from rodent brains. Curr Protoc Stem Cell Biol; Chapter 2:Unit 2D.5
52. Roedter A, Winkler C, Samii M, Walter GF, Brandis A, Nikkhah G (2001) Comparison of unilateral and bilateral intrastriatal 6-hydroxydopamine-induced axon terminal lesions: evidence for interhemispheric functional coupling of the two nigrostriatal pathways. J Comp Neurol 432:217–229
53. Nikkhah G, Cunningham MG, Jödicke A, Knappe U, Björklund A (1994) Improved graft survival and striatal reinnervation by microtransplantation of fetal nigral cell suspensions in the rat Parkinson model. Brain Res 633:133–143
54. Nikkhah G, Olsson M, Eberhard J, Bentlage C, Cunningham MG, Björklund A (1994) A microtransplantation approach for cell suspension grafting in the rat Parkinson model: a detailed account of the methodology. Neuroscience 63:57–72
55. Nikkhah G, Rosenthal C, Falkenstein G, Roedter A, Papazoglou A, Brandis A (2009) Microtransplantation of dopaminergic cell suspensions: further characterization and optimization of grafting parameters. Cell Transplant 18:119–133
56. Hahn M, Timmer M, Nikkhah G (2009) Survival and early functional integration of dopaminergic progenitor cells following transplantation in a rat model of Parkinson's disease. J Neurosci Res 87:2006–2019
57. Brandis A, Kuder H, Knappe U, Jödicke A, Schönmayr R, Samii M, Walter GF, Nikkhah G (1998) Time-dependent expression of donor- and host-specific major histocompatibility complex class I and II antigens in allogeneic dopamine-rich macro- and micrografts: comparison of two different grafting protocols. Acta Neuropathol (Berl) 95:85–97
58. Steiner B, Winter C, Blumensath S, Paul G, Harnack D, Nikkhah G, Kupsch A (2008) Survival and functional recovery of transplanted human dopaminergic neurons into hemiparkinsonian rats depend on the cannula size of the implantation instrument. J Neurosci Methods 169:128–134

Chapter 4

Rat Middle Cerebral Artery (MCA) Occlusion Models Which Involve a Frontotemporal Craniectomy

Hideaki Imai, Nobuhito Saito, and I. Mhairi Macrae

Abstract

In this chapter, we describe the technical approach for exposure and occlusion of the rat middle cerebral artery (MCA), focusing mainly on proximal electrocoagulation of the MCA, the Tamura model. This model requires training and expertise in microsurgical techniques so that the artery can be exposed and occluded without damaging the underlying brain tissue. However, once the required skills are acquired, a very reproducible ischemic insult can be produced with good recovery and low mortality.

Through extensive experience in the use of this model, we have modified the original Tamura model to make the surgery more straightforward and less invasive. In this chapter, we describe the MCAO procedure step by step, comprehensively noting the surgical preparation, body position, skin incision, craniotomy, dural incision, diathermy of the MCA, and the prevention of infection. We have also included a series of photographs of the surgical site at each step to facilitate training in the model.

Key words Rat, Middle cerebral artery occlusion, Focal cerebral ischemia, Rodent, Tamura model, Diathermy, Electrocoagulation, Endothelin-1, MCAO

1 Introduction

Animal stroke models are indispensable for both the investigation of the pathophysiology of cerebral ischemia and the evaluation of preclinical pharmacological intervention [1]. Most rodent models involve occlusion of the middle cerebral artery (MCAO) using either an intraluminal approach (e.g., Koizumi model) [2] or by exposure and direct surgical occlusion of the blood vessel (e.g., Tamura model) [3]. The most suitable experimental stroke model to use in a given study depends on the scientific question being investigated and each model has inherent advantages and disadvantages. The Tamura model, for example, has a robust advantage in terms of the reliability and reproducibility in the production of ischemic lesions in the ipsilateral middle cerebral artery (MCA) territory including the frontal cortex, dorsal parietal cortex, and the lateral part of the caudate–putamen. However, presumably due

Miroslaw Janowski (ed.), *Experimental Neurosurgery in Animal Models*, Neuromethods 116 vol. 116,
DOI 10.1007/978-1-4939-3730-1_4, © Springer Science+Business Media New York 2016

to the requirement for substantial microsurgical expertise, the use of this model, and others that require surgical exposure of the MCA, have become less prevalent compared to the intraluminal filament model. The Tamura model induces permanent MCAO while the intraluminal filament model can be adapted for either permanent or transient MCAO. However, because of the craniectomy, the Tamura model is associated with lower mortality than the intraluminal filament and other intact skull models.

In this chapter, we provide a full description of the neurosurgical procedures for direct exposure of the MCA in the rat including anesthesia, physiological monitoring, positioning, and neurovascular anatomy. Details are provided for MCA occlusion using electrocoagulation (proximal and distal MCA occlusion), endothelin-1 (ET-1), and mechanical devices (Fig. 1).

2 Animals

Sprague-Dawley rats are the strain of choice in the authors' laboratories. Adult male rats (e.g., from Charles River, Tsukuba, Japan) are maintained in controlled temperature (24 ± 1 °C) and humidity (55 ± 5 %) under a 12-h light: 12-h dark cycle with free access to rat chow and water (*see* **Note 1**).

3 Methods

3.1 General Surgical Preparation

For all surgical procedures, anesthesia is induced with isoflurane (4–5 %) and then subsequently maintained with 1.5–2 % isoflurane in nitrous oxide:oxygen (70:30) on a facemask. The rats are then intubated transorally (16 gauge intubation tube is suitable for rats of 250–350 g) and artificially ventilated by a small animal respirator pump with a tidal volume of 3–4 ml and respiration rate from 45 (surgical tracheostomy) to 60 (nonsurgical, oral intubation) breaths per minute. The right femoral artery is cannulated for continuous physiologic monitoring (Fig. 2). Arterial pressure is monitored throughout the experiment and arterial blood samples are taken at regular intervals for assessment of respiratory status using a direct-reading electrode system (Bayer, Newbury, Berkshire, UK). Rats are maintained normotensive (MABP > 80 mmHg), normocapnic ($36 < PaCO_2 < 44$ mmHg), and adequately oxygenated ($PO_2 > 100$ mmHg) while under anesthesia. Rectal temperature is maintained at 37 °C with a heating lamp or homeothermic blanket during the operation.

3.2 Surgical Exposure of the MCA

Good aseptic technique is essential in experimental stroke surgery and efforts should be made to ensure the operating area is as clean and sterile as possible. All surgical instruments should be sterilized

a. Proximal electrocoagulation

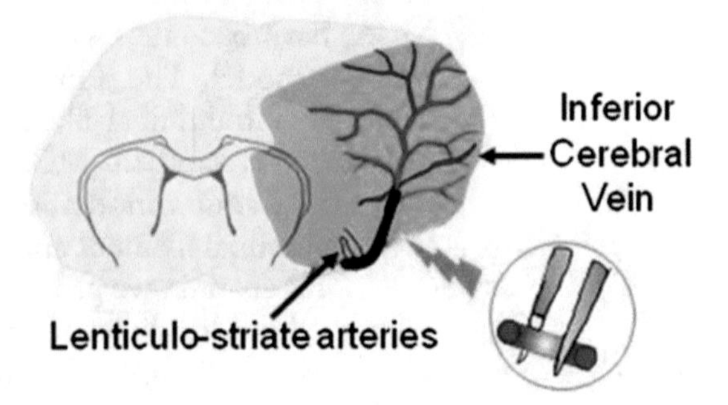

b. Distal electrocoagulation

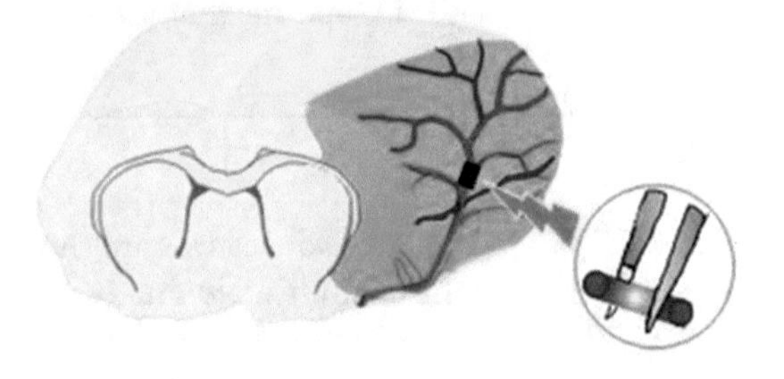

c. Topical application of endothelin-1

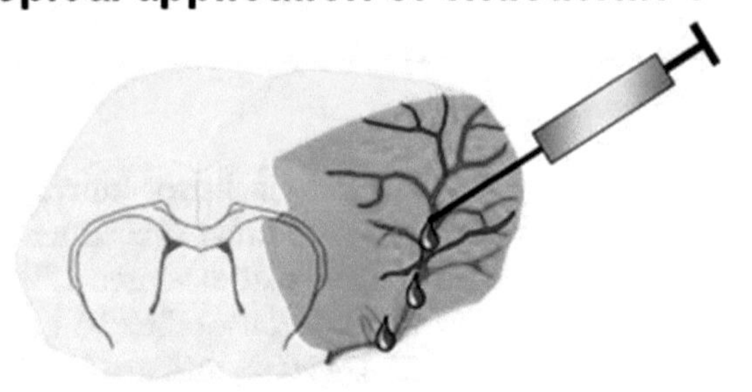

d. Clip or ligature occlusion

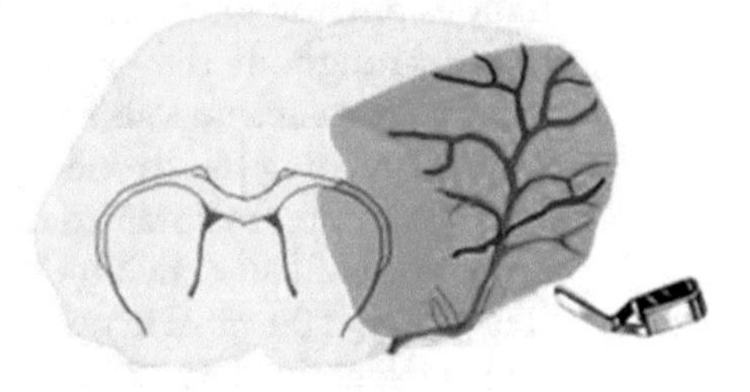

Fig. 1 There are several ways to occlude the MCA—by using neurosurgical procedures such as electrocoagulation (proximal (**a**) and distal (**b**) MCA occlusion), endothelin-1 (**c**), and mechanical devices (**d**)

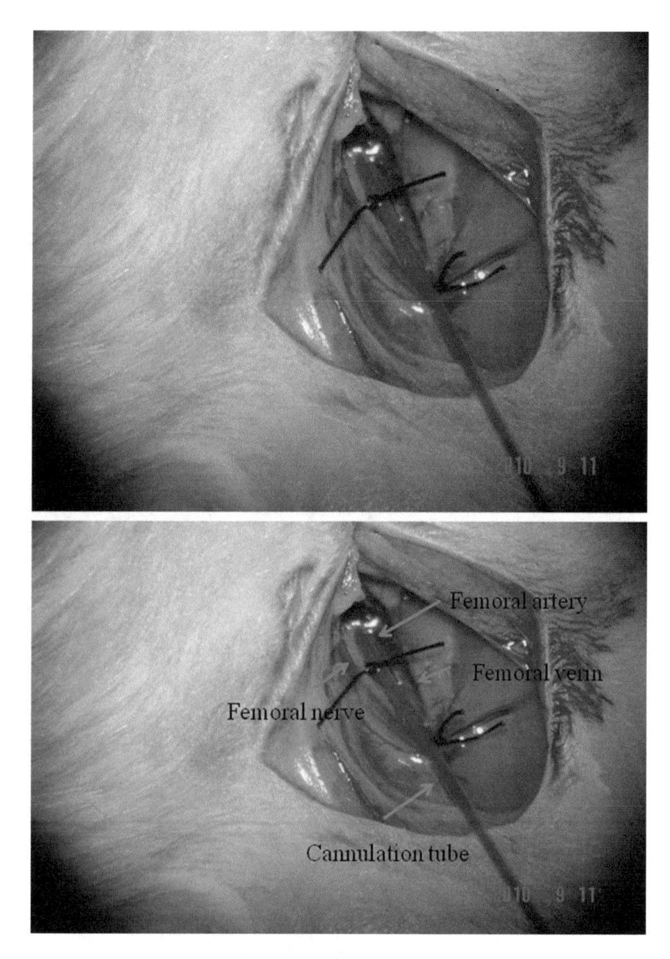

Fig. 2 The right femoral artery is cannulated for continuous physiologic monitoring. If needed, the femoral vein is also available for intravenous injection

prior to use and laid out on a sterile drape beside the animal. Sterile swabs, sutures, saline, etc., are employed. The use of a surgical operating microscope is recommended for the entire surgical procedure and the craniectomy site should be irrigated frequently with sterile saline or artificial CSF (*see* **Note 2**).

3.3 Position of the rat on the operating table

The animal is placed in the right lateral position (Fig. 3), raising the head 20° from the surgery table to allow a left frontotemporal approach (*see* **Note 3**). In order to protect the left eye, the eyelid should be closed by a suture or tape (Fig. 4). Fur around the planned skin incision should be shaved using an electric hair clipper and then a skin antiseptic solution applied (Fig. 4). Administration of a local anesthetic (1–2 mg/kg ropivicaine, bupivacaine) subcutaneously (line block) to the wound site prior to skin incision is recommended.

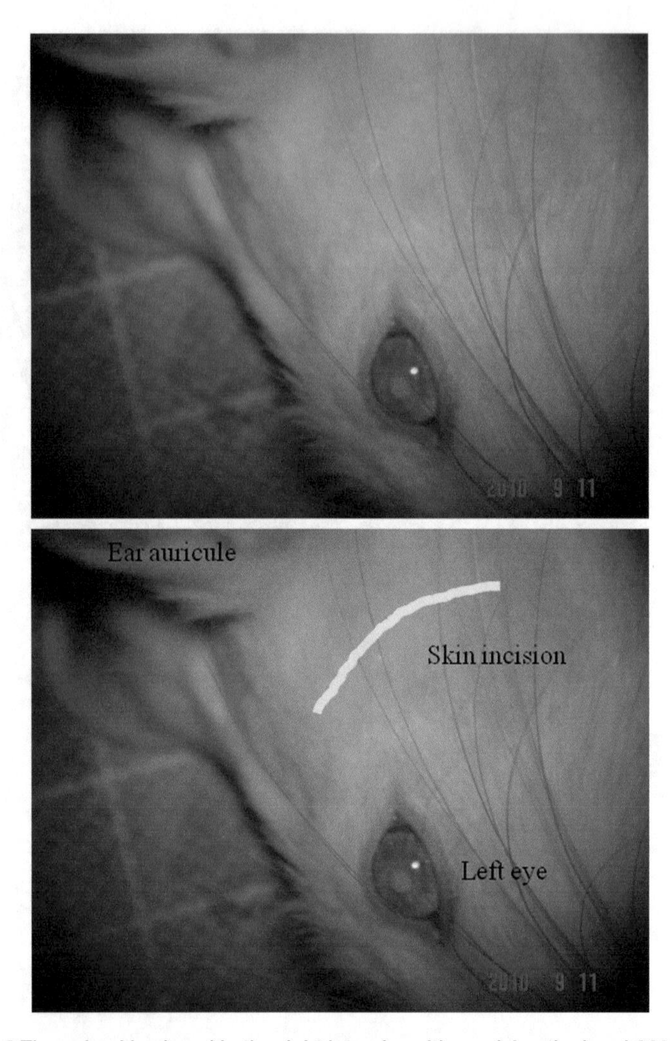

Fig. 3 The animal is placed in the right lateral position, raising the head 20° from the surgery table to allow a left frontotemporal approach

3.4 Skin Incision and the Way to Approach the Skull Base

A 1.5 cm vertical skin incision is performed between the left eyeball and ear auricule, using electrocoagulation to stem any bleeding (Fig. 5). The temporal fascia and muscle are incised just under the skin incision from the zygomatic arch to the linear tempolaris (top of the temporal muscle) using bipolar forceps for cutting with coagulation. If the surgical approach is correct, the coronoid process of the mandible (Fig. 6) is a good landmark and should emerge surrounded by the temporal muscle. After dissecting the temporal muscle, the coronoid process is taken away. The zygoma is not rongeured away. After splitting the temporal muscle and reflecting it to the rostral and caudal side, the surgical field is open to observe the temporal skull, root of zygoma, and zygomatic arch.

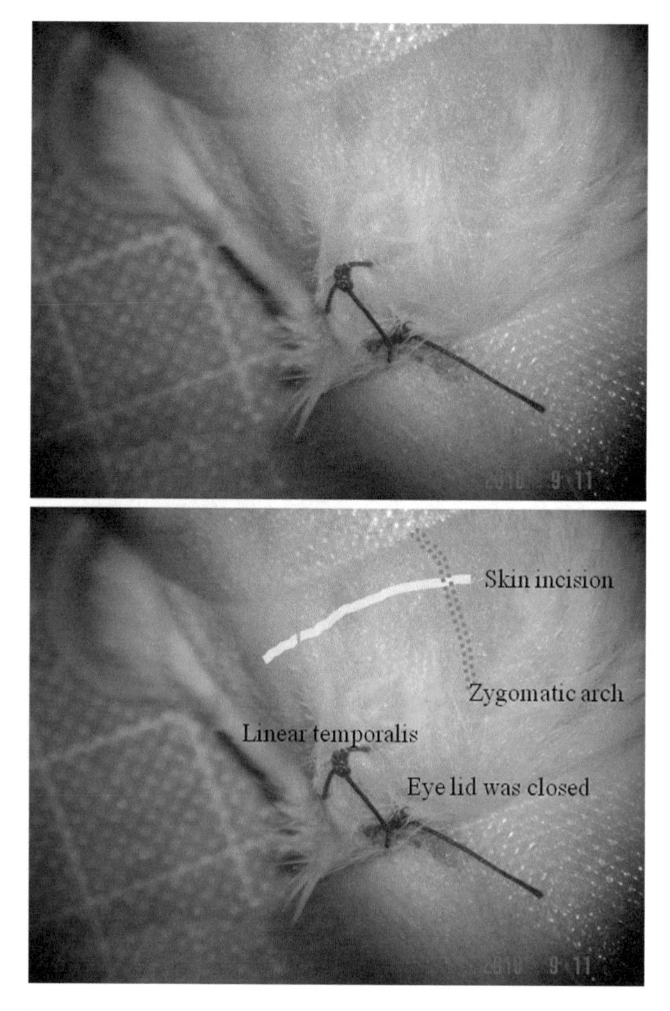

Fig. 4 The eyelid should be closed by a suture or tape in order to protect the left eye. Fur around the planned skin incision should be shaved using an electric hair clipper and then a skin antiseptic solution applied

3.5 Approach to the Middle Fossa: Expose the Frontotemporal Skull

The thin membrane is penetrated and dissected from the temporal skull base. Then just along the temporal muscle under the zygomatic arch, the mandibular nerve is visible from the foramen ovale. The craniotomy point of the temporal skull base between the foramen ovale and orbital fissure can now be identified (Fig. 7).

3.6 Craniectomy

A single entry burr hole is made with a dental drill. The temperature is controlled by irrigation with sterile saline solution to keep the dura matter intact. Micro forceps are used to expand the burr hole and to perform the frontotemporal craniectomy (Fig. 8). The lateral part of the temporal bone and temporal skull are removed with rongeurs. Complete removal of the bone ridge facilitates access from the proximal end of the MCA in the basal cistern to

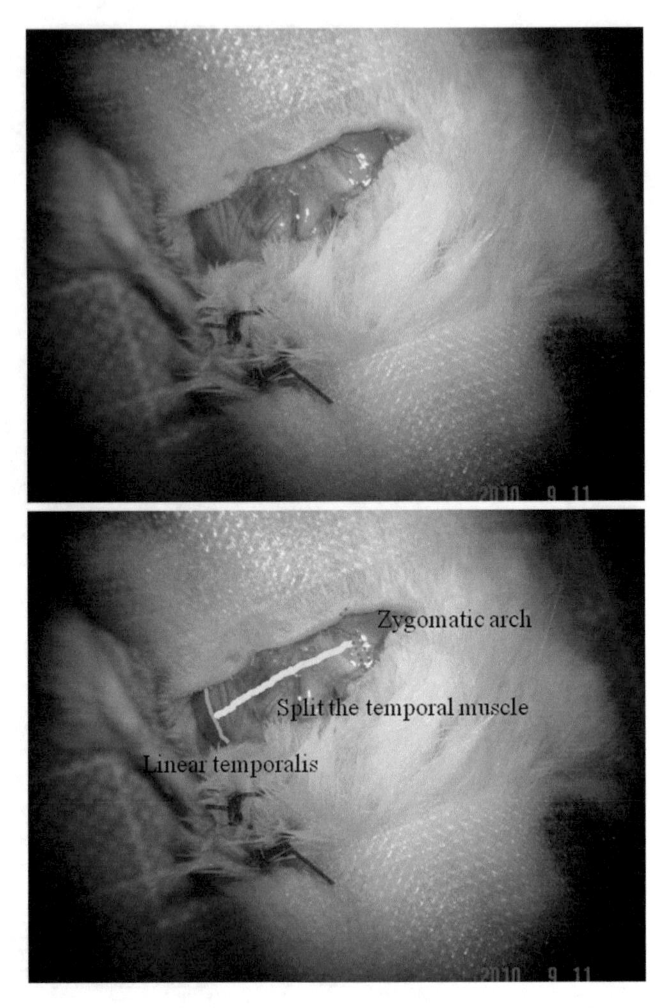

Fig. 5 A 1.5-cm vertical skin incision is performed between the left eyeball and ear auricule, using electrocoagulation to stem any bleeding. Then, the temporal fascia and muscle are incised just under the skin incision from the zygomatic arch to the linear tempolaris (top of the temporal muscle) using bipolar forceps for cutting with coagulation

the distal end of the MCA where it crosses the inferior cerebral vein (ICV).

The dura matter and arachnoid membrane are opened by perforating with a fine needle and retraction, exposing the full visualized brain within the craniectomy. CSF is released, thereby producing further brain exposure.

3.7 Exposure of MCA from Proximal to Distal Extent

The olfactory tract and ICV must be exposed as landmarks to definitively identify the MCA. Distally, the ICV crosses the MCA [at an angle of ~90°] and proximally the branching arteries of the MCA such as lenticulo striate artery (LSA) (Fig. 9).

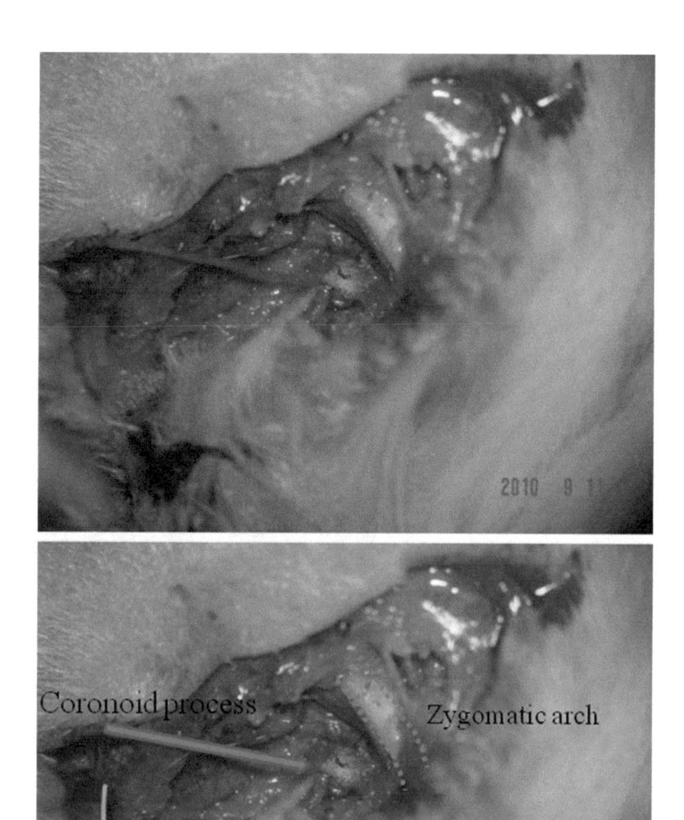

Fig. 6 After splitting the temporal muscle, the coronoid process of the mandible emerges surrounded by the temporal muscle and is a good landmark. After dissecting the temporal muscle, the coronoid process is taken away

3.8 Permanent Focal Cerebral Ischemia

3.8.1 Electrocoagulation of the MCA: Proximal MCAO (Fig. 1a)

Permanent focal cerebral ischemia is accomplished by occlusion of the MCA (*see* **Note 4**), as introduced by Tamura et al. [3] with some modification [4] as follows: For electrocoagulation of the main trunk of the MCA, the forceps are delicately inserted under the MCA at the olfactory tract. To facilitate this, you should first electrocoagulate the proximal perforating arteries and lateral striate arteries of the MCA with short bursts of current using the diathermy forceps. This will provide space to then place one side of the diathermy forceps between the MCA stem and brain surface, the other side on the surface of the MCA stem (*see* **Notes 5** and **6**).

After gently clamping the MCA, to reduce the blood flow through the blood vessel, electrocoagulation is exerted for a few seconds. This procedure is repeated several times until the blood flow is completely stopped. Then, using a segmental approach, the

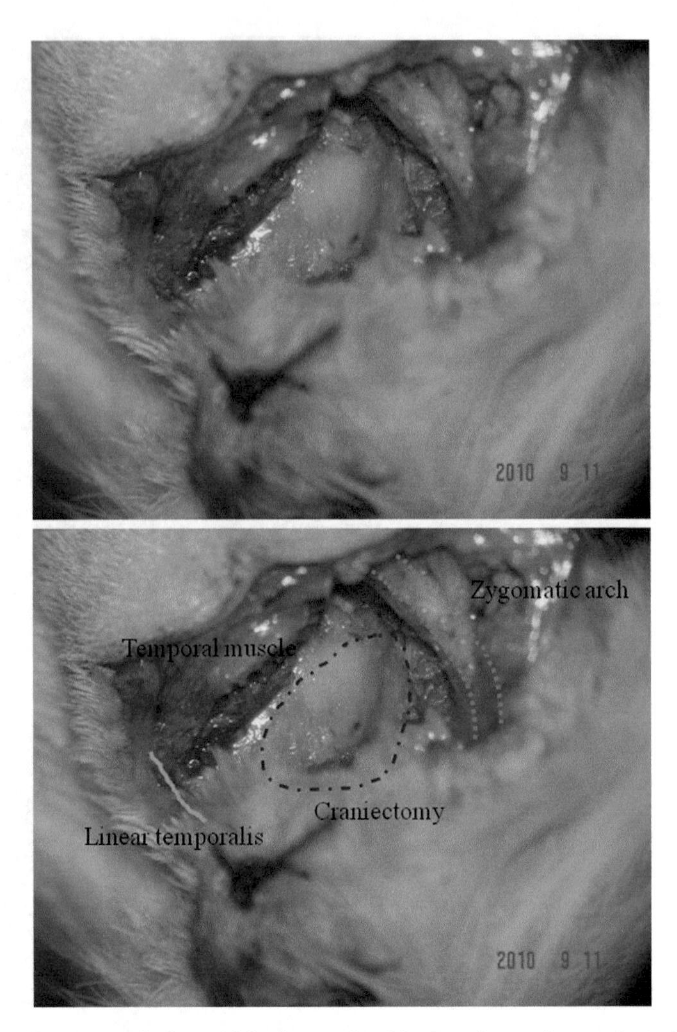

Fig. 7 To approach the middle fossa, the thin membrane is penetrated and dissected from the temporal skull base. The craniotomy point of the temporal skull base between the foramen ovale and orbital fissure can now be identified

main trunk of the MCA is electrocoagulated starting from the proximal MCA and working your way distally till you reach the intersection of the MCA with the ICV (Figs. 9, 10, and 11). The MCA is then transected at the level of the olfactory tract to ensure the completeness of the occlusion. Sham-operated controls undergo the same procedure to expose the MCA but the artery and its branches are not occluded.

3.8.2 Electrocoagulation of the MCA: Distal MCAO (Fig. 1b)

The model can be modified to reduce the size and location of ischemic tissue by applying electrocoagulation to a small, more distal portion of the MCA. For example, a short (2 mm) occlusion and transection, just distal to the ICV, will spare the caudate–putamen and confine ischemia to the cortex [5] (Fig. 1b). This variation is

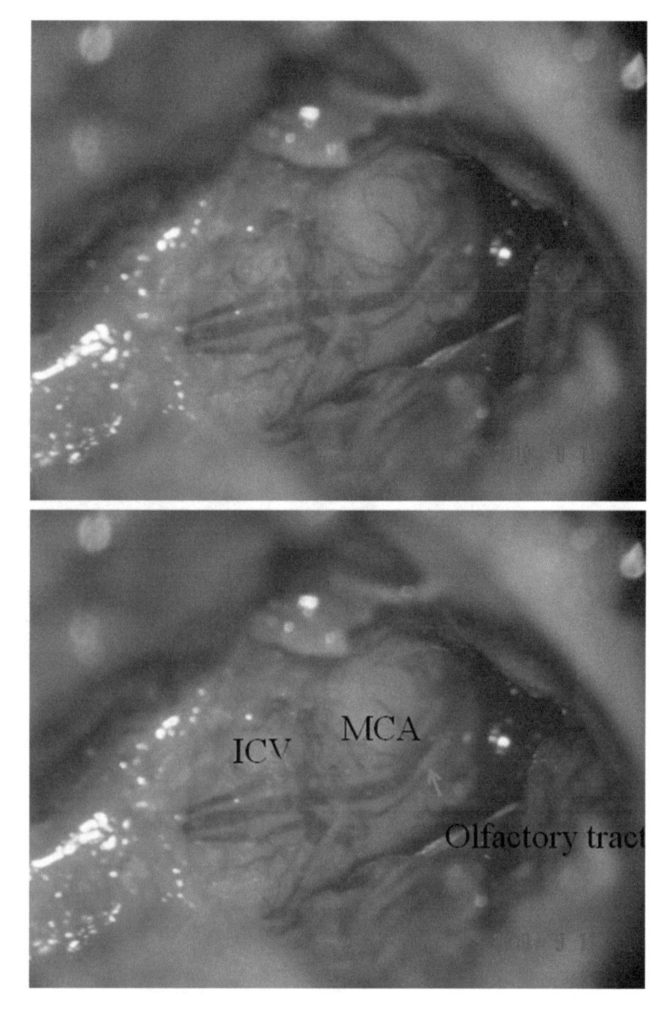

Fig. 8 For the craniectomy, firstly, a single entry burr hole is made with a dental drill. Secondly, microforceps are used to expand the burr hole and to perform the frontotemporal craniectomy. At this stage, the MCA, ICV, and olfactory tract can be visualized through the intact dura matter

less technically demanding and induces a milder neurological deficit than proximal MCAO.

3.9 Transient Focal Cerebral Ischemia

3.9.1 ET-1-Induced MCA Occlusion (Fig. 1c)

The peptide ET-1 is a potent vasoconstrictor of cerebral blood vessels with a prolonged duration of action [6] capable of blocking flow and inducing downstream ischemia. The model requires the same surgical approach as is used for the electrocoagulation models, and the exposed MCA can be transiently occluded by topical application of ET-1 (Fig. 1c). Once the dura has been opened and the MCA exposed, a fine (30 gauge) sterile needle is used to puncture the arachnoid membrane at several points on either side of the blood vessel to improve peptide access. ET-1 (25 μl of 10^{-7} to 10^{-4} M) is then topically applied to constrict the artery

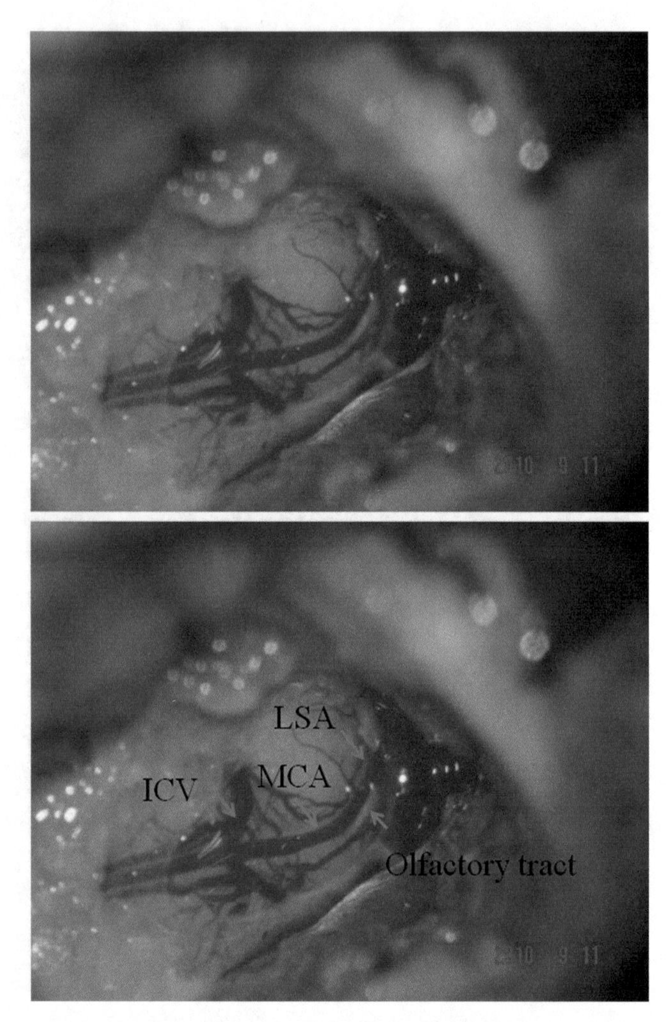

Fig. 9 After the dura matter and arachnoid membrane are opened, the MCA, inferior cerebral vein (ICV), which crosses the MCA, the olfactory tract and the branching arteries of the MCA such as lenticulo striate artery (LSA) are exposed

sufficiently to block blood flow (*see* **Note 7**). This can be confirmed visually using the operating microscope. The higher the concentration of ET-1 applied, the more severe and prolonged the ischemia and the larger the infarct, which has both a cortical and subcortical component, similar to proximal MCAO [7]. As the effect of the peptide wears off, the MCA diameter returns to normal and blood flow is gradually reestablished.

3.9.2 Mechanical Occlusion of the MCA (Fig. 1d)

Mechanical occlusion of the MCA provides the flexibility to induce permanent or transient occlusion of the main trunk of the MCA or its branches. Using mechanical devices such as microaneurysm clips [8, 9] (Fig. 1d), hooks [10], and ligature snares [11], models have been developed to induce focal ischemia (30 min to 2 h), followed by reperfusion (*see* **Note 8**). Mechanical occlusion at a

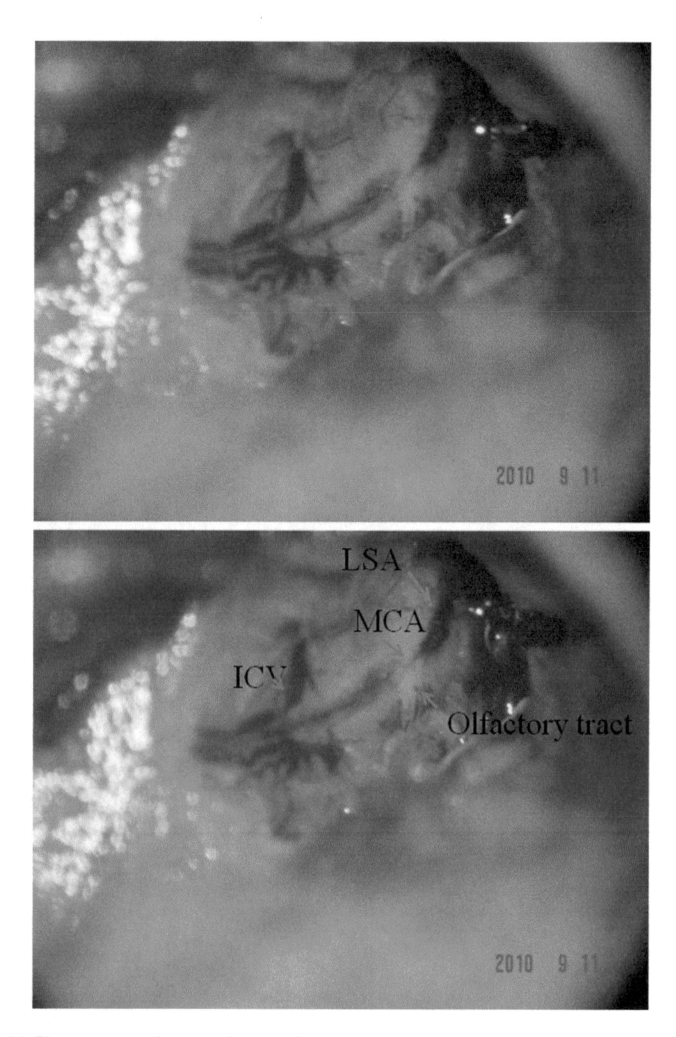

Fig. 10 Electrocoagulation of the MCA is performed over the olfactory tract. Then, using a segmental approach, the main trunk of the MCA is electrocoagulated starting from the proximal MCA and working your way distally till you reach the intersection of the MCA with the ICV

single point on the MCA can result in unacceptable variability in lesion size. In order to limit this, MCAO is often combined with uni- or bilateral common carotid artery occlusion or employed specifically in strains such as the spontaneously hypertensive and spontaneously hypertensive stroke-prone rat which have impaired cortical collateral blood flow. Microaneurysm clips (e.g., Codman, AVM micro clips with 10 g closing pressure) are small enough for use in the rat and are loaded into a special applicator for attachment to the MCA.

Mechanical occlusion can cause damage to the MCA with resultant impairment in reperfusion when the occluding device is removed. Recent modifications of the suture model have been published [12], which limit direct damage to the occluded vessels with the use of an occluding suture.

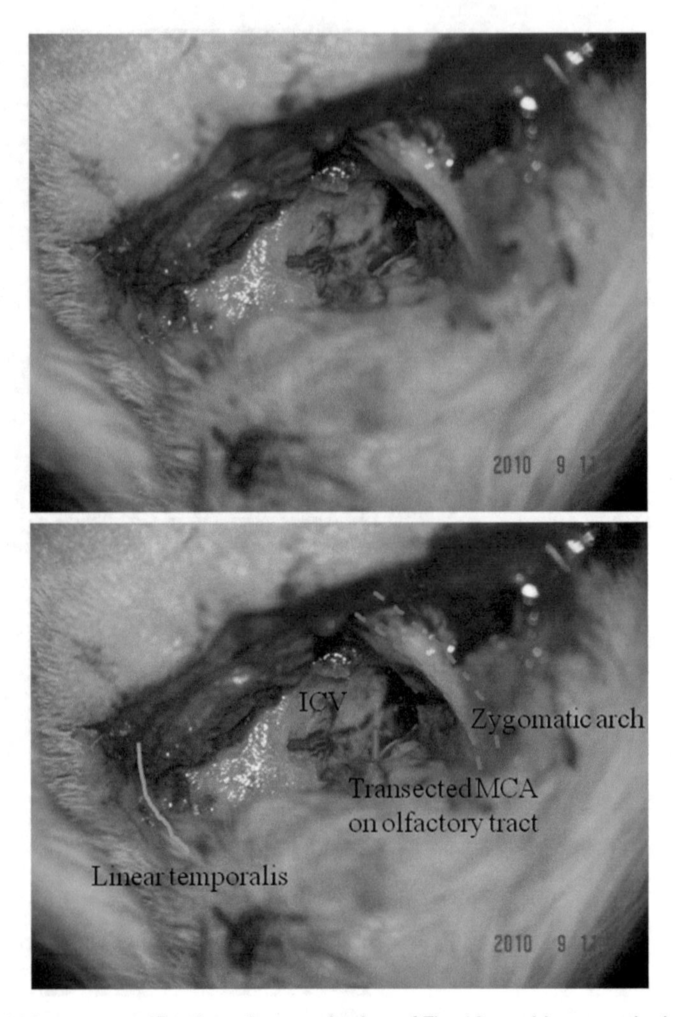

Fig. 11 Lower magnification microscopic view of Fig. 10 provides a good orientation of the surgical anatomy of the MCA. The transected MCA on the olfactory tract can be identified in the center of the craniectomy

3.10 Closure of the Surgical Wound

Before closing the wound, the surgical field should be washed with quantities of sterile saline to prevent infection. Muscle and skin layers are sutured with 4-0 Vicryl (Johnson & Johnson, New Brunswick, NJ) (Fig. 12). After surgery, a subcutaneous injection of sterile saline (2.5 ml into each of two sites) is administered to prevent post-anesthetic dehydration and should be repeated twice a day until the animal is drinking normally Analgesia should also be administered for the first 2–3 days after stroke surgery to limit postoperative pain (e.g., carprieve, buprenorphine, and paracetamol. Follow the recommendation of your local vet). The eyelid suture is removed and anesthesia withdrawn to allow the animal to recovery. Body temperature is maintained until the rat is fully conscious and when spontaneous respiration returns, the intubation tube is withdrawn and the rat returned to a clean cage with softened rat chow and water.

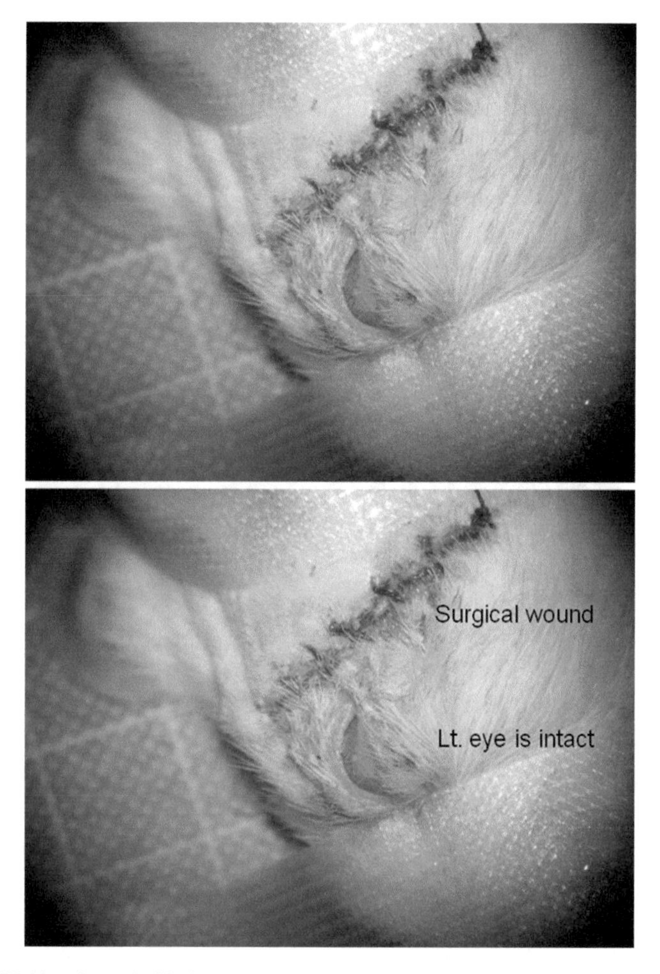

Fig. 12 Muscle and skin layers are sutured with 4-0 Vicryl. The eyelid suture is removed

3.11 Assessment of Ischemic Damage in the Rat

Quantitative histopathology, employing light microscopic examination of neuronal perikarya on multiple hematoxylin- and eosin-stained sections (Fig. 13a), is an established technique for the volumetric assessment of ischemic damage and drug efficacy [13]. Areas of infarct are transcribed onto line diagrams of coronal sections throughout the MCA territory (Fig. 13b) Alternatively, 2,3,5-triphenyltetrazolium chloride (TTC) staining can be used to detect ischemic damage [14]. TTC reacts with intact mitochondrial oxidative enzyme systems to produce a red colored formazan. The region where ischemia has damaged the tissue, mitochondria remain uncolored and easily distinguishable (Fig. 13c). More recently, magnetic resonance imaging is increasingly used for the assessment of ischemia and ischemic damage. Diffusion-weighted imaging is a sensitive and reliable modality for detection of the ischemic injury in the acute phase (Fig. 13d) and T2 scans provide images of the final infarct.

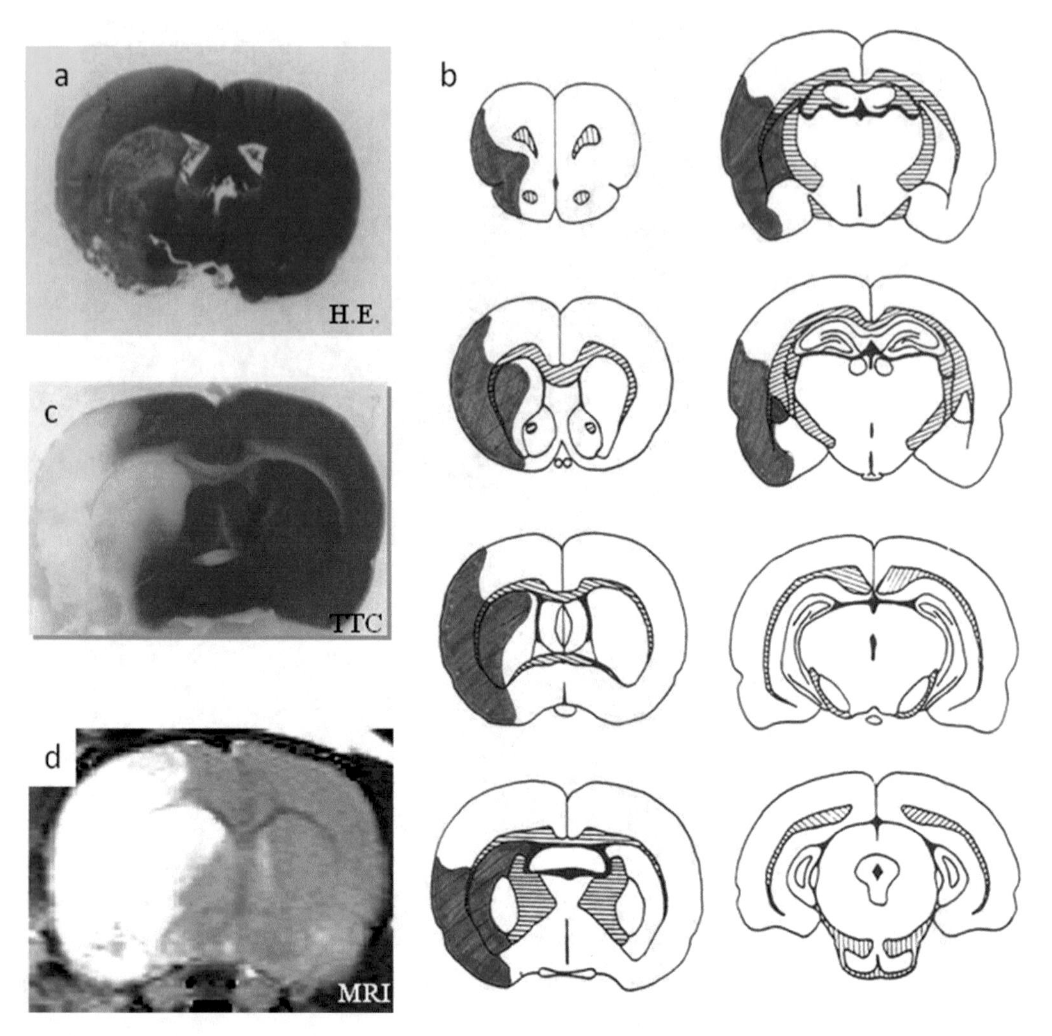

Fig. 13 Assessment of ischemic damage after MCA occlusion can be achieved on hematoxylin and eosin (H.E.) stained sections (**a**), 2,3,5-triphenyltetrazolium chloride (TTC) brain slices (**c**), and MR imaging (**d**). Line diagrams from a stereotaxic atlas of the rat brain can be used for the volumetric assessment of ischemic damage (**b**). Infarct is represented by black shading

4 Notes

1. Selection of the strain of rat: MCA occlusion can be carried out on any strain of rat but reproducibility in outcome measures such as infarct size varies in different strains [15]. The authors prefer to use Sprague-Dawley rats of 300–350 g and aged 10–12 weeks. In animals below 300 g the surgical field is relatively small, making the approach more difficult and in animals above 350 g the skull is noticeably thicker making the exposure of the MCA more challenging. Reproducibility can also be improved by using animals of a defined age and bodyweight.

2. To avoid any postsurgical infection, the surgical field should be washed with copious amounts of sterile saline or artificial cerebral spinal fluid. In our experience, no cases of infection have occurred when this procedure has been followed.
3. The complexity of the Tamura electrocoagulation model is almost exclusively attributed to achieving the correct anatomical orientation so that the craniectomy is made to reveal the MCA with minimum invasiveness. Achieving the correct orientation of the head for the surgical approach is crucial because of the deep surgical field needed to expose the proximal MCA at the skull base. It takes time and practice for the researcher to learn the correct position of the rat, angle of the microscope, and surgical approach.
4. Occlude the MCA on the same side of the brain in each animal to limit variability in outcome measures.
5. For the Tamura model, the most critical point in the procedure is learning how to electrocoagulate the MCA without rupturing the blood vessel or injuring the underlying brain tissue. This requires a fine and controlled technique that is more difficult than electrocoagulation techniques used in human surgery where an equivalent size of microvessel is easily coagulated. The blood vessel walls of the rat MCA are much thinner than those of the human and are electrocoagulated under mean arterial pressures of approximately 80–90 mmHg.
6. Electrocoagulation of the rat MCA: (a) To ensure reproducible ischemic lesions, the perforating artery from the MCA and the LSA must be coagulated first when carrying out proximal MCA occlusion. This ensures that the ischemic lesion includes the dorsolateral caudate nucleus. Moreover, once this procedure is complete, it enables the diathermy forceps to be inserted between the MCA and cerebral surface to clamp the main trunk of the MCA thereby facilitating a successful MCA occlusion by reducing the blood flow through the artery; (b) keep the tips of the diathermy forceps clean and polished so that the forceps do not adhere to the MCA when the blood vessel is being electrocoagulated; (c) once electrocoagulation has stopped the blood flow through the MCA at the level of the olfactory tract, the subsequent proximal to distal electrocoagulation of the main MRC trunk is much easier to achieve.
7. For ET-1 induced transient MCAO, consistency in the potency of ET-1 is very important to control variability. The peptide is purchased in lyophilized form and should be made up to the required concentration, aliquoted out into single use vials and frozen at −80 °C. A fresh aliquot should be thawed for each experiment and not refrozen.

8. Microaneurysm clips, ligatures, and sutures are applied to the MCA under the magnification of an operating microscope. This also facilitates visual inspection of the artery and/or its branches after removal of the occluding device to ensure the return of blood flow.

References

1. Macrae IM (1992) New models of focal cerebral ischaemia. Br J Clin Pharmacol 34:302–308
2. Koizumi J, Yoshida Y, Nakazawa T, Ooneda G (1986) Experimental studies of ischemic brain edema, I: a new experimental model of cerebral embolism in rats in which recirculation can be introduced in the ischemic area. Jpn J Stroke 8:1–8
3. Tamura A, Graham DI, McCulloch J, Teasdale GM (1981) Focal ischemia in the rat. Part I: description of technique and early neuropathological consequences following middle cerebral artery occlusion. J Cereb Blood Flow Metab 1:53–60
4. Imai H, McCulloch J, Graham DI, Masayasu H, Macrae IM (2002) New method for the quantitative assessment of axonal damage in focal cerebral ischemia. J Cereb Blood Flow Metab 22:1080–1089
5. Shigeno T, McCulloch J, Graham DI, Mendelow AD, Teasdale GM (1985) Pure cortical ischemia versus striatal ischaemia. Surg Neurol 24:47–51
6. Robinson MJ, McCulloch J (1990) Contractile responses to endothelin in feline cortical vessels in situ. J Cereb Blood Flow Metab 10:285–289
7. Macrae IM, Robinson MJ, Graham DI, Reid JL, McCulloch J (1993) Endothelin induced reductions in cerebral blood flow: dose-dependency, time course and neuropathological consequences. J Cereb Blood Flow Metab 13:276–284
8. Dietrich WD, Nakayama H, Watson BD, Kanemitsu H (1989) Morphological consequences of early reperfusion following thrombotic or mechanical occlusion of the rat middle cerebral artery. Acta Neuropathol 78:605–614
9. Buchan AM, Xue D, Slivka A (1992) A new model of temporary focal neocortical ischemia in the rat. Stroke 23:273–279
10. Kaplan B, Brint S, Tanabe J, Jacewicz M, Wang X-J, Pulsinelli W (1991) Temporal thresholds for neocortical infarction in rats subjected to reversible focal cerebral ischemia. Stroke 22:1032–1039
11. Shigeno T, Teasdale GM, McCulloch J, Graham DI (1985) Recirculation model following MCA occlusion in rats. Cerebral blood flow, cerebrovascular permeability and brain edema. J Neurosurg 63:272–277
12. Luo W, Wang Z, Li P, Zeng S, Luo Q (2008) A modified mini-stroke model with region-directed reperfusion in rat cortex. J Cereb Blood Flow Metab 28:973–983
13. Bederson JB, Pitts LH, Germano SM, Nishimura MC, Davis RL, Bartkowski HM (1986) Evaluation of 2,3,5-triphenyltetrazolium chloride as a stain for detection and quantification of experimental cerebral infarction in rats. Stroke 17:1304–1308
14. Osborne KA, Shigeno T, Balarsky AM, Ford I, McCulloch J, Teasdale GM, Graham DI (1987) Quantitative assessment of early brain damage in a rat model of focal cerebral ischaemia. J Neurol Neurosurg Psychiatry 50:402–410
15. Duverger D, MacKenzie ET (1988) The quantification of cerebral infarction following focal ischemia in the rat: influence of strain, arterial pressure, blood glucose concentration, and age. J Cereb Blood Flow Metab 8:449–461

Chapter 5

Inferior Colliculus Approach in a Rat

Dennis T.T. Plachta

Abstract

The inferior colliculus (IC) of the rat is a well-investigated and understood model for topographical mapping of the frequency domain (Clopton et al., Exp Neurol 42(3):532–540, 1974; Kelly and Masterton, J Comp Physiol Psychol 91(4): 930–936, 1977; Borg, Hear Res 8(2) 101–115, 1982; Ryan et al., Hear Res 36(2–3): 181–189, 1988; Zhang et al., Hear Res 117(1–2):1–12, 1998). As a central hub for binaural auditory processing in the midbrain (Du et al., Eur J Neurosci 30(9): 1779–1789, 2009) it shows a variety of response patterns to a given complex auditory stimulation (Kelly et al., Hear Res 56(1–2):273–280, 1991; Kelly and Li, Hearing Res 104:112–126, 1997). It is therefore a major target for neuroscientific approaches of the ascending and descending auditory pathway. Approaching the IC is, however, not only valuable for scientists interested in auditory processing, but also for students learning the proceedings of standard electrophysiological experimentation. In addition, engineers of biomedical devices (e.g., flexible penetrating electrodes) can take benefit from the IC approach (Kisban et al., Conference proceedings: annual international conference of the IEEE Engineering in Medicine and Biology Society IEEE Engineering in Medicine and Biology Society Conference, 2007:175–178). A critical test of the suitability of shaft electrodes is their successful implantation in vivo. The steady tonotopic structure of the IC and its three subdivisions provides an almost perfect anatomic testing ground (Saldana and Merchan, J Comp Neurol 319(3):417–437, 1992). Additionally, the anatomical procedure to access the IC requires only a medium level of surgical skills and the testing apparatus can be kept relatively small and manageable. The current study describes the necessary anatomical steps and materials needed for the aforementioned scenarios.

Key words Inferior colliculus, IC approach, Auditory, Mid-brain, Rat, Tonotopy

The inferior colliculus (IC) of the rat is a well-investigated and understood model for topographical mapping of the frequency domain [1–5]. As a central hub for binaural auditory processing in the midbrain [6] it shows a variety of response patterns to a given complex auditory stimulation [7, 8]. It is therefore a major target for neuroscientific approaches of the ascending and descending auditory pathway. Approaching the IC is, however, not only valuable for scientists interested in auditory processing, but also for students learning the proceedings of standard electrophysiological experimentation. In addition, engineers of biomedical devices (e.g., flexible penetrating electrodes) can take benefit from the IC approach [9]. A critical test of the suitability of shaft electrodes is their successful implantation in vivo. The steady tonotopic

Miroslaw Janowski (ed.), *Experimental Neurosurgery in Animal Models*, Neuromethods, vol. 116,
DOI 10.1007/978-1-4939-3730-1_5,

structure of the IC and its three subdivisions provides an almost perfect anatomic testing ground [10]. Additionally, the anatomical procedure to access the IC requires only a medium level of surgical skills and the testing apparatus can be kept relatively small and manageable. The current study describes the necessary anatomical steps and materials needed for the aforementioned scenarios.

1 Materials

Female Sprague-Dawley rats weighing between 250 and 300 g are used. For acute experiments, Ketamine (10%, 0.6 ml/kg bodyweight (BW)) and Medetomidine (0.3 mg/kg) are used for initial anesthesia. If the experiments last more than 4 h, additional anesthesia must be applied: use the same concentrations but at 1/5th of the initial dose. Four hours after the initial application of the anesthetics, a refresh has to be applied at least every second hour. Both medications can be applied subcutaneousely (s.c.) using a single syringe.

Especially for students and beginners in animal experimentation, a plexiglass tube (diameter: 8 cm, height: 20 cm, one end closed and perforated for air flow) can be helpful to safely apply the initial anesthetics. The rats like to crawl into the tube. The experimenter can then hold the tail and apply the anesthetic while the rat is safely immobilized in the tube.

The anesthetized rat will cool down and dehydrate. To prevent this and to maintain stable physiological parameters, ringer solution (Braun, Ringerlösung 0.9% NaCl) has to be applied every other hour (1 ml s.c.). Additionally the body temperature has to be supported using either a heat-lamp (e.g., Thermolux Pet-Mat 10 W, for experiments <4 h) or a closed-loop heat pad (e.g., Animal Temperature Controller ATC, WPI). For the opening of the skull a drill is necessary (Proxxon, Mini-Mot 40). For optimal results, use a drilling head

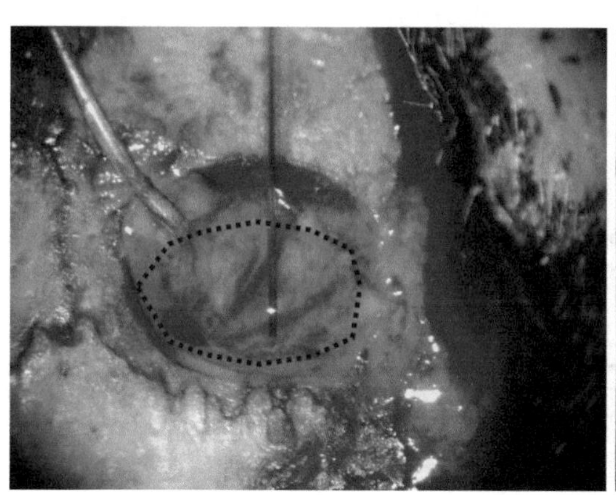
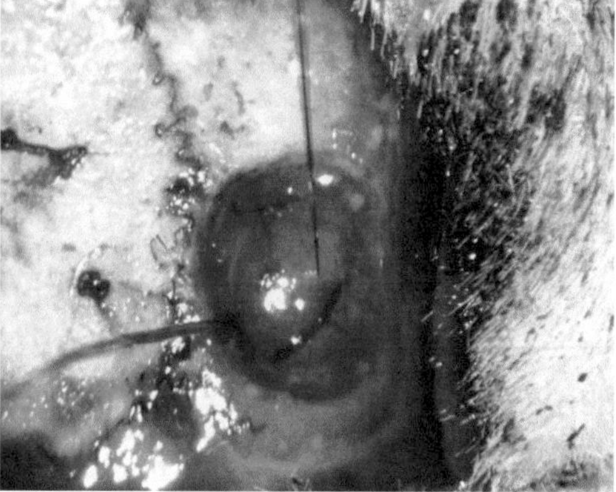

Fig. 1 REM image of the head of a flexible polyimide electrode with a large application tip. This tip is used to retract the insertion guide. Note the electrode contacts on the surface. Those are ideal for recording from different recording depth and in this approach will deliver different best frequencies. The bar indicates 100 μm

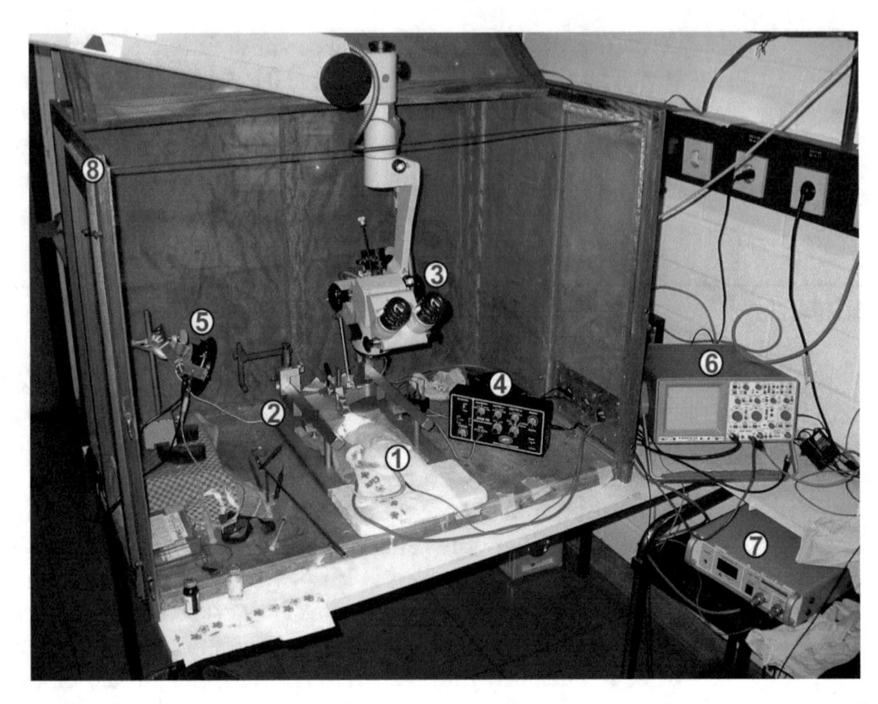

Fig. 2 Image of the rat attached to the stereotactic device. Note that no metal is close to the opening wound and the head is fixed to prevent respiration artifacts

with a ball (0.5 mm ball head). Opening of the skull with the drill and positioning of the microelectrode should be performed using a dissecting microscope (e.g., Leica M60). For postsurgical fixation of the head and for precise positioning of electrodes a stereotactic device is strongly recommended (e.g., Stoelting, AnyAngle Stereotaxic Instrument). Finally, the penetration of the electrode requires a micromanipulator (e.g., WPI, KITE-R micromanipulator).

Our setup including the auditory stimulation and electrophysiological recording instrumentation is shown in Figs. 1 and 2. The complete list of devices for surgery and experimentation is provided in Table 1.

2 Methods

2.1 General Anatomy and Considerations

The inferior colliculus (IC) is a mesencephalic nucleus, primarily part of the ascending auditory pathway. It is roughly oval in shape with a diameter of around 3 mm and is situated some 2–4 mm beneath the lambda point (see Fig. 3 for saggital section). The IC is part of the corpora quadrigemina and separated into three subdivisions: the dorsal cortex of the inferior nuclei (DCIC), the external cortex of the inferior nuclei (ECIC), and the core, the central nucleus of the IC (CNIC; DCIC and ECIC are not separated in Fig. 3). The CNIC in particular shows a highly linear tonotopy with low frequency responsiveness in the dorsal area and high frequency responses in ventral regions. The hearing range of a rat extends far into the ultrasound

Table 1
Instruments for surgery and devices for electrophysiology

Surgical equipment
Scalpel, scalpel blade #10, WPI
Noyes scissors, stainless steel, 14 cm, straight, WPI
Vannas scissors superfine, stainless steel, 8 cm, straight, WPI
Dissecting scissors, 10 cm, straight, WPI
2× Dumont #5, 12 cm, straight, WPI
Dumont #5, 11 cm, curved, WPI
Gillies dissecting forceps, 15 cm, 1 × 2 teeth
ALM retractor, 7 cm, 2 × 2 or 4 × 4 prongs, WPI
Cutting needles, ½ circle, size 0, WPI
Probe 14 cm, round tip, 0.25 mm, WPI
Drill, Mini-Mot 40, Proxxon
Drill head, 0.5 mm, round, Proxxon
Longhair shaver, Exact Power EP50, Braun
Cauterizer, Cautery Kit, FST
Q-tip, local dealer
Syringe, 0.1 ml, short needle
Syringe, 1 ml, short needle
Syringe, 5 ml, short needle
Accessory surgical equipment
Animal heat pad, animal temperature controller ATC, WPI
Dissecting microscope, M60, Leica
Anesthetic tube for rat, self-made
Stereotactic device, AnyAngle, Stoelting
Surgical consumables
Ringer, 0.9% NaCl, Braun
Ketamine, 10%, Dormitor
Medetomidine, Dormitor
Analgetics, Ketoprofen, Sigma
Dermal glue, Nexaband, Abbot
Dental cement, Promedica
Collagenase, Sigma
Vaseline, local dealer

(continued)

Table 1 (continued)

Agarose, Sigma
Nembutal, Dormitor
Skin disinfectant, Kodan, local CVS
Electrophysiological equipment
Micromanipulator, Kite-R, WPI
Micro stepper, SM325, WPI
Faraday cage, self-made
Preamplifier, Medusa, TDT
AD-Board, Medusa, TDT
DA-Board, RP2, TDT
Oscilloscope, TDS1012, Tektronix
Software, brainware, TDT
Audio amplifier, 7600, KronHite
Loudspeaker, MT22, Morel
Sound-level meter, 2238 Mediator, Bruel&Kjaer

domain. Frequencies from a few Hz up to 38 kHz can be detected [11]. A comprehensive tonotopical mapping of the IC requires therefore quite sophisticated stimulation hardware.

2.2 Preparing the Skull

Depending on whether the planned approach will be an acute one or not, the operation place and the instruments have to be sterilized and prepared. By default the following steps in this chapter describe acute implantation.

The head of the anesthetized rat is first shaved using an electric shaver (e.g., Braun, long hair shaver). Even though the approach targets only the IC contralateral to the stimulation side, the coat between the neck and both eyes should be removed above both hemispheres. This is a matter of precaution. If the remaining side has to be opened up later, this additional shaving avoids hair particles from contaminating the already open wound. If the experiment is not acute, the skin has to be disinfected locally prior to using a scalpel (Kodan®). The next step is a 1.5 cm midline incision between the eyes and the ears (see Fig. 2). The skin is clamped back with a retractor (Alm self-retaining retractor, 7 cm, WPI). Alternatively add two transversal terminal incisions (T-shaped incisions) to both ends of the midline cut; this allows the two skin flaps to be folded inwards. The attached subcutaneous connective tissue has to be removed from the scull using a pair of fine forceps, fine scissors, and a scalpel. Do not hesitate to make extensive use of the scalpel to scratch away the remaining and strongly adhesive

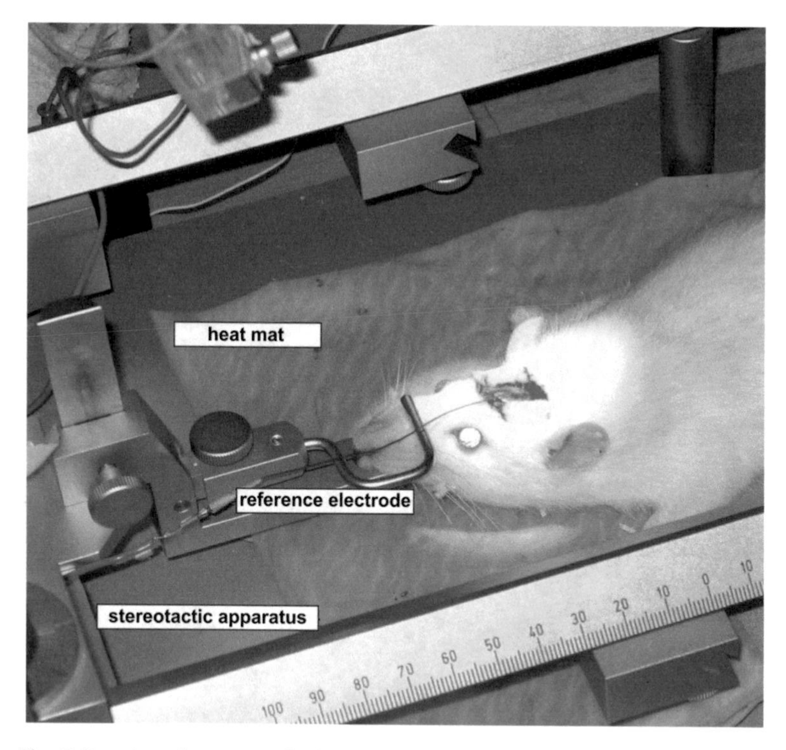

Fig. 3 Drawing of a parasagittal section of the rat brain. Attached are the scale bars (in mm) given by the stereotactic device. (*1*) skull, (*2*) telencephalon, (*3*) cerebellum, (*4*) brainstem, (*5*) DCIC and ECIC, (*6*) CNIC, and (*7*) "lambda" point. The *red bar* indicates the position of the opening of the skull the *arrow* represents an electrode penetrating through the brain in an angle of 30°

periosteum; otherwise the identification of the sagittal and the lambdoidal suture is difficult (see Fig. 4). There will be only very mild bleeding caused by this latter procedure.

2.3 Opening of the skull

Now the skull should be exposed and the sagittal and the lambdoidal suture should be identifiable (see Fig. 4), respectively. Even though the target nucleus is directly underneath the sagittal suture, it is not wise to penetrate the skull at this spot. The superior sagittal sinus is just below the suture and penetrating this major blood vessel will bring any experiment to a swift end. Instead take the drill and force a 4–5 mm diameter whole between the sagittal and lamboidal suture as shown in Figs. 5a, b and 6. This drilling is one of two crucial steps in this approach and requires a steady hand and a dissection binocular. Take the head of the drill and gently drill a circular channel in the bone until the blood vessels on top of the brain become visible through the thinned out bone structure. The channel the drill carves into the bone should be cleaned with ringer solution every now and then to wash away the bone particles and facilitate optically control of the depth. If the drill is used with too much force it will penetrate through the remaining thin layer of bone and damage the brain tissue beneath.

A raw egg presents a very suitable exercise to find the right moment to stop drilling. Just let the drill do the job and dig in

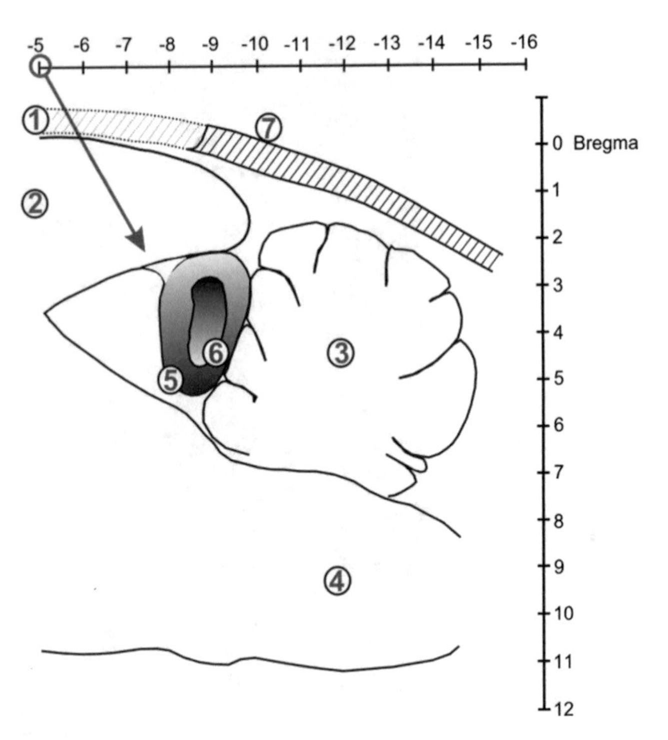

Fig. 4 Drawing of a rat skull with attached opening for IC approach. (*1*) sutura coronalis, (*2*) bregma point, (*3*) sutura sagittalis, (*4*) sutura lamboidea, and (*5*) lambda point. The large wire in the *right insert* is the reference electrode. This wire is formed like a hook to be fixed underneath the bone

deeper and deeper to the shell with every new circle. Once the resistance drops, one is generally close to inserting the drill into the egg and the right moment to stop drilling has been reached.

As soon as the blood vessels are visible through the thinned out bone in the channel (Fig. 5c) and there are no major bony bridges left which connect the area within the circle with the remaining skull, the encircled bone material can be removed using a pair of thin but strong forceps (see Fig. 5d). Use one end of the forceps to penetrate between the encircled bone and the dura mater on top of the brain in a steep angle. Then carefully liftoff the encircled material. This step is the second critical one since the penetration must be done very precisely in order to not damage the dura and the underlying brain surface as well. The liftoff step might require quite some force depending on how much bone structure is left after the drilling. Again this liftoff can be best practiced using a raw egg.

2.4 Preparing the Penetration

After the liftoff step the opening has to be prepared for the penetration of the electrode. If this approach is not acute, it is reasonable to keep the removed bone lid in a cooled ringer dish until the craniotomy is subsequently closed.

Due to the drilling the fringe of the opening may possess sharp edges, which even under magnified vision might be difficult to identify. As a matter of precaution it is conducive to use a tiny round probe

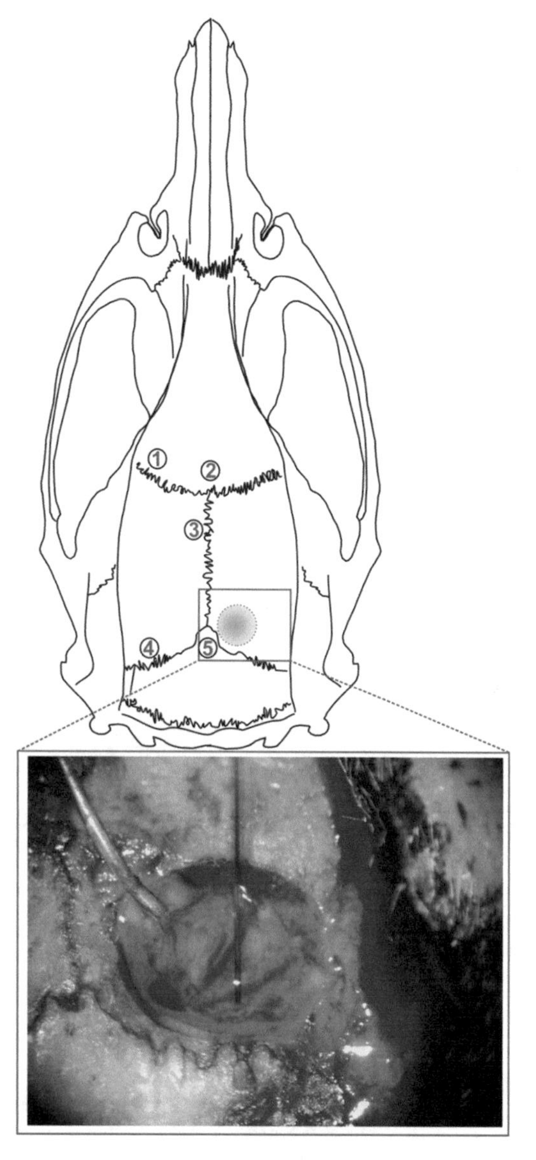

Fig. 5 The two critical steps of the IC approach are shown in a sketch. The bone of the skull is shown in *black*, the dura mater in *blue*, the drill in *red*, and the tips of the forceps and the rounded tip of the probe in *gray*. (**a**) Positioning of the drill, (**b**) movement of the drill in circles in order to carve a circular channel (**c**) stopping of drilling before penetration, (**d**) use of forceps to liftoff the circular bone piece, (**e**) use of round-tipped probe to remove sharp edges, and (**f**) removal of all debris and blood from the wound

to detect possible sharp edges and break them away from the dura (Fig. 5e, f). This way, the risk of electrodes being bent and damaged while coming too close to these edges is reduced to a minimum.

Modern electrode designs are quite flexible and therefore tricky to penetrate through intact dura such as the one generally found in the rat (e.g., flexible array electrode Fig. 7). Even stiff glass electrodes tend to buckle, bend, and finally break if

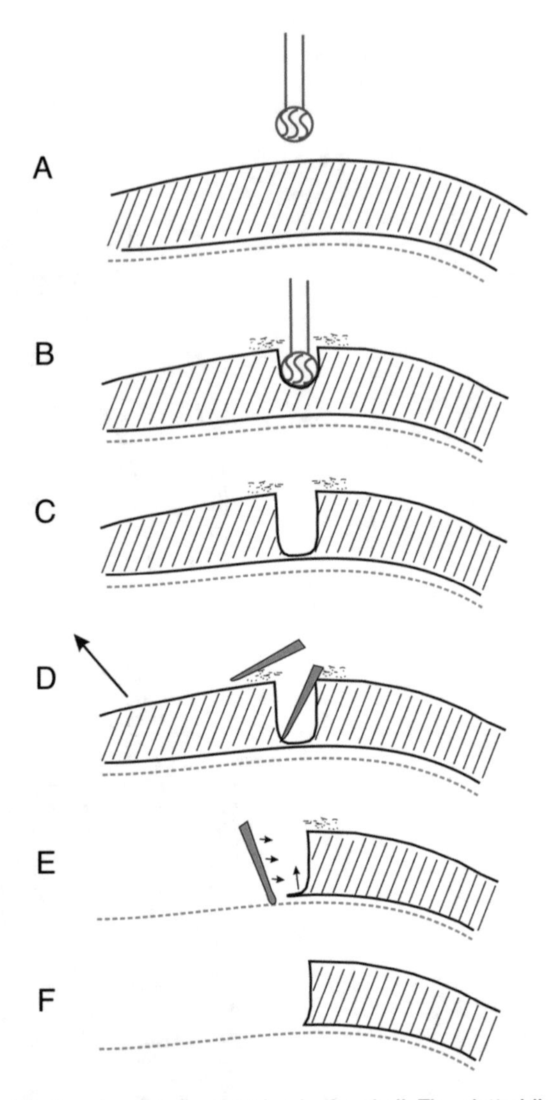

Fig. 6 Two images showing the opening in the skull. The *dotted line* in the *left* image shows a recommended area of penetration for a successful approach toward the IC

penetrated straight through the dura. Except where stiff metal electrodes will be used it is therefore preferable in most cases to remove the dura above this round window before penetration. This can be realized in two ways: either remove the dura using fine forceps and micro scissors (WPI) or use collagenase (Sigma) to degrade the collagen fibers and "digest" the dura. The first approach might result in some bleeding, which can be stopped using a fine cauterizer (FST Cautery Kit). The latter approach is "safer" in terms of excess wounds, but might not weaken the dura enough to allow very sensitive electrodes to penetrate without buckling (see Fig. 6 for final view of the prepared recording site). In addition, the digestion of the dura requires some 20–30 min incubation time.

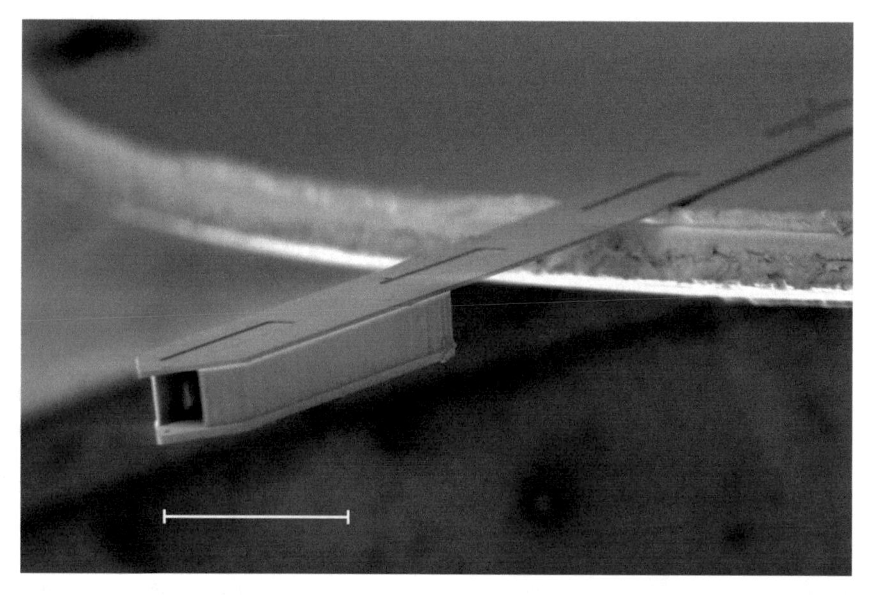

Fig. 7 Setup used for acute IC-recordings. (*1*) Animal with heat pillow underneath, (*2*) stereotactic device, (*3*) dissection microscope for electrode positioning and general control of skull opening, (*4*) preamplifier, (*5*) loudspeaker, (*6*) oscilloscope, (*7*) closed-loop thermo element, and (*8*) Faraday cage. Note that parts of the recording and stimulation chain are not shown (like PC, attenuator, AD-board, headphone)

2.5 Recording Track Through the IC

Once the opening window for the electrodes is prepared the animal can be placed in the stereotactic device. Since the first three steps of the IC approach require full access to and flexibility of the head of the rat it is not conducive to have the rat fixed in the stereotactic device prior to step 4. The primary functions of the stereotactic device are to allow the fixation of the animal and to reduce respiration artifacts. If cortical pulsation becomes an issue during recording, 3% agar or Vaseline can be used to fill the hole, dampening movement of the cortex and diminishing this artifact.

Since the IC has an offset toward the opening of the skull, the penetration has to be applied at an angle of 30° (see Fig. 3). For reproducible tracks, a micromanipulator is crucial. If the penetration depth is of importance an adjustable or programmable micro stepper is a welcome but expensive auxiliary piece of equipment, which delivers precise mapping coordinates.

For a frequency dependent recording, the following equipment is necessary on the recording and the stimulation side of the setup. The recording requires an electrode and an appropriate head stage. The signal is then fed into a bio-amplifier and finally into an AD-board (analog-digital-converter). If the software does not present the recording in a window it is convenient to have an extra oscilloscope at hand to monitor the activity at the electrode.

For auditory stimulation, a voltage controlled function generator of a programmable DA-board is necessary to generate analog signals. These signals have to be adjusted in amplitude using an attenuator and fed into a loudspeaker using an audio amplifier. If calibrated signals are desired, use a hand calibration tool, e.g., from Bruel&Kjaer.

Details on the setup devices we used can be found in Table 1. There are three major approaches to record frequency dependent signals from the IC:

1. Use sinusoidal signals of defined duration and amplitude.
2. Use sweeps of different frequencies and defined amplitude.
3. Use white noise and perform a RevCor (reverse correlation) analysis [12].

A broadband search stimulus can be used to control the activity at the electrodes. In the same context is it most convenient to have at least one channel of the recording streamed to a headphone to listen for the background "hash," which typically precedes or accompanies the presence of clear spiking activity at a given recording site.

2.6 Termination of the Experiment or Closing of the Wound

After the experiment is conducted, there are now two possible courses of action, depending on whether the experiment was an acute one, or whether for example a tracer was applied and the animal must survive for a period of days or weeks in order for the tracer to be distributed throughout the brain. To terminate an experiment, use sodium pentobarbital (Nembutal®, 40–50 ml/kg) and apply it i.p. If the animal is subject to a tracer study, pick up the lid of bone material lifted off during trepanation and place it back so the craniotomy is closed again. Now use dental cement (Promedica) to fixate it. Take care that the cement covers the lid as well as the surrounding area of the skull. The wound can be closed using a suture. Try to place the stitches underneath the skin since rats are extremely adept at ripping out surgical suture material during recover. If the use of a suture is not applicable the wound can also be closed using dermal glue like Nexaband s/c®.

For the post surgery phase, the animal should be administered with analgesics (Ketoprofen, 5 mg/kg s.c.).

3 Notes

- Especially during the surgical process, ensure that the mouth of the rat is open and the nose as well as the mouth are not blocked by the tongue or a surgical drape.
- Monitor the respiration of the animal every half hour. The frequency and depth of the respiration is an important vital indicator regarding longer experimental sessions.
- Use a syringe and the 0.9% NaCl solution to frequently clean the wound. This prevents it from drying. It is especially important that the exposed surface of the brain does not dry out during experimentation. This will result in recording artifacts and eventually, by means of clogging, lead to damage of the recording electrode.

- Keep an eye out for micturition of the experimental animal. If this does not happen after the first 2–3 h the animal might suffer from dehydration or renal failure. Immediately apply additional 1 or 2 ml 0.9% NaCl.
- With ongoing experimental duration the animal will fall into deeper relaxation, i.e., the animal will "flat out" and the eyes will pop out of the skull. Therefore, use Vaseline to prevent the eyes of the animal from drying out.
- Do not leave any metal containing parts attached to the wound (e.g., retractor and forceps). This will result in increased levels of noise, since the metal will pick up electrical fields like an antenna.
- A Faraday cage is not mandatory but generally helps to get stable results and a decent level of background noise during the recording. In the same context, do not forget to turn off cell phones and to ground yourself while positioning the electrode with open amplifiers.
- If you purchase a heating mat for an electrophysiological experiment, make sure the device is appropriate for this since some heat mats are strong sources of pulsed electromagnetic radiation, which then will spit in your recording.

References

1. Clopton BM, Winfield JA (1974) Unit responses in the inferior colliculus of rat to temporal auditory patterns of tone sweeps and noise bursts. Exp Neurol 42(3):532–540
2. Kelly JB, Masterton B (1977) Auditory sensitivity of the albino rat. J Comp Physiol Psychol 91(4):930–936. doi:10.1037/h0077356
3. Borg E (1982) Auditory thresholds in rats of different age and strain. A behavioral and electrophysiological study. Hear Res 8(2):101–115
4. Ryan AF, Furlow Z, Woolf NK, Keithley EM (1988) The spatial representation of frequency in the rat dorsal cochlear nucleus and inferior colliculus. Hear Res 36(2–3):181–189
5. Zhang DX, Li L, Kelly JB, Wu SH (1998) GABAergic projections from the lateral lemniscus to the inferior colliculus of the rat. Hear Res 117(1–2):1–12
6. Du Y, Ma T, Wang Q, Wu X, Li L (2009) Two crossed axonal projections contribute to binaural unmasking of frequency-following responses in rat inferior colliculus. Eur J Neurosci 30(9):1779–1789. doi:10.1111/j.1460-9568.2009.06947.x
7. Kelly JB, Glenn SL, Beaver CJ (1991) Sound frequency and binaural response properties of single neurons in rat inferior colliculus. Hear Res 56(1–2):273–280
8. Kelly JB, Li L (1997) Hearing Research: two sources of inhibition affecting binaural evoked responses in the rat's inferior colliculus: the dorsal nucleus of the lateral lemniscus and the superior olivary complex. Hear Res 104:112–126
9. Kisban S, Herwik S, Seidl K, Rubehn B, Jezzini A, Umiltà MA, Fogassi L, et al (2007) Microprobe array with low impedance electrodes and highly flexible polyimide cables for acute neural recording. Conference proceedings: annual international conference of the IEEE Engineering in Medicine and Biology Society IEEE Engineering in Medicine and Biology Society Conference, 2007, 175–178. doi:10.1109/IEMBS.2007.4352251
10. Saldana E, Merchan M (1992) Intrinsic and commissural connections of the rat inferior colliculus. J Comp Neurol 319(3):417–437
11. Heffner HE, Heffner RS, Contos C, Ott T (1994) Audiogram of the hooded Norway rat. Hear Res 73(2):244–247
12. Klein DJ, Depireux DA, Simon JZ, Shamma SA (2000) Robust spectrotemporal reverse correlation for the auditory system: optimizing stimulus design. J Comput Neurosci 9(1):85–111. doi:10.1023/A:1008990412183

Chapter 6

Why Robots Entered Neurosurgery

Jason W. Motkoski and Garnette R. Sutherland

Abstract

Progress in neurosurgery has paralleled technological innovation. Image-guided surgical robotic systems have emerged as a potential hub for integration of the complex sensory, pathologic, and imaging data sets that are available to contemporary neurosurgeons. These systems couple the executive capacity of surgeons with the technical capabilities of machines and have the potential to improve surgical care as neurosurgery progresses towards the cellular level. Surgery is often performed in animal models prior to clinical application, representing a very important safety step in regulatory approval. As the capital investment for surgical robotic systems decreases, robotic systems may be specifically designed for animal application. In this chapter, we review neurosurgical robotic systems used in humans and animals; present the development, preclinical testing, and early clinical use of a unique image guided MR-compatible neurosurgical robot called neuroArm; and review the strengths and limitations of using surgical robotic systems in animal models.

Key words neuroArm, Image guidance, Robotics, Clinical integration, Stereotaxy, Microsurgery, Neurosurgery

1 Introduction

Historically, progress in neurosurgery has paralleled technological innovation. This trend began with improved neurosurgical instrumentation. In 1927, Bovie and Cushing revolutionized neurosurgery with the introduction of electrocautery [1]. For the first time, neurosurgeons were provided a technology that allowed control of bleeding, resulting in a substantial reduction of operative morbidity and mortality. The ability to achieve hemostasis also allowed surgeons to begin surgically managing lesions that were previously considered inoperable.

During the same timeframe, visualization of the surgical site evolved from the incandescent headlamp, progressing into today's counterbalanced operating microscope [2]. Magnification of the surgical field initiated the microsurgical paradigm, with narrower surgical corridors and greatly enhanced visual differentiation between normal and pathological tissue. The resulting

Miroslaw Janowski (ed.), *Experimental Neurosurgery in Animal Models*, Neuromethods, vol. 116,
DOI 10.1007/978-1-4939-3730-1_6, © Springer Science+Business Media New York 2016

microsurgical revolution necessitated the creation of increasingly small, precise surgical instrumentation [3].

Prior to the 20th century, surgical localization was based on the principles of Paul Broca and contemporaries, who established concepts of cortical compartmentalization of function based on clinical pathological correlation [4]. In 1895, Wilhelm Conrad Roentgen revolutionized diagnostic medicine with the discovery of X-rays [5]. This was improved upon by the serendipitous discovery of pneumoencephalography, or air injection, in 1917, which visualized brain shift through displacement of ventricles [6]. Pneumoencephalography remained the primary neurological imaging modality until the mid 1950s, when contrast angiography became widespread because of decreased toxicity of contrast agents. The invention of the microprocessor allowed for the explosive growth of computer technology, which when coupled with X-ray imaging, created computerized tomographic (CT) imaging in the early 1970s [7]. Characterization of electron spin in the hydrogen atom was also coupled to computer technology to create magnetic resonance (MR) imaging [8, 9]. These technologies allowed serial 2D imaging of the human body, and accurate lesion localization within a particular 2D plane. Volumetric reconstruction followed, presenting interactive 3D virtual models from which additive or destructive lesions on the brain could be observed. In addition, MR technology evolved to allow imaging of brain function and metabolism [10, 11].

The explosive growth of imaging technology contributed to a new paradigm in surgery: one of minimalistic technique. Technologies rapidly emerged in response, equipping the surgeon with an array of new instrumentation including endoscopy, high-definition cameras, computer displays, and elongated tools capable of accessing body compartments through small portals. It was very quickly shown that minimalistic surgical technique resulted in decreased length of hospital stay, lower rates of surgical complication, improved patient outcome, and increased patient satisfaction [12, 13].

In addition, as intracranial operations involve fixed anatomy in a contained volume, investigators began to exploit triangular geometry to link preoperative images with intraoperative surgical navigation [14, 15]. As a result, cranial openings became smaller and surgeons were able to accurately target deep brain pathology using patient-specific imaging. Unfortunately, the act of surgical dissection results in brain shift, whether through anesthetic management, patient positioning for surgery, drainage of cerebral spinal fluid, progressive excision of pathology, or brain edema. In response to this challenge, several investigators began to integrate various imaging technologies into surgical procedure [16–19].

Surgeons were soon provided with exquisite lesion localization before and during each operation. Neurosurgical corridors became smaller, pushing surgeons toward their physical limits of precision, accuracy, and coordination. Investigators from around the world began to integrate robotic technology into the increasingly complex surgical environment. Robotics allowed the executive capacity of the human brain to be coupled with the increasingly precise and accurate technology of machines. The modern neurosurgical paradigm of informatic surgery was thus born [20].

2 Neurosurgical Robotics

The initial concept of integrating a robotic system into neurosurgery was a daunting proposition. In addition to the technical requirements of the system and its surgical objectives, investigators needed to resolve challenges associated with sterility, patient safety, ethical and regulatory approval, considerable financial cost, integration with imaging technology, and introduction into established surgical procedure and processes. Multidisciplinary cooperation between science, medicine, engineering, and industry was crucial in overcoming the complexity of these challenges. Global awareness and networking allowed incorporation of diverse technologies as they were developed and became available (Fig. 1).

Neurosurgical robotics (Table 1) began with the introduction of an industrial robot, Programmable Universal Machine for

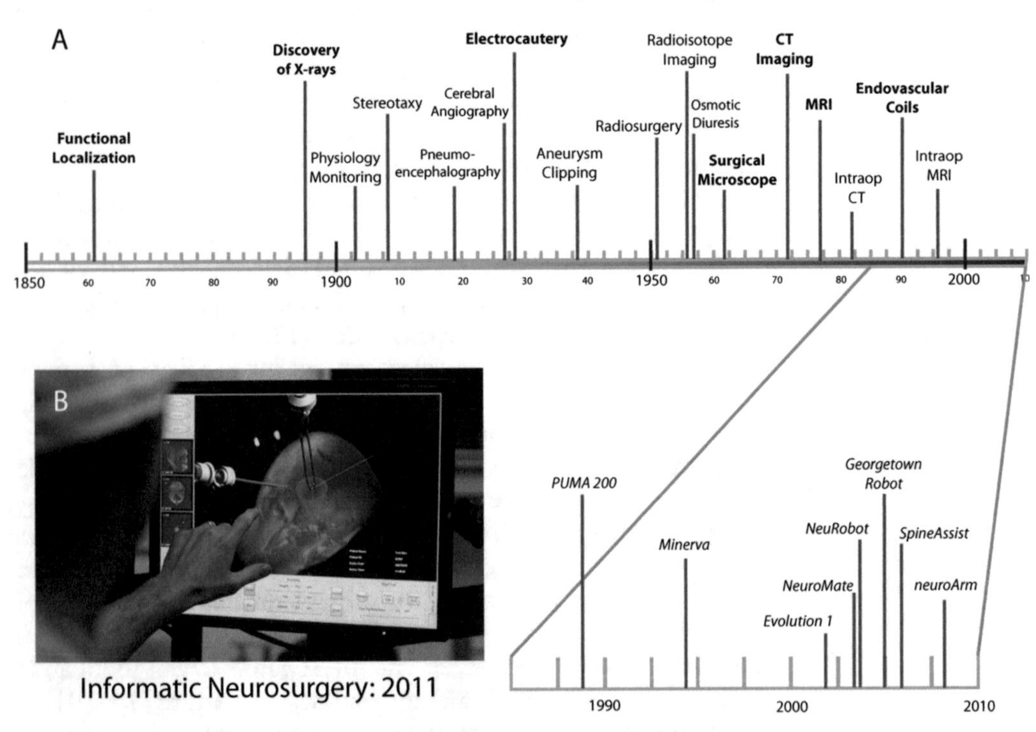

Fig. 1 (**a**) Timeline showing the chronological introduction of technologies into clinical neurosurgery. Over the past 20 years, robotic systems have been developed to couple the executive decision-making capacity of the surgeon with the accuracy of imaging technology and the precision of advanced robotics. (**b**) A screenshot of the neuroArm human–machine interface showing the integration of 3D magnetic resonance images with surgical planning (*blue cone*) and robotic tools (*blue* and *off-white*) to target patient pathology

Table 1
The evolution of neurosurgical robotics

Year	Project name	DOF	# EE	Navigation	Imaging	Purpose	Documented use	Accomplishments	Limitations
1988	PUMA[1,2]	6	1	BRW frame	CT	Retractor	Glioma surgery	First neurosurgical robot Demonstration of safety	No FDA approval Safety concerns with industrial robot in OR
1994	Minerva[3–5]	5	1	BRW frame	CT	Image guided surgery	Stereotactic biopsy Stereotactic implantation	Mounted inside CT machine	Only 5 DOF Requires dedicated CT scanner Obstructive within CT gantry
1997	CyberKnife[6]	6	1	Frameless	X-ray	Radiosurgery	Neurosurgical radiation	Focused radiosurgery FDA approved	Expensive
1999	RAMS[7,8]	6	1	N/A	N/A	Microsurgery	Rat carotid endarterectomy	Microsurgical ability	No haptic feedback Increased length of procedure
1999	Steady Hand[9]	6	1	N/A	N/A	Microsurgery	Tremor filter and motion stabilizer	Increased surgical precision Haptic feedback	No image guidance No clinical application
2001	Harvard MRI Robot[10,11]	5	1	Frameless	MRI	Navigation and tool placement	Position needle holder	Nonmagnetic actuators	Pointing device only

2002	Evolution 1[12–14]	6	1	Frameless	MRI	Endoscopy	Neuroendoscopy Endoscopic ventriculostomy Transphenoidal skull base surgery	Endoscopic application	Narrow working envelope Single arm No haptic feedback
2003	NeuroMate[15–18]	5	1	Frame based or Frameless	CT	Stereotaxy and lesion localization	Stereotaxy Functional neurosurgery Drilling at the skull base	Commercially available First FDA-approved neurosurg robot Diverse clinical application	No microsurgical ability No tool actuation
2003	NeuRobot[19–21]	3	3	Frameless	N/A	Microsurgery	Tumor resection	Partial resection of meningioma Telesurgery on rat from 40 km away	Limited to 3 DOF for each arm Low payload
2005	Georgetown[22,23]	6	1	N/A	Fluoroscopy	Stereotaxy	Percutaneous facet blocks	Accuracy comparable to manual technique	Movement occurs in 1 DOF at a time Requires fluoroscopy suite
2006	Pathfinder[24]	6	1	Frameless	CT	Stereotaxy	Epilepsy surgery	Highly accurate navigation	CT required for feducial placement Surgical ergonomics

(continued)

Table 1 (continued)

Year	Project name	DOF	# EE	Navigation	Imaging	Purpose	Documented use	Accomplishments	Limitations
2006	SpineAssist[25–27]	6	1	Frame based	CT	Spinal instrumentation	Guide for tool positioning Guide for screw placement	FDA approval for spinal surgery	Limited range of application
2009	NISS[28]	5	1	CT	CT	CT-guided navigation	Image-guided implantation	In vivo and in vitro surgical implantation	Ionizing radiation
2009	neuroArm[29–33]	7	2	Frameless	MRI	Presurgical planning Microsurgery and stereotaxy	Various intracranial pathology Intracranial tumor resection	Microsurgery and stereotaxy MRI-compatible robot and tools Haptic feedback	Expensive

DOF degrees of freedom, *#EE* number of end effectors, *PUMA* programmable universal machine for assembly, *BRW* Brown–Roberts–Well, *CT* computer tomographic, *FDA* Food and Drug Administration, *OR* operating room, *MRI* magnetic resonance imaging, ***km*** kilometer, *NISS* Neuroscience Institute Surgical System

References: [1]Kwoh YS, et al. (1988) IEEE Trans Biomed Engg 35(2):153-160. [2]Drake JM, et al. (1991) Neurosurgery 29(1):27-33. [3]Glauser D, et al. (1995) J Image Guid Surg 1(5):266-72. [4]Hefti J-L, et al. (1998) Comp Aid Surg 3:1-10. [5]Frankhauser H, et al. (1994) Stereotact Funct Neurosurg 63(1-4):93-8. [6]Adler JR, et al. (1997) Stereotact Funct Neurosurg 69(1-4 Pt 2):124-8. [7]Das H et al. (1999) Comp Aided Surg 4:15-25. [8]Le Roux PD, et al. (2001) Neurosurgery 48(3):584-9. [9]Taylor R, et al. (1999) Int J Rob Res 18:1201-1210. [10]Chinzei K, et al. (2001) Med Sci Monit 7(1):153-63. [11]Chinzei K, et al. (2003) Min Invas Ther & Allied Technol 12(1-2):59-64. [12]Zimmerman M, et al. (2002) Neurosurgery 51:1446-52. [13]Zimmerman M, et al. (2004) Acta Neurochir (Wein) 146:697-704. [14]Nimsky CH, et al. (2004) Minim Invasive Neurosurg 47(1):41-6. [15]Li QH, et al. (2002) Comp Aided Surg 7(2):90-98. [16]Varma TRK, et al. (2006) Int J Med Rob Comp Assist Surg 2(2):107-113. [17]Varma TRK, et al. (2003) Stereotact Funct Neurosurg 80(1-4):132-5. [18]Xia T, et al. (2008) Int J Med Robot 4(4):321-330. [19]Hongo K, et al. (2002) Neurosurgery 51(4)985-988. [20]Goto T, et al. (2003) J Neurosurg 99(6),1082-4. [21]Hongo K, et al. (2006) Acta Neurochir Suppl (Wien) 98:63-66. [22]Cleary K, et al. (2002) Acad Radiol 9(7):821-5. [23]Cleary K, et al. (2005) Int J Med Robot 1(2):40-47. [24]Eljamel MS, et al. (2006) Int J Med Rob Comp Assist Surg 2:233-7. [25]Lieberman IH, et al. (2006) Neurosurgery 59(3):641-50. [26]Sukovich W, et al. (2006) Int J Med Robot 2(2):114-122. [27]Barzilay Y, et al. (2006) Int J Med Robot 2(2): 181-193. [28]Chan F, et al. (2009) Surg Neurol 71:640-8. [29]Louw DF, et al. (2004) Neurosurgery 54(3):525-36. [30]Sutherland GR, et al. (2008) Neurosurgery 62(2):286-93. [31]Pandya S, et al. (2009) Neurosurgery 111(6):1141-9. [32]Sutherland GR, et al. (2008) IEEE Eng Med Biol Mag 27(3):59-65. [33]Greer AD, et al. (2008) IEEE/ASME Trans Mech 13(3):306-315

Assembly (PUMA), into the operating theater in 1985 [21]. In 1987, Benabid et al. coupled the PUMA 200 robot to a Brown–Roberts–Wells (BRW) head frame for frame-based stereotaxy [22]. The robotic arm had six degrees of freedom (DOF), and each joint was equipped with spring-applied, solenoid-release brakes that would immediately stop motion should any system defect arise. Using CT imaging for navigation, the system was able to orient a cannula for needle insertion. The robot was modified to hold a retractor for the resection of multiple pediatric thalamic astrocytomas in 1991 [23]. These developments provided proof-of-concept of robotic neurosurgery, allowing multidisciplinary research teams to begin development of robotic systems designed for specific neurosurgical applications.

In 1994, Frankenhauser et al. introduced the Minerva robot [24]. The system was coupled to the same BRW headframe used by Benabid, and mounted inside a CT scanner for image-guided biopsy and implantation. Unfortunately, the requirement of a dedicated CT scanner limited applicability of Minerva to the neurosurgical community at large. Two systems, the Robot Assisted Microsurgical System (RAMS) and Steady Hand system, were developed in 1999 to enhance microsurgery [25, 26]. While these projects were shown to improve microsurgical precision and provided haptic feedback, respectively, they have not yet been clinically applied.

By the year 2000, advances in computer technology, the ubiquity of neurological imaging, and increasing adaptation of robotic technology across the manufacturing and aerospace industries provided the potential for development of an image-guided neurosurgical robotic system. The Harvard MRI robot was developed and integrated with intraoperative MR imaging (iMRI) [27]. The robot manipulator was MR compatible, and used for intraoperative tool orientation. NeuRobot emerged as a microsurgical robot in 2003 [28]. It was telecapable, and tested in an experimental model with the surgeon located 40 km away from the surgical site.

In addition to developments in microneurosurgery, robotic systems were designed for endoscopic and stereotactic application. The Evolution I robot was developed for endoscopic applications, and used in 2002 for the transphenoidal removal of pituitary adenoma [29]. The CT-based NeuroMate system became the first neurosurgical robot to receive FDA approval for clinical use [30]. The Pathfinder system followed in 2006, with highly accurate CT-guided stereotactic application [31]. Finally, the NISS collaboration was published in 2009 with direct application in CT-guided implantation procedures [32].

Two important robotic systems have been developed for spinal surgery. In 2005, the Georgetown robot was used for fluoroscopy-guided percutaneous facet blocks in cadaveric studies and clinical patients [33]. Accuracy was deemed comparable to manual technique, but movement was only available in 1 DOF at a time. The

SpineAssist robot has been used for tool positioning and pedicle screw placement [34]. It received FDA approval and is commercially available to this day.

3 neuroArm

Advances in technology provided the opportunity to develop a robotic system capable of both microsurgery and stereotaxy [35]. Furthermore, due to the increasing acceptance of intraoperative MR imaging, it was also desirable to construct a system that could operate within this imaging environment [36]. While this presented a significant challenge, it would resolve the problem of disrupting surgical rhythm for intraoperative image acquisition. In 2002, investigators at the University of Calgary (Calgary, Alberta, Canada), in collaboration with MacDonald, Dettweiller and Associates, began the development of such a robot.

4 neuroArm: Design and Manufacture

The initial requirements document for neuroArm included the ability to perform both microsurgery and stereotaxy within the bore of a 1.5 T magnet. At the time of preliminary design review, it became evident that stereoscopic vision, a necessary requirement for microsurgery, could not practically be captured within the magnet bore. Furthermore, due to requirements of payload (750 g) and speed (200 mm/s), as well as the size of then existing position encoders, the manipulators needed to be relatively large. Both could not be placed within the 70-cm working aperture of the intraoperative 1.5 T magnet. For these reasons, scope was changed such that image-guided microsurgery would be performed outside of the bore of the magnet, and stereotaxy, using a single arm, within the bore of the magnet.

4.1 Manipulators

neuroArm consists of two anthropomorphic manipulators, each with a total of seven DOF: six spatial and one degree of tool actuation [37]. Each manipulator was designed to reflect the limbs and joints of a human surgeon: shoulder joint to allow for rotation in both the horizontal and vertical plane, elbow joint to allow for flexion/extension of the arm, and a wrist joint to allow adjustment of rotation and pitch. Tool actuation is accomplished through motion of one tool holder relative to the other. The arms are mounted on a mobile base, which allows height adjustment to accommodate the position of the operating table.

To achieve MR compatibility, the manipulators were manufactured from titanium, polyetheretherketone (PEEK), and polyoxymethylene (Delrin; Dupont, Wilmington, DE). Motion of the arms is accomplished using ultrasonic piezoelectric actuators (Nanomotion, Yokneam, Israel) that have a 20,000-h lifetime, 1-nm resolution, and inherent braking characteristics if power is lost. Absolute 16-bit sine/cosine encoders provided 0.01-degree accuracy at each joint and retained positional information when powered off for imaging. Haptic feedback was provided in three translational DOF from six-axis force/torque sensors (ATI Industrial Automation, Apex, NC) that were specifically manufactured for neuroArm. The end effectors were designed to hold a variety of tools using a standardized interface that allows for tool roll and tool actuation (Fig. 2).

4.2 Mobile Base

The manipulators are moved in and out of the operating room on a height-adjustable mobile base. A digitizing arm, mounted on the mobile base, allows registration of the manipulators to the radio frequency (RF) coil. The information, transferred to the computerized human–machine interface, allows 3D MR image display with tool-overlay. The field camera, mounted on the mobile base, provides overall visual feed of the surgical field. For stereotactic procedures, the mobile base is used to transport the manipulators to a platform inserted into the MR magnet (Fig. 3).

4.3 Main System Controller

The main system controller consists of four main software applications, each operating on an individual computer: (1) The command and status display provides the main graphical control interface for the neuroArm end effectors. (2) The MRI display provides 2D and 3D volumetric images of patient pathology with tool overlay. (3) The hand controller interfaces to the left and right human–machine interface hand controllers process kinematic motion at the human–machine interface. (4) The controller interfaces to the manipulator arms and other hardware.

4.4 Human–Machine Interface

The human–machine interface recreates the sight, sound, and touch of surgery, while facilitating integration of advanced imaging and surgical planning technologies [38] (Fig. 4). Two high-definition cameras (Ikegami Tsushinki Co, Tokyo, Japan) mounted on the surgical microscope provide a stereoscopic image at 1000 TV lines horizontal resolution to a 3D computer monitor (Alienware, USA). The MRI display can be manipulated by touch with on-screen controls to view patient-specific MR images in 2D or 3D, with real-time tool overlay. The surgeon is thus able to see the tools as they are manipulated down the surgical corridor, and their spatial relationship to the pathology. During stereotaxy, the command status display provides real-time feedback of end effector orientation relative to the RF coil and magnet bore.

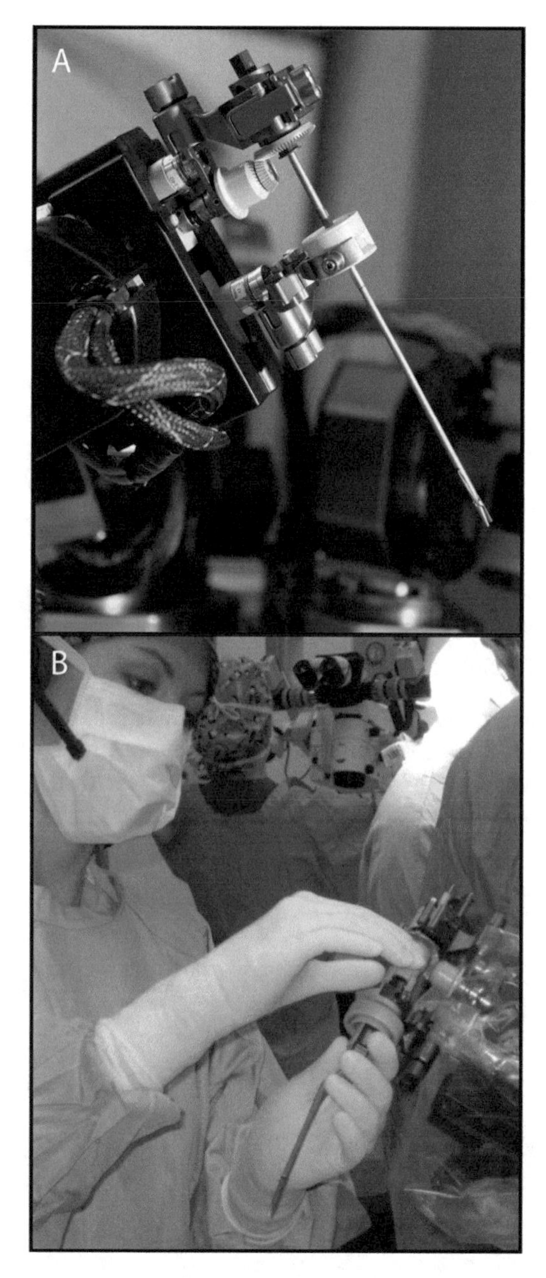

Fig. 2 (**a**) The neuroArm end effector uses two standardized connectors (*blue*) to hold a surgical tool. The upper connection includes a gear to control tool roll, while the lower connection moves upward and downward to allow for tool actuation. (**b**) During surgery, the neuroArm manipulators are draped for sterility, while the two standardized tool connectors penetrate the drape. The scrub nurse is able to exchange all the neuroArm tools with the standardized tool connectors

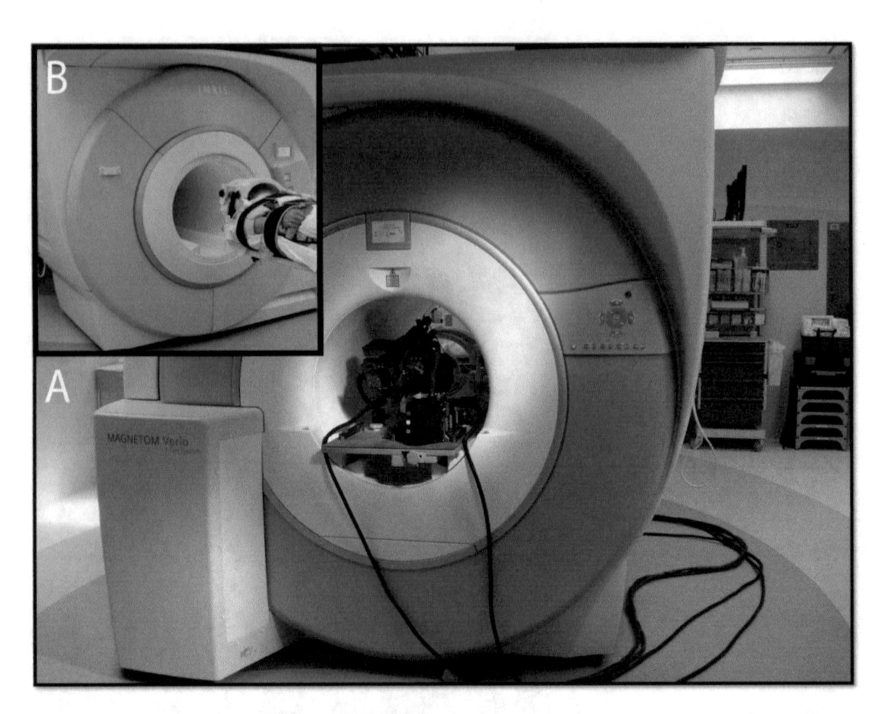

Fig. 3 (**a**) During stereotaxy, one neuroArm manipulator is placed on a specialized board within the iMRI magnet bore, opposite the patient. (**b**) The iMRI machine moves to the operating table, so the patient and neuroArm end effector meet at the magnet isocenter. Stereotaxy near the magnet isocenter allows for simplified registration and optimized image quality throughout the procedure

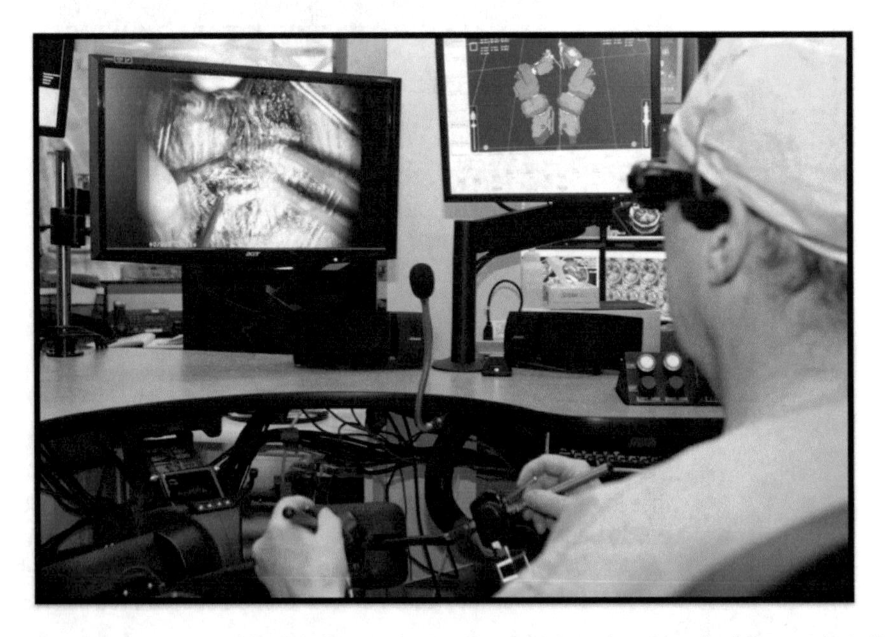

Fig. 4 The neuroArm human–machine interface recreates the sight, sound, and touch of the surgical site for the surgeon. The surgeon is provided with a 3D stereoscopic view of the surgical corridor. The command status display (*right* of the stereoscopic display) shows the position of the neuroArm manipulators in relation to the radio frequency coil. The surgeon controls neuroArm using two modified PHANTOM hand controllers, providing 7 DOF control with 3 DOF force feedback. The surgeon communicates with the surgical team using a wireless headset

The human–machine interface is height adjustable, and two modified PHANTOM hand controllers (SensAble Technologies, Inc., Woburn, MA), which provide haptic feedback in three translational DOF, are mounted at 45° to optimize ergonomics. The surgeon activates each of the manipulators by continuously depressing a corresponding foot pedal. Each pedal acts as a *live-man switch*: if disengaged, all motion input to the corresponding manipulator is immediately stopped, bypassing the main system controller. This design is not dependent on the computational capacity of the main system controller, providing the surgeon ultimate control over any adverse motion of the manipulators. Motion scaling and tremor filtration can be applied from the command status display.

As the neuroArm human–machine interface is telecapable, precise and reliable audio communication with surgical staff is of critical importance. Each member of the surgical team wears a wireless headset with a microphone to maintain smooth surgical rhythm. This alone seems to enhance communication between all members of the surgical team, empowering each member of the surgical team to optimize their role in the procedure

4.5 Safety

The creation of a telecapable human–machine interface provides an ergonomic platform from which the surgeon can interact with complex imaging and sensory data sets. For the surgeon to interact within this complex data set, and at the surgical site, increasingly sophisticated robotic manipulators have been developed and provided. This creates some unique challenges relative to patient safety, as the surgeon is no longer mechanically engaged with the surgical site. Thus, motion of the robotic manipulators passes through the main system controller, where all of the input and output from the surgical site is being analyzed and interpreted. Computational complexity is further increased by redundancy in all manipulator position encoders, force measurements, computer power supplies, and redundant hardware wiring. This introduces the potential for uncontrolled motion of the robotic manipulators at the surgical site, should any individual element within the computational processing malfunction. While the software is designed to localize any malfunction in the system, this can take time due to the complex nature of the computation. To overcome this, the *live-man switch* was introduced to provide the surgeon with the absolute ability to stop robotic motion at any time of the robotic procedure. In addition, all members of the surgical team are trained on robotic safety, in particular, on the potential for unexpected collision with object or patient. As stereotaxy with neuroArm occurs within the confines of the MR magnet bore, the surgeon is provided with visual feed of the tool by field cameras within the bore. Prior to executing the procedure, the surgeon must perform a verification intraoperative scan containing the tool and destination pathology to assess accuracy of graphical MR tool overlay.

5 neuroArm: Preclinical Studies

neuroArm and its components arrived and were installed in the intraoperative MR Operating Theatre at the Seaman Family MR Research Centre (Calgary, Alberta, Canada) in 2008. Regulatory and ethics approval were received in the same year. While such approval is required for clinical application, introduction, and acceptance into the international neurosurgical community requires suitable preclinical studies to demonstrate performance and capability of the system within the time-sensitive environment of surgery. A two-stage study was conceived to sequentially test (1) the microsurgical capabilities of neuroArm in a rodent model and (2) the stereotactic capabilities of neuroArm in a cadaveric model [39].

5.1 Animal Studies

Animal studies allowed for in situ testing of neuroArm and acclimatization to the novel neuroArm interface. Compared to conventional neurosurgery, when using neuroArm for microsurgery, surgeons were sitting rather than standing, viewing the surgical site through miniature cameras rather than an operating microscope, manipulating hand controllers rather than the tools directly, requesting for tool exchange from the human–machine interface, communicating via wireless headsets, and relying on the assistant surgeon for manipulations requiring increased dexterity at the surgical site. Surgeons adjusted to these changes over the course of the study, indicating a short learning curve.

A Sprague-Dawley rat model was selected to evaluate neuroArm in microsurgery mode. Each procedure involved three objectives (bilateral nephrectomy, splenectomy, and bilateral submandibular gland excision), selected to provide reasonable models of varying microsurgical landscapes. Procedures were completed using either neuroArm or conventional hand techniques, and with a common assistant surgeon. Total surgical time, blood loss, thermal injury, and vascular injury were recorded for each procedure, then weighted and combined into a surgical performance score to compare all measures with a single variable. Using neuroArm and hand techniques, each surgeon was allowed one complete procedure for familiarization with the equipment and surgical objectives. Results of the subsequent four procedures were recorded for evaluation [39].

For each procedure, the abdominal cavity and neck were opened with midline incisions. The renal vessels were exposed, coagulated with bipolar electrocautery and the kidneys removed from the abdominal cavity. Splenectomy required hemostasis of the splenic vessels, then dissection from abdominal contents and removal from the abdomen. Submandibular gland excision required more cutting and less hemostasis than the previous objectives given the fibrous nature of surrounding neck tissue. For all robotic trials, neuroArm was equipped with bipolar forceps in the right end effector and tissue forceps in the left. For nonrobotic trials, the surgeons were provided bipolar forceps and a standard selection of microsurgical instruments.

Results from procedures using neuroArm were compared to those of procedures using hand techniques. While the use of neuroArm increased the total surgical time, there was decreased blood loss compared to hand trials, resulting in equal overall surgical performance. Increased surgical time had been, in part, expected as the introduction of intraoperative technology has previously been shown to increase surgical time. However, surgical time is not the only predictor of surgical outcome, which is why blood loss and other performance measures had been recorded. The decreased blood loss when using neuroArm may have been a result of increased caution from the surgeon, who was no longer directly present at the surgical site for a rapid response to a vascular event should one occur. While decreased blood loss did not reach statistical significance due to small sample size, it remains an important measure of surgical performance.

5.2 Cadaveric Studies

Following animal studies, the neuroArm navigation system was tested by image-guided implantation of ferrous oxide coated nanoparticles in a cadaveric model. The head of the caudate nucleus and globus pallidus were selected as target implantation sites and identified on all trials by a senior resident neurosurgeon. This was an important preclinical study to evaluate the accuracy of the frameless neuroArm navigation software as compared to an already established navigation system.

Following bilateral frontal craniotomy with a pneumatic drill, the cadaveric specimens were placed in a head clamp. T1-weighted MR images were acquired at 2 mm slice thickness. For neuroArm trials, one neuroArm end effector was placed inside the MR magnet bore and registered to the head clamp and images. The senior resident neurosurgeon identified the targets, then planned implantation trajectory using the tool tip extension feature and the Z-lock feature, which restricts end effector motion to only the direction of the tool axis. These features, coupled with 3D tool overlay at the neuroArm human–machine interface, greatly simplified the implantation procedure. Nonrobotic implantation was completed on the contralateral side using the VectoVision Sky system (BrainLAB). The specimens remained in the same head clamp, and the same preoperative T1-weighted MR images were loaded onto VectorVision. For this implantation, the surgeon was presented with sagittal, axial, and coronal images at the tool tip. Following all implantation procedures, the specimens were imaged using the same acquisition sequences to determine the final position of nanoparticles relative to the desired targets. The neuroArm system was more accurate than the VectorVision system, but did not reach statistical significance due to small sample size ($n = 4$ targets for each modality).

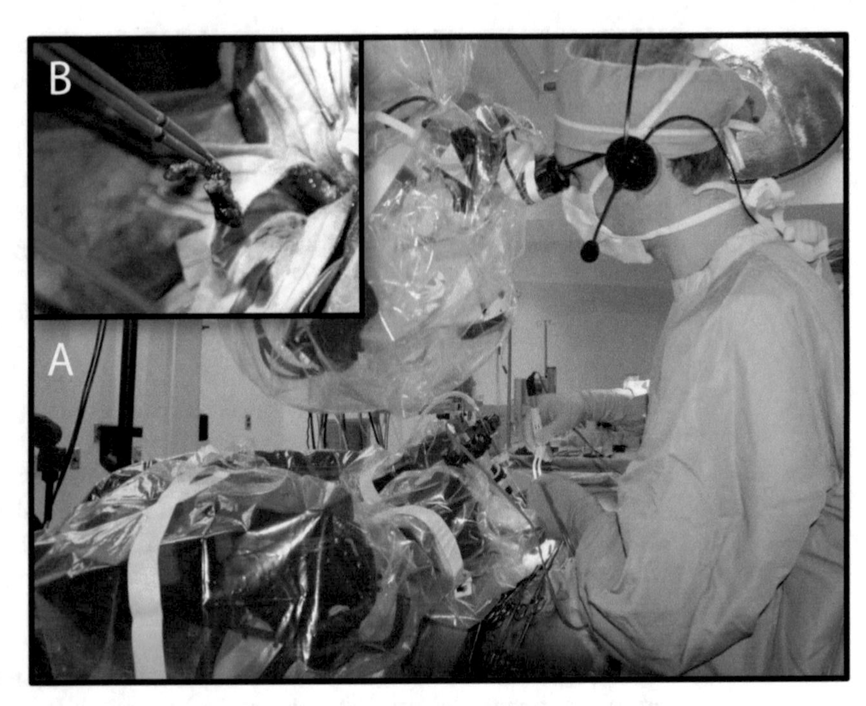

Fig. 5 (**a**) For microsurgical procedures, neuroArm is positioned at the operating table in the position of the primary surgeon. The assistant surgeon is able to operate in an ergonomic position relative to neuroArm and the operating microscope. (**b**) The neuroArm bipolar forceps can be used to coagulate, as well as remove pathological tissue

6 neuroArm: Clinical Studies

neuroArm represents a novel paradigm for neurosurgery, which created a number of practical considerations for implementation into established neurosurgical procedure. Wireless headsets were implemented for all members of the surgical team to allow communication with the surgeon at the human–machine interface. The scrub nurse became responsible for exchanging the neuroArm tools. Draping of the robot, and the time of draping relative to preoperative MR imaging and registration to patient anatomy required careful planning into existing nursing and MR technician protocols (Fig. 5). For these reasons, clinical integration of neuroArm was accomplished in a step-wise fashion.

Among the first 22 cases, 10 were meningioma, 9 glioma, 2 acoustic schwannoma, and 1 brain abscess. Each of these procedures required general anesthetic and craniotomy. For four cases, neuroArm was registered to the intraoperatively acquired surgical planning MR images, and used to target the pathology and determine craniotomy placement (Fig. 6). In all cases, neuroArm was draped during craniotomy, and brought into the surgical field after partial dissection of the pathology. Working at the human–machine

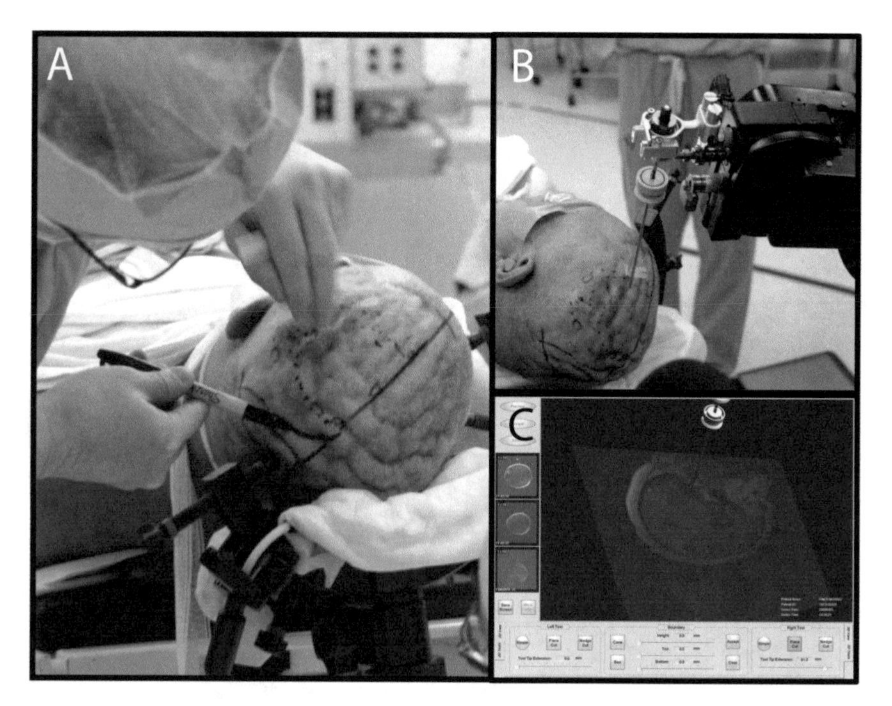

Fig. 6 (**a**) Conventional presurgical planning involves marking of the surgical site following anesthetic and fixation with pins in a head clamp. (**b**) Prior to craniotomy, neuroArm navigation software can be used to confirm and refine craniotomy placement based on intracranial pathology. (**c**) Patient-specific MR images are loaded into the MRI display at the neuroArm human–machine interface

interface, the surgeon was able to manipulate tools within the surgical corridors, coagulate vessels to control bleeding, and aspirate (Fig. 7). For the brain abscess case, the bipolar forceps, mounted in the right arm, was used to open the tumor capsule and allow drainage of pus.

There was a disruption in the ongoing integration of neuroArm into neurosurgery in 2009, as the 1.5 T iMRI environment with local RF shielding was upgraded to a 3.0 T iMRI suite that included whole room RF shielding. This upgrade required a 10-month interval, during which the operating room was shut down to patient cases. The upgrade to whole room shielding allowed dramatic improvements in practical aspects for stereotactic procedures. The manipulator is now able to be attached directly to the magnet bore, rather than being mounted on an extension board from the OR table. Cables are now run through the backside of the magnet, rather than along the OR table, which was previously required to prevent penetration of the RF shielding. Finally, the registration procedure is much simpler as the location of the manipulator is always constant relative to the magnet's isocenter, and thus the patient's pathology (Fig. 3).

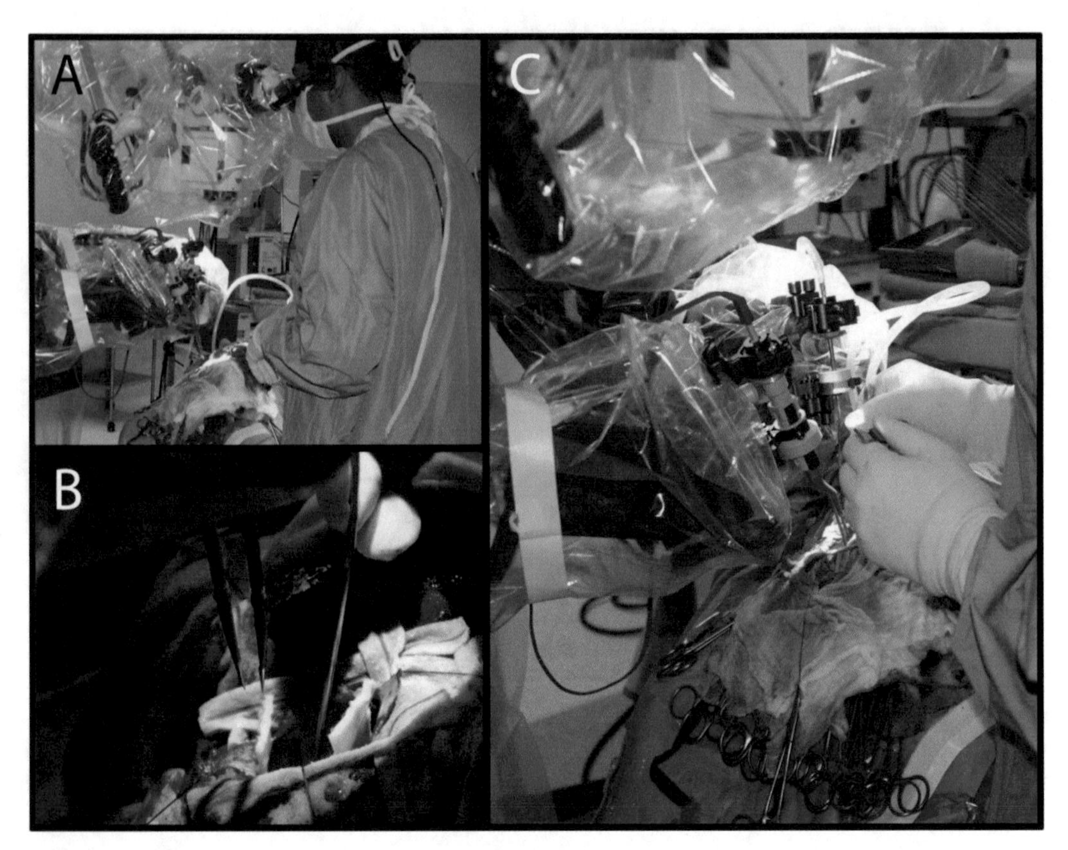

Fig. 7 (**a**) Positioning neuroArm into surgical procedure is important so that ergonomics of the surgical assistant and scrub nurse are not compromised. (**b**) At the surgical site, both neuroArm and the assistant surgeon are able to manipulate tools within the surgical corridor. (**c**) Sterile drapes are placed over the neuroArm manipulators, while the tool holders (*blue*) are able to penetrate the sterile drapes and hold the tools

7 Use of Surgical Robotics in Animal Models

Robotic surgery in animal models occurs primarily as preclinical testing of systems intended for clinical use. Animal studies provide optimal simulation of the surgical environment, recreating the dynamic landscape of the surgical field and time-sensitive events such as hemorrhage. The surgeon becomes puppeteer of the surgical corridor; manipulating its characteristics to achieve the surgical objective. To date, there is no superior simulation of clinical surgery than surgery in animal models. This fact has driven regulatory requirements for surgical technology across North America, and all surgical robotic systems require testing in animal surgery. The primary investigator and colleagues often complete these validation studies, but once regulatory approval has been achieved, it is no longer necessary to sacrifice animal life to validate robotic performance.

There does, however, remain a need to train individual surgeons on the use of any surgical robotic system. For neuroArm, a detailed training paradigm has been developed. It begins with familiarization of equipment using an interactive, internet-based computerized tutorial. The surgeon manipulates a virtual neuroArm through space to accomplish very simplistic objectives including positioning of equipment, stacking of objects, and tracing of convoluted tool tip trajectories. The next step involves the use of the real neuroArm robot in simple tasks: moving rings on pegs or sponges between containers. These stages provide excellent familiarization of the neuroArm equipment, but do not replicate the dynamic complexities of the surgical site. For this reason, it remains customary for new surgeons to complete minor procedures in a rodent model prior to using neuroArm in clinical application. There is simply no adequate substitution for this experience at the present time. It is hoped, that through patient-specific virtual surgery simulators, dynamic environments will be created that allow surgical rehearsal without the expense of animal life.

When neuroArm is used in the animal setting, it offers the surgeon certain advantages over conventional techniques. These advantages are similar to the benefits of using neuroArm in the clinical setting, and include:

- Increased kinetic precision of the distal tool tip to the order of micrometers
- Increased tool tip accuracy from sine/cosine absolute position encoders
- Improved human–machine interface including real-time 3D tool overlay, visualization of the surgical site and the intuitive Z-lock feature for stereotaxy-biopsy

The present challenges of robotic surgery in animals are being rapidly achieved through advances in robotics and human–machine interface technology. Robotic systems of the future will have increased dexterity and push the limits of precision and accuracy toward the cellular level. Human–machine interfaces will become more intuitive to operate and increasingly integrated with intraoperative imaging. Finally, developments in low and high fidelity haptic technology will improve the sense of touch that is recreated at the human–machine interface. This factor may be the most important in driving international adaptation of robotic systems. Present limitations of robotic surgery with neuroArm include:

- The dexterity of a human surgeon is greater than 6 DOF because a tool may be held in a single position with multiple positions of proximal joints in the arms. Due to present computer algorithms, neuroArm dexterity is limited to 6 DOF: any specific tool position can only be achieved with a unique combination of proximal joint positions. The greater robotics industry has solved this challenge already, and algorithms are presently being evaluated for suitability with neuroArm.

- While neuroArm offers low frequency haptic feedback in 3 DOF, it is not able to completely recreate a surgeon's sense of kinesthetic touch at the surgical site. Ongoing international research in haptics is advancing this science and will improve future haptic interfaces.
- Since the preclinical animal studies, the human–machine interface visualization system was upgraded from Ikegami cameras inside a microscope-style viewfinder to a single high-definition 3D display. This has overcome limitations at the time of preclinical studies of decreased picture quality compared to the microscope view at the surgical site.

It is very likely that the future of animal surgery will involve an increasing number of robotic systems. Perhaps the greatest issue preventing the widespread development of robotic systems for animal applications is the high cost of surgical robotics. Robotics intended for clinical use have benefited from the greater funding that has been required to initiate their development. However, as robotic technology becomes more universally applied in animal procedures, costs of development and implementation will drop, allowing creation of systems for specific therapeutic purposes in animal applications.

8 Conclusion

The future of neurosurgery lies in the realm of multidisciplinary teamwork. As surgical staff gather increasing clinical experience with neurosurgical robotic systems, robotic technology will continue to be advanced by teams of engineers, scientists, and technologists around the world. Mechanically, advances in MR-compatible robotics will allow miniaturization without sacrifices in surgical performance. This will overcome present spatial limitations and allow movement toward real-time, image-guided microsurgery within the physical constraints of intraoperative imaging devices.

Perhaps of more impact to the surgeon will be the rapid upgrades in human–machine interface technology. In the future, the surgeon will be presented with pre- and intraoperative images in a manner that is surgically relevant and intuitive to manipulate. Improved computer processing will provide relevant data related to anatomy, function, and metabolism. The surgeon will be provided with realistic recreation of touch through ongoing developments in haptic feedback and hand controller design. Measurement of surgical forces and their relationship to tissue deformation will open new areas of research in basic science toward the understanding of tactile perception and its relation to surgical decision-making.

Neurosurgical robotics will become the hub of technology in the operating room, allowing an interface for imaging and surgical management, advanced tool design, image-guided biopsy and implantation, and realization of individualized therapy through cell-specific intervention. From this, patients will receive better, more accurate, and increasingly precise neurosurgical care.

Acknowledgments

Supported by grants from the Canada Foundation for Innovation, Western Economic Diversification Canada, Alberta Advanced Education and Technology, Alberta Heritage Foundation for Medical Research, and the Canadian Institute for Health Research.

References

1. Cushing H, Bovie WT (1928) Electrosurgery as an aid to the removal of intracranial tumors, with a preliminary note on a new surgical-current generator by W.T. Bovie. Surg Gynecol Obstet 27:751–785
2. Kriss TC, Kriss VM (1998) History of the operating microscope: from magnifying glass to microneurosurgery. Neurosurgery 42(4):899–907
3. Yasargil MG (1969) Microsurgery applied to neurosurgery. Academic, New York
4. Broca P (1861) Nouvelle observation d'aphe´mie produite par une le´sion de la troisie`me circonvolution frontale. Bulletins de la Socie´te´ d'anatomie 2e serie 6:398–407
5. Roentgen WC (1895) On a new kind of rays. Proc Phys-Med Soc, Wurzburg
6. Dandy WE (1919) Roentgenography of the brain after injection of air into the spinal canal. Ann Surg 70:397
7. Hounsfield GN (1973) Computerized transverse axial scanning (tomography): part 1. Description of system. Br J Radiol 46:1016
8. Lauterbur PC (1980) Progress in n.m.r. zeugmatogrpahy imaging. Philos Trans R Soc Lond B Biol Sci 289(1037):483–487
9. Mansfield P, Maudsley AA (1977) Medical imaging by NMR. Br J Radiol 50:188–194
10. Ogawa S, Tso-Ming L, Nayak AS et al (1990) Oxygenation-sensitive contrast in magnetic resonance image of rodent brain at high magnetic fields. Magn Reson Med 14:68–78
11. Peeling J, Sutherland GR (1992) High-resolution 1H NMR spectroscopy studies of extracts of human cerebral neoplasms. Magn Reson Med 24:123–136
12. Lacy AM, Garcia-Valdecasas JC, Delgado S et al (2002) Laparoscopy-assisted colectomy versus open colectomy for treatment of non-metastatic colon cancer: a randomized trial. Lancet 359(9325):2224–2229
13. Liem MS, van der Graaf Y, van Steensel CJ et al (1997) Comparison of conventional anterior surgery and laparoscopic surgery for inguinal-hernia repair. N Engl J Med 336(22):1541–1547
14. Kelly PJ, Kall B, Goerss S (1983) Stereotactic CT scanning for the biopsy of intracranial lesions and functional neurosurgery. Appl Neurophysiol 46:193–199
15. Kanner AA, Vogelbaum MA, Mayberg MR et al (2002) Intracranial navigation by using low-field intraoperative magnetic resonance imaging: preliminary experience. J Neurosurg 97:1115–1124
16. Chandler WF, Knake JE, McGillicuddy JE et al (1982) Intraoperative use of real-time ultrasonography in neurosurgery. J Neurosurg 57(2):157–163
17. Lunsford LD (1982) A dedicated CT system for the stereotactic operating room. Appl Neurophysiol 45(4–5):374–378
18. Black PM, Moriarty T, Alexander E III et al (1997) Development and implementation of intraoperative magnetic resonance imaging and its neurosurgical applications. Neurosurgery 41(4):831–835
19. Sutherland GR, Kaibara T, Louw D et al (1999) A mobile high-field magnetic resonance system for neurosurgery. J Neurosurg 91(5):804–813
20. Lang MJ, Sutherland GR (2010) Informatic surgery: the union of surgeon and machine. World Neurosurg 74(1):118–120
21. Kwoh YS, Hou J, Jonckheere EA et al (1988) A robot with improved absolute positioning

accuracy for CT guided stereotactic brain surgery. IEEE Trans Biomed Eng 35(2):153–160
22. Benabid AL, Cinquin P, Lavaile S et al (1987) Computer-driven robot for stereotactic surgery connected to CT scan and magnetic resonance imaging: technological design and preliminary results. Appl Neurophysiol 50(1–6):153–154
23. Drake JM, Joy M, Goldenberg A et al (1991) Computer- and robot-assisted resection of thalamic astrocytomas in children. Neurosurgery 29(1):27–33
24. Fankhauser H, Glauser D, Flury P et al (1994) Robot for CT-guided stereotactic neurosurgery. Stereotact Funct Neurosurg 63(1–4):93–98
25. Le Roux PD, Das H, Esquenazi S et al (2001) Robot-assisted microsurgery: a feasibility study in the rat. Neurosurgery 48(3):584–589
26. Taylor R, Jensen P, Whitcomb L et al (1999) A steady-hand robotic system for microsurgical augmentation. Int J Robot Res 18(12):1201–1210
27. Chinzei K, Miller K (2001) Towards MRI guided surgical manipulator. Med Sci Monit 7(1):153–163
28. Hongo K, Kobayashi S, Kakizawa Y et al (2002) NeuRobot: telecontrolled micromanipulator system for minimally invasive microneurosurgery-preliminary results. Neurosurgery 51(4):985–988
29. Zimmermann M, Krishnan R, Raabe A et al (2004) Robot-assisted navigated endoscopic ventriculostomy: implementation of a new technology and first clinical results. Acta Neurochir (Wien) 146(7):697–704
30. Varma TR, Eldridge PR, Forster A et al (2003) Use of the NeuroMate stereotactic robot in frameless mode for movement disorder surgery. Stereotact Funct Neurosurg 80(1-4):132–135
31. Eljamel MS (2006) Robotic application in epilepsy surgery. Int J Med Robot 2:233–237
32. Chan F, Kassim I, Lo C et al (2009) Image-guided robotic neurosurgery—an in vitro and in vivo point accuracy evaluation experimental study. Surg Neurol 71(6):640–647
33. Cleary K, Watson V, Lindisch D et al (2005) Precision placement of instruments for minimally invasive procedures using a "needle driver" robot. Int J Med Robot 1(2):40–47
34. Lieberman IH, Togawa D, Kayanja MM et al (2006) Bone-mounted miniature robotic guidance for pedicle screw and translaminar facet screw placement: part I—technical development and a test case result. Neurosurgery 59(3):641–650
35. Louw DF, Fielding T, McBeth PB et al (2004) Surgical robotics: a review and neurosurgical prototype development. Neurosurgery 54(3):525–536, discussion 536–537
36. Sutherland GR, Latour I, Greer AD (2008) Integrating an image-guided robot with intraoperative MRI: a review of the design and construction of neuroArm. IEEE Eng Med Biol 27(3):59–65
37. Sutherland GR, Latour I, Greer AD et al (2008) An image-guided magnetic resonance-compatible surgical robot. Neurosurgery 62(2):286–292, discussion 292–293
38. Greer AD, Newhook P, Sutherland GR (2008) Human-machine interface for robotic surgery and stereotaxy. IEEE/ASME Trans Mech 13(3):355–361
39. Pandya S, Motkoski JW, Serrano-Almeida C et al (2009) Advancing neurosurgery with image-guided robotics. J Neurosurg 111(6):1141–1149

Chapter 7

Impact Model of Spinal Cord Injury

Dorothée Cantinieaux, Rachelle Franzen, and Jean Schoenen

Abstract

Spinal cord injury (SCI) is a frequent disorder with effective treatment still to be developed. Amongst the various experimental models of SCI, the impact model at the thoracic level is one of the most useful as it mimics the contusion injury, which represents 33 % of all spinal cord injuries encountered in the clinic.

Key words Spinal cord injury, Laminectomy

1 Material

Lesions are realized on adult female Wistar rats weighing approximately 200 g. The whole surgical procedure is performed under general anaesthesia, with a continuous delivery of a mixture of O_2 (N25) with 5 % isoflurane for induction and 2–3 % isoflurane for maintenance.

« Infinite Horizons Spinal Cord Impactor » was acquired from *Precision Systems and Instrumentation*, LLC. Version 5.0.

Mini-osmotic pumps are from Alzet, model 1007D with flow rate of 0.5 μl/h, during 7 days.

1.1 Preparation of Rat for the Lesion

Lesion is realized at the T10 level. This level is chosen because it is rostrally far enough from the central pattern generator (situated at the T13-L1 levels and whose lesion would prevent any locomotor recovery). The T10 level can be located by feeling the thorax of the rat to identify the final floating ribs: T13 vertebra is at this level, so T10 is three vertebras rostrally further. It is easier to count vertebras by feeling them and by feeling spaces between them with a scalpel. Moreover, the T10 vertebra can be spotted by the fact that (1) T9-10-11 vertebras are very squeezed, (2) T9 spinous process is directed backwards (caudal side), T10 spinous process is almost vertical and T11 spinous process is oriented towards rostral part, (3) the T8 vertebra is situated just underneath neck fat tissue, and

Miroslaw Janowski (ed.), *Experimental Neurosurgery in Animal Models*, Neuromethods, vol. 116,
DOI 10.1007/978-1-4939-3730-1_7,

(4) there is a larger space lightly sloping between T8 and T9 compared to other vertebras.

After careful laminectomy at the T10 level, bared spinal cord is protected with sponge material. Spinous process of the following caudal vertebra (11th) is smoothed away by mechanical abrasion, up to obtain a plane surface or even slightly bended, to create a gutter directed towards the bared spinal cord (for placing catheter later).

The sponge material can then be removed.

1.2 Positioning of Rat on Impactor and Lesion

Rat is placed on the impactor table (see Fig. 1). Position of the forceps on rachis determines the success of the lesion. They must be fixed to the vertebras directly adjacent to the laminectomy (9th and 11th), and must penetrate the muscular tissue deeply enough on both sides (3–4 mm) in order to firmly immobilize the rachis and stabilize it during impact (see Fig. 2). Forceps must pinch vertebras on both sides between accessory and lateral processes (see Fig. 3). Jointed arms carrying forceps must be also tightly screwed. It's important the bared spinal cord to be horizontal for the impactor tip, so that all of the spinal cord surface to be injured is reached with the same impact force.

With micrometric screws allowing to adjust impactor table, rat is placed in such a way that the center of the laminectomized region is just below impactor tip (to check the exact position of the tip, bring it down next to the spinal cord).

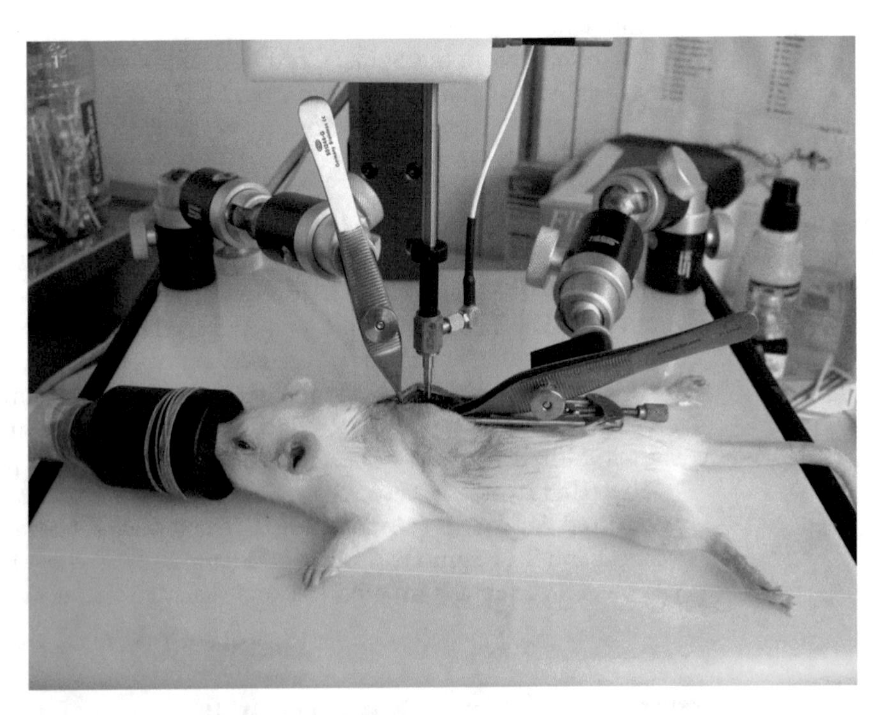

Fig. 1 Positioning of the rat on the impactor table

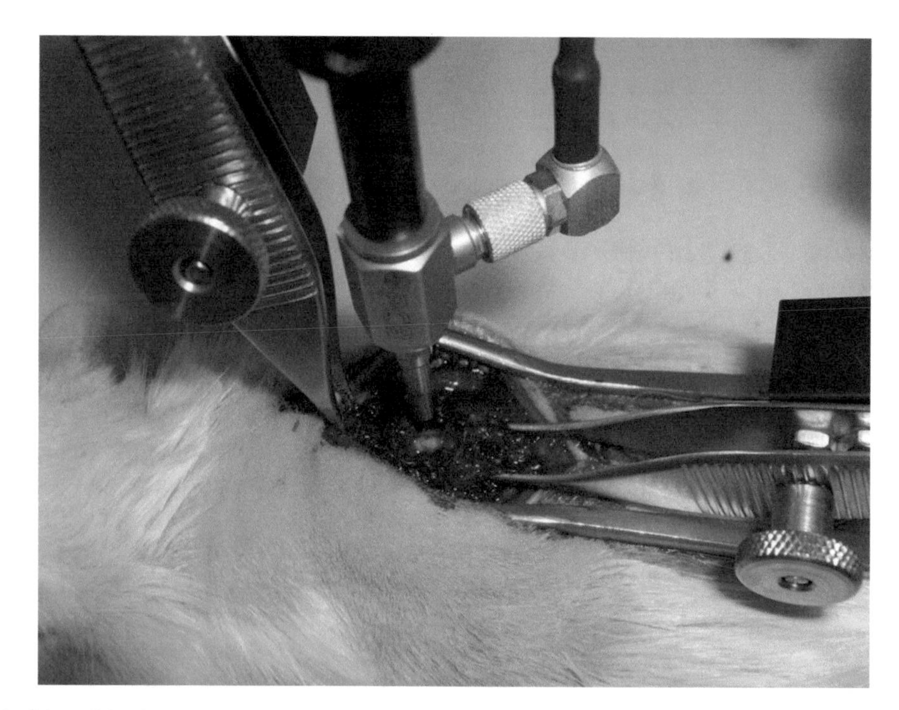

Fig. 2 Position of the forceps on the rachis

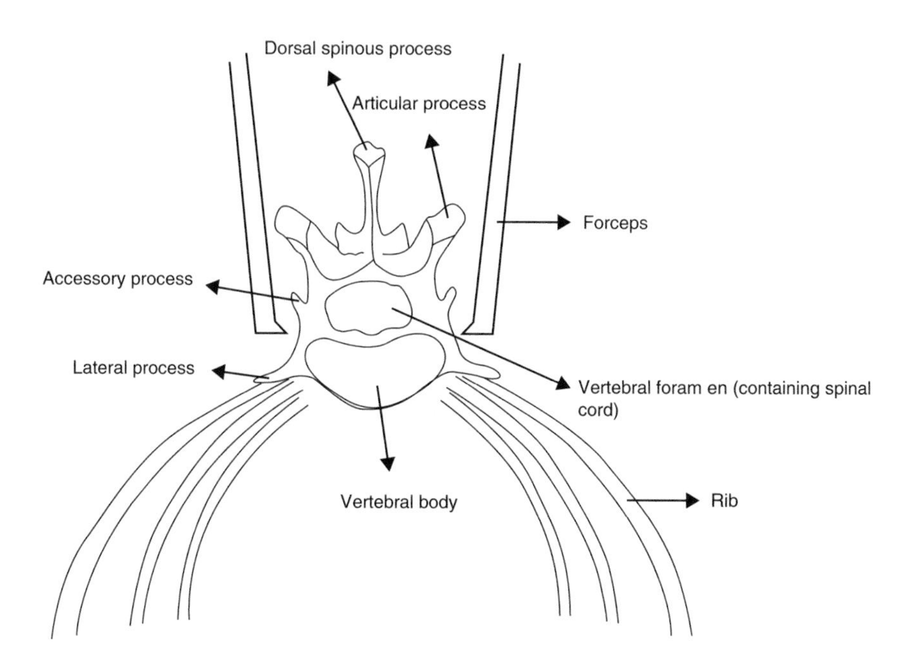

Fig. 3 Outline of position of the forceps on the thoracic vertebras (*transversal view*)

Impactor tip must not touch any bone or muscle pieces at the time of impact. Before performing the impact, the impactor tip is carefully brought against spinal cord, and then it is raised up by four micrometric screw turns (this is for the rod to reach a sufficient speed). The system is now ready, and the lesion is realized at 250 kilodynes.

On adult female Wistar rats weighting 200 g, this impact force allows to reach a displacement in the spinal cord of approximately 1400–1500 μm.

Following the impact, the tip is raised up, forceps are removed, and the rat is replaced on a hot carpet maintained at 37 °C.

1.3 Treatment Administration

In order to allow the treatment solution to penetrate the injured spinal cord tissue, dura must be opened. Under the microscope, and using a very thin needle, a hole is made in the dura (without touching the spinal cord tissue), allowing a yellowish clear liquid to escape. The hole is enlarged by gently pulling dura on 4–6 mm^2 out. With a Hamilton syringe, 10 μl of treatment or control solution is delivered on the top of the lesion. Then, the reservoir of a mini-osmotic pump linked to a catheter and containing the treatment or control solution is placed subcutaneously in the back, caudally to the lesion. A first stitch with a Vycril 5-0 thread is realized around tank with adjacent muscles, to fix the tank to the muscles. The next stitches are made all along the catheter, always with muscles, in order to fasten and direct the catheter towards the lesion (to place it into the gutter made at the beginning) (see Fig. 4). Catheter is next cut just above the lesion to deliver solution at the correct place (see Fig. 4). Muscles between stitches are sutured above the catheter and the skin is sutured with a Vycril 3-0 thread.

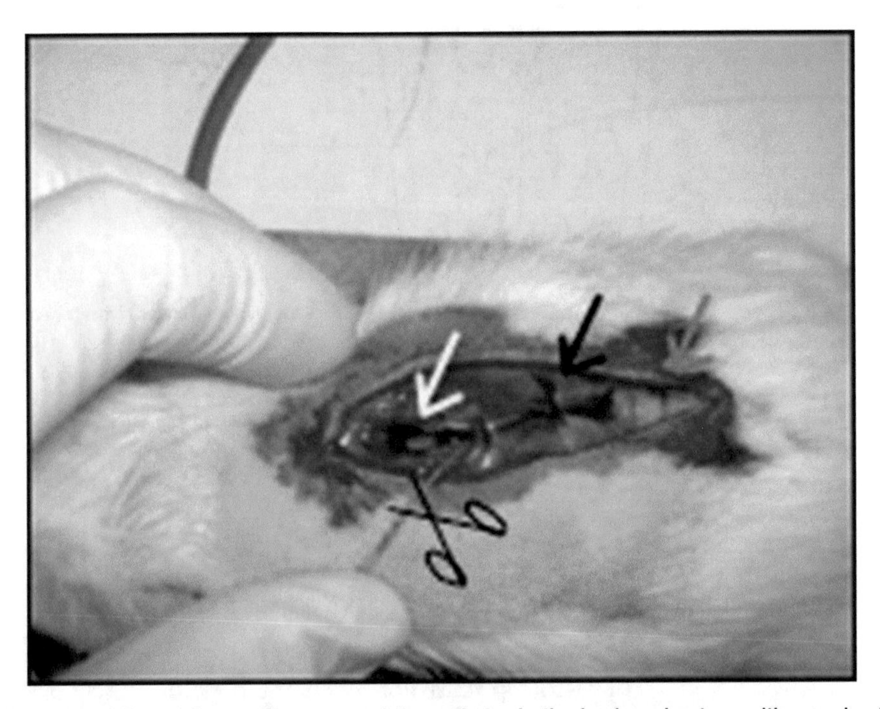

Fig. 4 Placement of the mini-osmotic pump and the catheter in the back and suture with muscles to direct catheter towards lesioned spinal cord. *Grey arrow* shows a stitch around the tank of the minipump, *black arrows* show stitches made along the catheter, and *white arrow* shows the bared spinal cord above which catheter must be cut

Chapter 8

Acute Clip Compression Model of SCI

Jared T. Wilcox and Michael G. Fehlings

Abstract

Developing effective therapeutics to treat spinal cord injury (SCI) requires robust preclinical animal models with substantive clinical relevance. To extrapolate preclinical studies of SCI to human medicine, the animal model must exhibit the proper pathophysiology processes, including hypoxia-ischemia, neuroinflammation, cell death, excitotoxicity, myelin disruption, axonal degradation, astrogliosis, and glial scarring. The modified aneurysm clip compression SCI model has been established and characterized over three decades of use (Dolan and Tator, J Neurosurg 51:229–233, 1979; Rivlin and Tator, Surg Neurol 10:38–43, 1978; Joshi and Fehlings, J Neurotrauma 19:175–203, 2002). We present the cervical clip compression SCI model that delivers a bilateral, dorsoventral lesion. The clip compression model is one of the first non-transection models of SCI (Dolan and Tator, J Neurosurg 51:229–233, 1979), and remains the only well-characterized SCI model incorporating dorsoventral compression. This approach requires concentric access to the dura, observant avoidance of spinal roots, and consistent application of force. The cervical clip compression SCI model has high translatability, and is considered to be one of the most highly relevant models of experimental SCI from a translational perspective (Kwon et al., J Neurotrauma 28:1525–1543, 2010). The clip compression model provides a robust and highly translational lesion for evaluation and assessment of cutting-edge and combinatorial therapeutics (Karimi-Abdolrezaee et al., J Neurosci 30:1657–1676, 2010).

Key words Spinal cord injury, Rat, Modified aneurysm clip, Compression, Cervical

1 Materials

Rivlin-Tator modified Kerr-Lougheed 1/4 slightly curved aneurysm clip and custom clip applicator (Rivlin-Tator rat clip and FEJOTA™ mouse clip available from the Fehlings Laboratory) are required (Fig. 1) [1, 2]. Lesions are realized on Wistar rats of 290 ± 20 g body weight anesthetized with isoflurane at 5 % induction and 2 % maintenance delivered with 1:1 NO_2:O_2 mixture. Standard surgical instrumentation is employed including aim screw retractor, angled offset bone nippers, and dull spinal hook (Fine Science Tools). Procedure is performed with visualization through surgical microscope with built-in illumination and zoom/focus capability.

Miroslaw Janowski (ed.), *Experimental Neurosurgery in Animal Models*, Neuromethods, vol. 116,
DOI 10.1007/978-1-4939-3730-1_8, © Springer Science+Business Media New York 2016

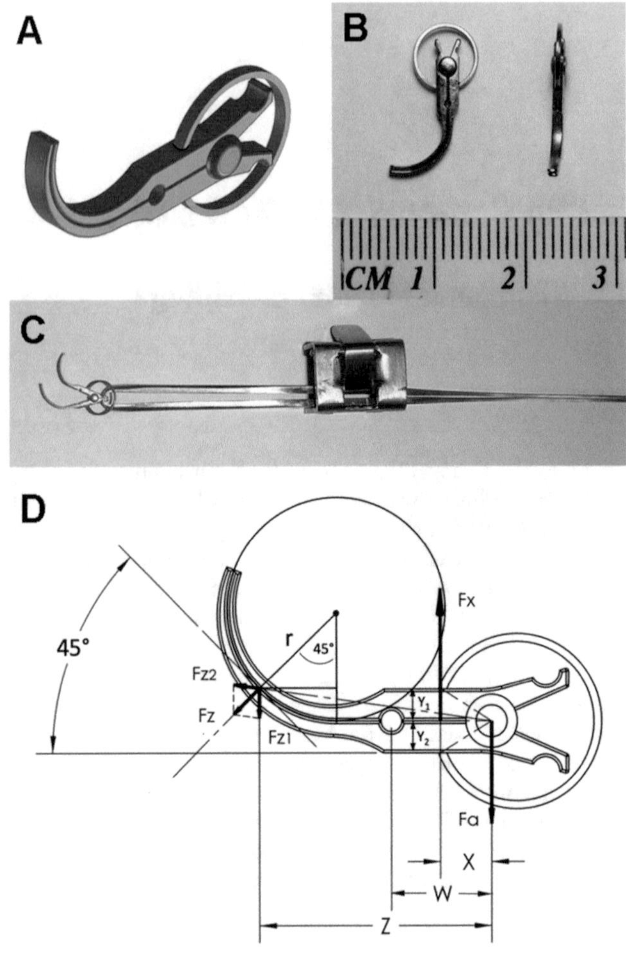

Fig. 1 Calibrated clip and schematic for force measurement. (**a**) Schematic drawing of the 1/4 slightly curved Fehlings modified aneurysm clip for rat models of SCI. (**b**) The rat clip is 1.5 × 1.0 × 0.1 cm with four components assembled. (**c**) The clip is applied using an applicator modified with a triggered quick-release mechanism. (**d**) Clips are calibrated using physical measurements of the clip and forces at a 45° tangential line of force at 1-mm intervals. Computer-generated graphics created by Nikolai Goncharenko and used with permission

2 Methods

2.1 Landmarks and Laminectomy

Standard care for animals, instrumentation, and safety are observed throughout the procedure. We recommend gaseous anaesthesia with a mixture of isoflurane and nitrous oxide to allow rapid induction, smooth anaesthesia, rapid awakening postoperatively, and avoidance of intraoperative hypotension. To realize a lesion with involvement of the muscles of the paw and not the shoulder, the calibrated clip (Fig. 1) is applied to the cord at the C7 vertebral level while avoiding the corresponding C8 nerve roots. The

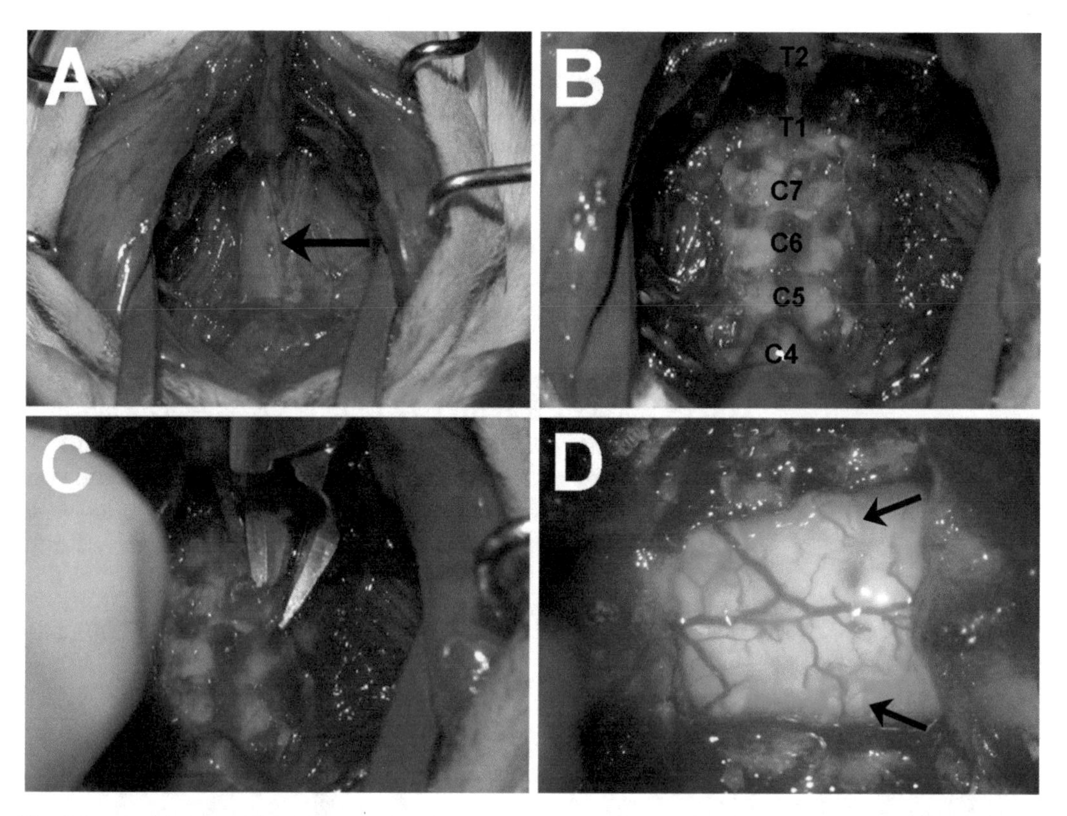

Fig. 2 Landmarking and laminectomy. (**a**) A dorsal approach is taken to access the cervical spinal column, retracting superficial and intermediate muscles, with the spinalis muscle layer incised along the midline (*arrow*) and retracted. (**b**, **c**) With the clear exposure of vertebral laminae immediately medial to facet joints, angled bone nippers are used to remove the laminae at the most lateral aspect without underlying tissue damage. (**d**) The location of tendon-like polygonal tissue (*arrows*) demarcates the interradicular space for approach

prominent spinal process of T2 is evident with palpation, and the splenius capitis muscle completely overlies the C4 vertebrae. Counting rostrally or caudally from these positions clarifies vertebral levels. The procedure is initiated by a skin incision made from occipital protuberance to T2. The cervical spinal column is accessed (Fig. 2): (1) latissimus and rhomboid muscles are cut along the midline raphe from the nuchal ligament caudally to the muscular origins at T2, (2) superficial muscles are retracted with a spring retractor, (3) incision is made at the midline raphe of spinalis muscle extended rostrally to end before the splenius capitis and residing artery, (4) attachments of spinalis and deep muscles are removed from the vertebrae to be laminectomized (C7, T1) and partially from one additional lamina rostral and caudal (C6, T2), (5) muscles are removed laterally to the articular facets by scraping these vertebrae with a scalpel, and (6) spinalis is then retracted with an aim screw retractor. The ligamenta flava of the junctions corresponding to the injury level (i.e., C6/C7 and C7/T1) are then removed with fine forceps. Angled offset bone nippers elicit the laminectomy by cutting the laminae as close to the left and right articular facets as possible. Very little bleeding occurs with proper avoidance of vessels residing adjacent to the T2 spinal process, and within the splenius capitis, ligamenta flava, and lateral posterior vertebral processes.

2.2 Extradural Microdissection of the Cervical Spinal Cord

Ease of entry to the cervical cord—with clarity provided by C7/T1 laminectomy and muscle retraction—determines the success of clip application (Fig. 3). Posterior vertebral processes are laminectomized to an extent allowing the surgeon to visualize the C8 and T1 dorsal spinal roots at the C7/T1 intervertebral bony junction. Extension of laminectomy is made using Friedman-Pearson rongeurs with 0.5 mm cup. Entrance to the ventral aspect of the cord is made at, or immediately rostral to, the C7/T1 bony junction to preserve the dura and ventral spinal vessels. Explicitly, the guiding

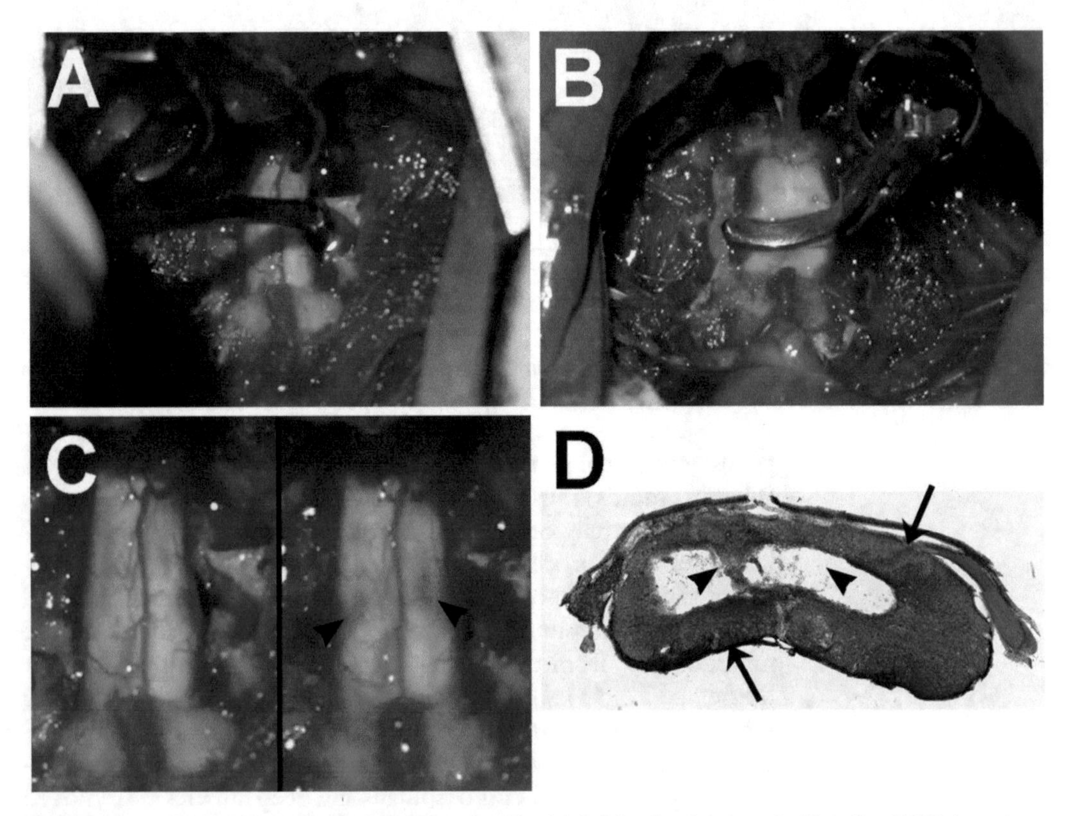

Fig. 3 Extradural microdissection and applying the clip. (**a**) Guiding hook is inserted into the C7/T1 bony junction to form a patent canal for the clip along the ventral spinal cord. (**b**) Once the clip is inserted between the C8 and T1 nerve roots until the distal tip is visible, the lesion is generated as with the contusive impact of the top clip blade and subsequent 60-s compression by clip-specified closing force. (**c**) Bruising of the cord occurs within seconds, resultant from the contusive force of the clip closing (*arrowheads*) indicating proper level of tissue damage. (**d**) Lesion epicenter stained with LFB/H&E displays a central cavitation and scarring (*arrowheads*), significant grey matter loss, and subpial white matter sparing (*arrows*)

hook and clip are inserted bilaterally in the space between the C8 and T1 spinal roots. Visualizing the triangular tendon-like appearance on the dorsolateral surface of the cord reveals the interradicular space. Landmarking this space for dural entry is imperative to avoid distraction, contusion, and avulsion of the spinal roots, which greatly impair the paw and upper limb. The guiding hook is (1) inserted into the space caudal to the C8 roots, (2) gently rotated around the ventral aspect of the cord using (3) small rostrocaudal angular sweeping motions to loosen the natural adhesion of the dural to the posterior face of the C6 vertebral body (4) until the hook tip is seen on the contralateral space.

2.3 Applying the Clip

The modified aneurysm clip applies a bilateral contusion followed by sustained dorsoventral compression (Fig. 3b) [3]. Clips are inexpensive and are calibrated using common instrumentation with highly reliable closing strengths (Fz, Fig. 1d) determined by mathematical reduction of empirical measurement. Stated strengths refer to the closure of blades at 5° displacement regressed linearly using sine law under assumptions of Hooke's law (calculations can be found in Ref. 1 using Fig. 1d schematic). Meticulous care of the instrumentation is required. Prior to application, the clip is opened using an applicator with locking and quick-release mechanism (Figs. 1d and 3a). The hook is used to guide the bottom blade of the clip through the space between C8 and T1 spinal roots until the blade tip is visualized in the contralateral space. Any restriction in movement while advancing the clip blade is overcome with gentle rostrocaudal pivoting. Continued resistance cannot be overcome by advancing the clip, and thorough ventral dissection with a fine nerve hook is instead performed slowly to avoid cord manipulation. Following smooth insertion of the clip around the cord, the applicator is held so the arm contacting the bottom clip blade is immobile and the other arm able to move freely. It is imperative that the bottom blade of the clip does not roam upon clip release, and the tips of the blades do not contact any bone or muscle. The release mechanism is then triggered so the clip snaps closed with a standardized contusive impact [4, 5]. The clip is fully released without any contact of the applicator for 60 s. Duration of compression can be varied; however, we recommend a standardized 1-min period of compression to reliably produce posttraumatic ischemia [6]. The applicator is then used to fully open the clip and remove it from the cord. The contusive impact initiates the primary physical insult while the compressive force realizes the secondary pathological processes essential to modelling the clinical presentation of traumatic SCI. Consistent application of the clip provides a resultant lesion exhibiting robust and highly reproducible central cavitation, astrogliosis, neuroinflammation, and spared subpial axonal rim (Fig. 3d) [3, 5].

2.4 Wound Closure and Monitoring

Following application and removal of the clip, retractors are released and the surgical field is monitored until bleeding and CSF leakage cease. A rectangular piece of surgifoam is placed over the exposed dura to provide a transient barrier. Muscles superficial to the spinalis muscle are closed in layers using 5-0 silk semicontinuous sutures. Skin incision is closed using Michaels wound clips or discontinuous absorbable sutures. Animals are administered 0.05 mg/kg buprenorphine and 5 cc saline sq, removed from isoflurane, and followed until verification of shoulder abduction and elbow extension. Animals are fed until competent of self-care, and bladders are expressed until urinary continence returns. Contracture and weakness of the forepaw should be evident, but shoulder involvement, cervical kyphosis, and spastic rigidity are unfavorable outcomes.

2.5 Quality Assurance and Outcome Measures

Grey matter destruction, central cavitation, and white matter preservation in the subpial rim are expected and consistent (Fig. 3d). Confirmation of reproducibility is performed using lesional dimension analysis with Abercrombie equation, Cavalieri method, or StereoInvestigator software (MBF Bioscience). The advantage of cervical clip compression model (C7 and rostral) is the utility of neurobehavioral outcomes to assess return of function to the forelimb. These include, but are not limited to, the use of grip strength meters, staircase reaching/grasping test, ladder/grid walk, inclined plane, catwalk gait assessment, IBB Forelimb Scale, and electrophysiological measure of motor-, sensory-, and spinal cord-evoked potentials, H reflex, and H/M Spasticity Index.

3 Notes

- While clip-spring combinations are reliable for over 900 open-close cycles, routine cleaning and calibration should be performed.
- Meticulous instrument care is required, including periodic clip re-calibration before and after each set of experiments or 40 applications.
- Care should be taken to avoid the large subcutaneous vein diving into the thoracic muscles at T4 with tributaries adjacent to T2 spinal process.
- If surgical field needs widening, spinalis can be cut just caudal to C5.
- Excessive distraction of the spinal roots is evident by a reduced grip strength or increased paw contracture on the side of clip/hook insertion.
- Radiculopathy due to the hook/clip is easily reduced by using a narrow (<0.8 mm) and dulled hook, and carefully visualizing the ventral roots.

- Fine forceps can be used to carefully move the lateral edge of the cord (padded with gauze) towards the midline to visualize nerve roots.
- Damage to the dorsal nerve roots will cause rats to chew their forepaws.
- Changes in the depth of the clip around the cord are evident as this causes cord tissue to be "pinched" between the blades close to the roller.
- Placement of the clip around the cord is the cornerstone of the model, as this determines all impact and forces and the resulting lesion.
- To eliminate all movement of the clip, insert the hook into the contralateral side and place it firmly underneath the bottom blade of the clip.

References

1. Dolan EJ, Tator CH (1979) A new method for testing the force of clips for aneurysms or experimental spinal cord compression. J Neurosurg 51(2):229–233
2. Rivlin AS, Tator CH (1978) Effect of duration of acute spinal cord compression in a new acute cord injury model in the rat. Surg Neurol 10(1):38–43
3. Joshi M, Fehlings MG (2002) Development and characterization of a novel, graded model of clip compressive spinal cord injury in the mouse: Part 1. Clip design, behavioral outcomes, and histopathology, and; Part 2. Quantitative neuroanatomical assessment and analysis of the relationships between axonal tracts, residual tissue, and locomotor recovery. J Neurotrauma 19(2):175–203
4. Kwon BK, Okon EB, Tsai E et al (2010) A grading system to evaluate objectively the strength of pre-clinical data of acute neuroprotective therapies for clinical translation in spinal cord injury. J Neurotrauma 28:1525–1543
5. Karimi-Abdolrezaee S, Eftekharpour E, Wang J et al (2010) Synergistic effects of transplanted adult neural stem/progenitor cells, chondroitinase, and growth factors promote functional repair and plasticity of the chronically injured spinal cord. J Neurosci 30(5):1657–1676
6. Fehlings MG, Tator CH, Linden RD (1989) The relationships among the severity of spinal cord injury, motor and somatosensory evoked potentials and spinal cord blood flow. EEG Clin Neurophysiol 74(4):241–259

Chapter 9

Microsurgical Approach to Spinal Canal in Rats

Mortimer Gierthmuehlen and Jan Kaminsky

Abstract

The rodent spine is used for a variety of models, including spinal instability (de Medinaceli and Wyatt, J Neural Transplant Plast 4:39–52, 1993), neuronal regeneration (Kwon et al., Spine 27:1504–1510, 2002), infection studies (Ofluoglu et al., Arch Orthop Trauma Surg 127:391–396, 2007), and studies about the cauda-equina-syndrome (Kobayashi et al., J Orthop Res 22:180–188, 2004). It is an interdisciplinary target for urologic (Hoang et al., J Neurosci 26:8672–8679, 2006), orthopedic (Iwamoto et al., Spine 20:2750–2757, 1995; Spine 22:2636–2640, 1997), neurologic (Takenobu et al., J Neurosci Methods 104:191–198, 2001), and neurosurgical (Xiao and Godec, Paraplegia 32:300–307, 1994) questions. However, no standard procedure to approach the spinal cord in rats has been published in detail. We present a description of a dorsal approach to the spine, spinal canal and myelon of the rat. This approach provides sufficient exposure of the neural structures to perform extended microsurgery at the spinal nerve roots, the lateral and dorsal myelon and vertebral structures under a surgical microscope. Perioperative management, anesthesia, and anatomical landmarks are discussed and common pitfalls are described.

Key words Dorsal approach, Spine, Spinal cord, Rat, Nerve root

1 Introduction

The rodent spine is used for a variety of models, including spinal instability [1], neuronal regeneration [2], infection studies [3], and studies about the cauda-equina-syndrome [4]. It is an interdisciplinary target for urologic [5], orthopedic [6, 7], neurologic [8], and neurosurgical [9] questions. However, no standard procedure to approach the spinal cord in rats has been published in detail. We present a description of a dorsal approach to the spine, spinal canal and myelon of the rat. This approach provides sufficient exposure of the neural structures to perform extended microsurgery at the spinal nerve roots, the lateral and dorsal myelon and vertebral structures under a surgical microscope. Perioperative management, anesthesia, and anatomical landmarks are discussed and common pitfalls are described.

Miroslaw Janowski (ed.), *Experimental Neurosurgery in Animal Models*, Neuromethods, vol. 116,
DOI 10.1007/978-1-4939-3730-1_9, © Springer Science+Business Media New York 2016

2 Materials

The procedure is performed in female Wistar rats weighing between 250 and 300 g. Anesthesia is done with intramuscular administration of ketamine 10% (0.75 ml/kg bodyweight (BW)) and medetomidine (0.15 mg/kg). The instruments we used for surgery are listed in Table 1. An operating microscope is necessary once the laminectomy is done.

A self-designed OR table (Fig. 1a) allows stable positioning of the rat with moderate kyphosis for optimal exposure of the spine. It also allows the surgeon to position the hands comfortably placed on each side of the animal. A heat lamp (standard infrared lamp for human use) and a warming pad (Thermolux Pet-Mat 10 W attached to a continuously variable dimmer switch) provide constant temperature of the animal during surgery. Sufficient hydration of the animal is ensured by intermittent subcutaneous injection of 0.9% NaCl (1 ml/h/100 g BW).

3 Methods

3.1 Anatomy

The lumbar spine of the rat consists of six vertebrae. The landmark for the caudal lumbar spine is the pelvis; the sixth lumbar vertebra can be identified between the two cristae iliacae (Fig. 1). This is the most caudal level of the spinal canal that can be opened safely since the sacral spinal roots leave the spinal canal dorsally and preparation in that area is extremely difficult.

Table 1
Instruments we used during surgery

Instrument	Manufacturer	Comments
OR table	Self-made	Providing kyphosis and warming
Scissors	Pfeilring	
Rongeur	Niegeloh	
Forceps (anat/surg.)	Aesculap	
Needle-holder	Aesculap	
Micro-forceps	Aesculap	
Micro-scissors	Aesculap	
Micro-needle holder	Aesculap	
Dura hook (sharp)	Aesculap	
Nerve dissector	Self-made	2-0 suture on top of a dental instrument (Fig. 15)

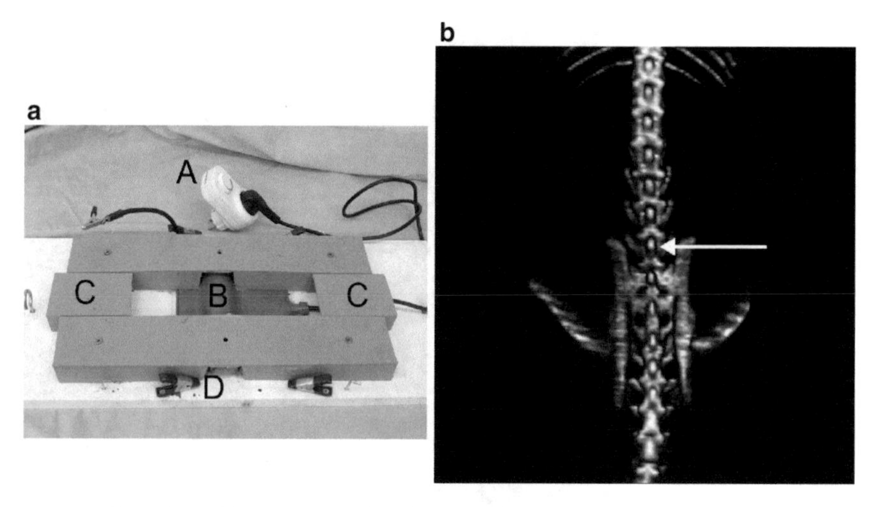

Fig. 1 (**a**) The operating table with a continuously adjustable dimmer (A) controls the heat of the warming pad (B). This pad is formed to provide a kyphosis. The hand rests on both sides (C) allow stable positioning of the hands. Small clamps (D) hold retracting sutures as necessary. (**b**) CT scan of the lumbar spine of a rat. The sixth lumbar vertebra is located between the cristae iliacae (*arrow*)

3.2 Approach to the Lumbar Spinal Cord

3.2.1 Skin Incision and Preparation of the Laminae and Spinal Processes

The coat at the operation field is removed with an electric shaver (a professional long-hair shaver is better suited than a cheap standard shaver). After palpating the cristae iliacae, the L6 spineous process is identified marking the most caudal process. After disinfection of the skin (Kodan®), a midline incision is done from 2 cm cranial to 1 cm caudal to the cristae iliacae. The skin is retracted with 2-0 suture, followed by blunt subcutaneous. After the subcutaneous connective tissue is carefully lifted with forceps and cut away (Fig. 2)—otherwise it might get entangled in the drill and cause trouble—the fascia of the paravertebral muscles becomes visible. The paravertebral tendons connecting to the L6 spineous process can be identified as the last white stripe before the muscles attach directly to the sacral bone (Fig. 3). The fascia is incised superficially and bilaterally to the spineous processes from L3 to L6 (Fig. 3), while a cranio-caudal direction allows cutting the paravertebral tendons. Of course, the fascia could also be incised in a caudo-cranial way, but the anatomy of the tendons can easily misguide the blade of the scalpel laterally. Again, blunt preparation is needed to separate the paravertebral muscles from the spineous processes; the remaining paravertebral tendons are dissected with scissors. Great care should be taken not to go too lateral but to dissect closely to the spineous processes. A lateral and deep preparation may damage the spinal roots emerging from the spinal canal. A small retractor is inserted to hold back the paravertebral muscles, and the interspineous tendons are dissected to make the spineous processes visible (Fig. 4).

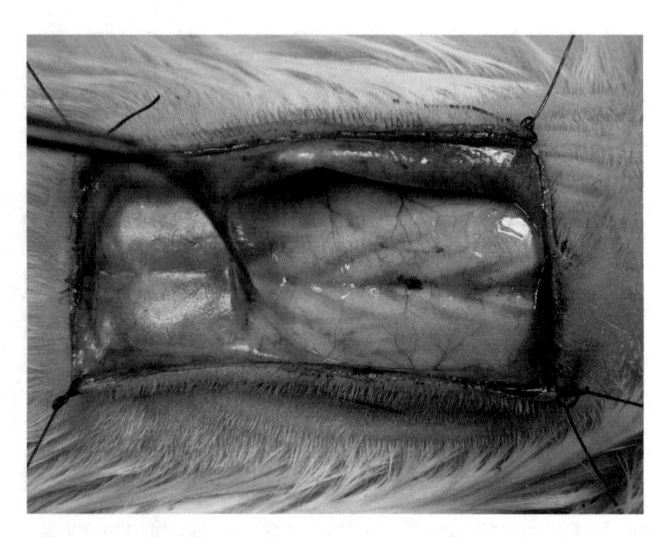

Fig. 2 The rat is positioned on the OR table - right side points cranially, left side caudally. The edges of the wound are retracted with 2-0 suture attached to small clamps on the operating table. The subcutaneous connective tissue (containing the blood vessels seen in the photo) should be removed as it may get entangled with the drill. The paravertebral muscles become visible. The spineous process of L6 can be localized as the most caudal insertion point for paravertebral tendons

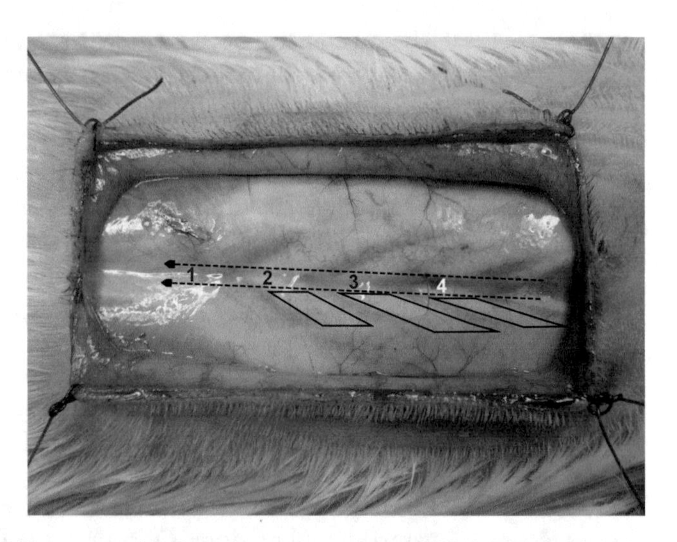

Fig. 3 The sacrum (*1*), the spineous processes L6 (*2*), L5 (*3*), and L4 (*2*) and the paravertebral tendons (*lines*) shine through the muscular and subcutaneous tissue. The tissue should be incised bilaterally in a cranio-caudal direction (*arrows*). This prevents the scalpel from being misguided laterally by the paravertebral tendons

3.2.2 Identification of the Correct Spinal Level

By holding the spinal process of L6 with forceps and moving the sacrum backward and forward with two fingers, a movement between the L6 and the S1 spineous process becomes obvious. This again ensures the correct level, as the spineous processes of S1 and S2 do not show any mobility against each other. After laminectomy, the level of L1/L2 can be confirmed if the caudal cone of

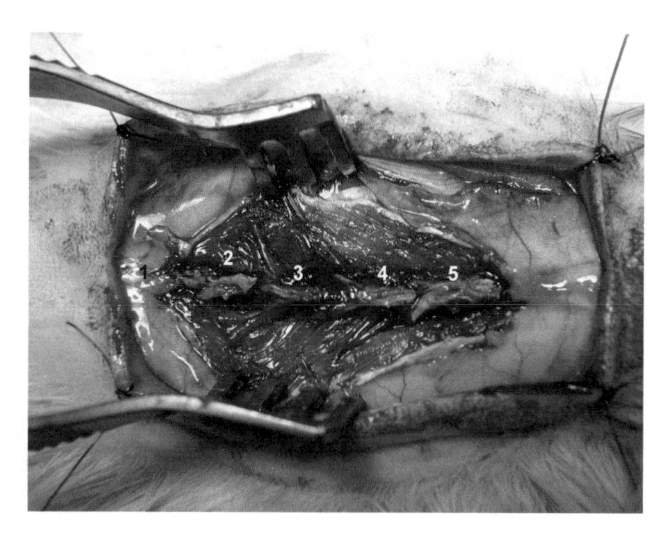

Fig. 4 The paravertebral and interprocessous tendons are dissected with scissors and the processes L6 (*2*), L5 (*3*), L4 (*4*), and L3 (*5*) and the sacrum (*1*) can be identified. It is helpful to hold the L6 process with the forceps and move the sacral bone to identify the correct level. Mobility is seen between L6 and S1, but absent between S1 and S2

the myelon is seen intradurally. It is easily identified by a small venous plexus on its dorsal surface.

The identification of the correct level is already a critical step in the lumbar preparation, but it becomes even more difficult when surgery at a specific thoracic level is necessary. In this case it is advisable to identify the lumbosacral junction and count the spineous processes cranially until the target region is found. As the thoracic spine consists of 13 vertebrae and anatomical variations are known, other ways to identify the correct region have been described (LITERATUR de Medinaceli).

3.2.3 Laminectomy

The spineous processes are now removed with a small rongeur. From this point on, an operation microscope is used.

A large drilling head is chosen to clean the operating site (Fig. 5). A combined irrigation-suction device (Hydroflow®) is helpful in providing good vision. Bleeding mostly occurs in the space between the facet joints and can be coagulated with the bipolar forceps, but since the spinal nerves emerge only a few millimeters below this area, the coagulation power is adjusted to the minimal possible. A cranial-to-caudal direction is chosen to open the spinal canal as its diameter decreases caudally. The drill is now held perpendicular to the spine; otherwise it could get stuck in osseous structures and damage the area lateral to the vertebrae. The lateral edges of the vertebrae and the facet joints are also reduced, making it more comfortable to cut them in a later stage of surgery (Fig. 6). Remaining tendons attached to the facet joints can also be cut now.

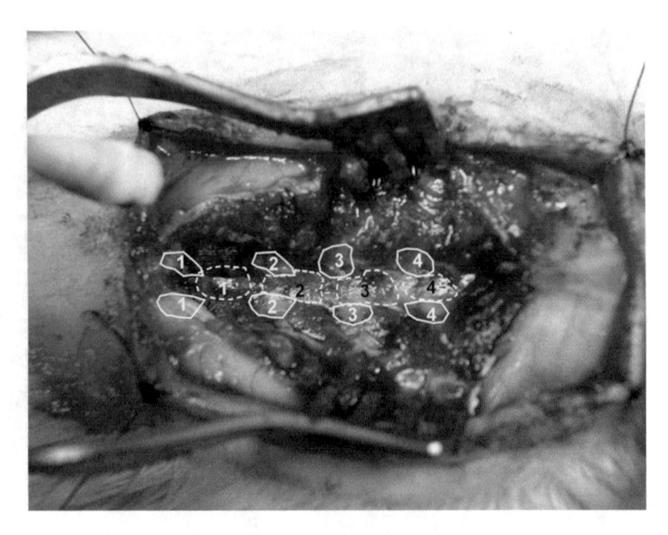

Fig. 5 The laminae L6 (*dashed line 1*), L5 (*dashed line 2*), L4 (*dashed line 3*), and L3 (*dashed line 4*) can be identified, also their respective facet joints (*bold lines 1–4*). At this stage, bleeding almost always occurs in the space between the facet joints and can easily be terminated with bipolar forceps. This is the last stage of surgery where the bipolar forceps can be used. Once the dura is visible, the bipolar should be avoided, as heat and electricity may damage the nerves or cause uncontrollable movements of the rat

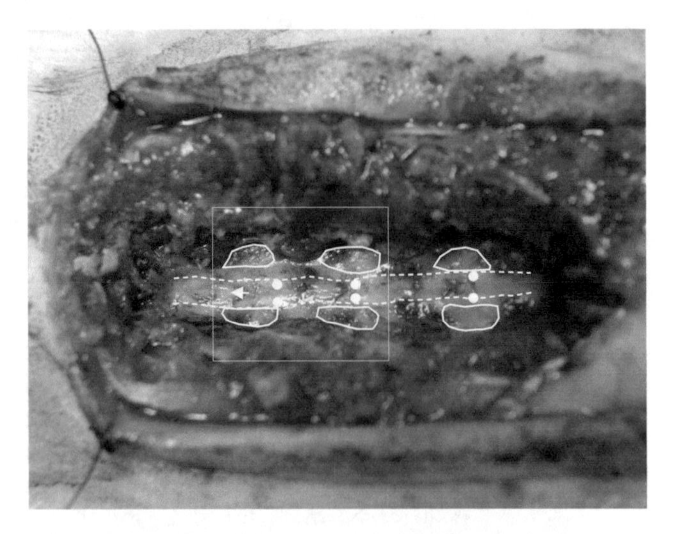

Fig. 6 With a smaller drilling head and much irrigation, the laminae are thinned out and the lateral edges, including the facet joints, are reduced. The spinal canal (*dashed line*) and the nerve roots (*arrow*) become visible. The spinal parts of the facet joint capsule appear as white stripes in the intervertebral region (*white dots*). Here, the bone layer is much thinner. The area framed by a white rectangle is shown in Fig. 7

The laminae are not entirely removed with the drill, but a small bone shell is left for safer manual removal. This layer is identified by small cracks reflecting the light of the OR microscope (Fig. 7). Drilling is stopped when these small cracks become visible over the entire length of the spinal canal.

3.2.4 Entering the Spinal Canal

At this step, the rat is positioned with the lower legs extended and hanging downwards, since the following manipulation at the spinal canal may provoke neural and muscular activity. The small cracks in the thin osseous layer are inspected with a sharp hook. The hook should therefore be slipped under the lateral edge of the osseous layer which is then carefully lifted (illustrated in Fig. 7). Starting this procedure at the level of a facet joint again reduces the risk of inadvertently penetrating the dura. Small fragments are removed with micro-forceps. Bleeding is stopped with warm water and Gelita®. The intraspinal, fibrous capsules of the facet joints (see Fig. 7) may sometimes mimic dissected nerve roots, but they are thicker and harder to remove than nerves. The rongeur is used to carefully cut away the remaining lateral edges of the laminae (Fig. 8), and the dural sack is prepared (Fig. 9).

3.2.5 Intra- and Extradural Approach

Extradural Preparation

By carefully resecting the vertebrae's pedicles it is possible to achieve a far lateral approach to the dural sack and the vertebral discs. In the lumbar spine this is relatively easy when the emerging spinal nerves are respected. In the thoracic region the attached ribs and the thorax make a lateral approach much more complicated.

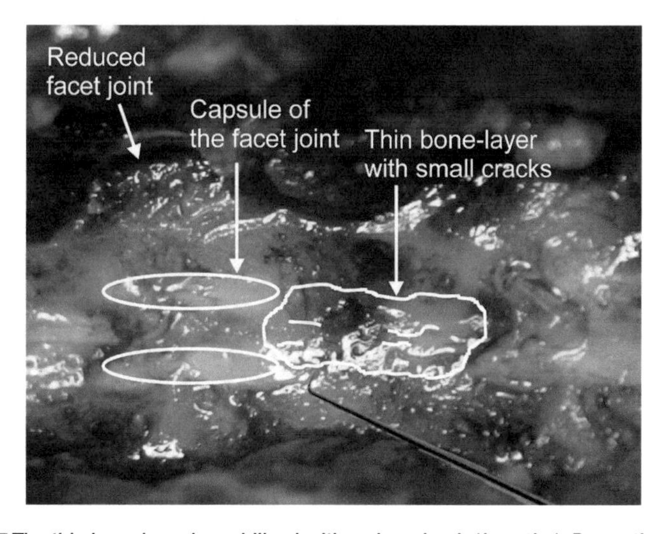

Fig. 7 The thin bone layer is mobilized with a sharp hook (*insertion*). By gently rotating the hook's tip, the layers can be lifted and removed with micro-forceps. Once the bone layer is gone, it is essential not to get confused by fragments of the facet joint capsule which may appear as damaged nerve stumps. It is advisable to remove these fragments, as they can interfere with the following steps of surgery

Fig. 8 Once all bone layers and facet joint fragments are removed, the entire spinal canal and the dura become visible. But the opening is still much too small to safely show nerve roots or even open the dura. It is necessary to reduce the lateral edges (*dashed lines*) with a rongeur. This should be done extremely carefully, since damage to the nerve roots should be avoided

Fig. 9 Once the spinal canal has been widely opened, the nerve roots L6 (*bold arrow*), L5 (*dashed arrow*), and L4 (*arrowhead*) are identified. The dura is incised with a sharp dura hook. It is advisable not to cut the dura open in the middle but to create a larger flap on the side where the dura can be retracted with a suture

Although the rongeur's size might look too big for the resection of the pedicles, injury to the spinal nerves is rare. Again, annoying venous bleeding from the bone is easily controlled with bone wax.

Intradural Preparation

If intradural surgery is necessary, a sharp dura hook can be used to open the dura. This should be started cranially, and once a small

Fig. 10 The dura is opened laterally and the flaps are retracted to the muscular wall with 7-0 dura suture (*arrowheads*)

distance (e.g., 5 mm) is opened, the edges should be attached to the paravertebral muscles with a 7-0 suture before continuing caudally (Fig. 10). This is much easier than opening the whole dura in one step since the dural edges retract laterally and are hard to find. A small piece of 2-0 prolene suture held by a dental instrument is used as an atraumatic nerve dissector (Fig. 15).

Identification of nerve roots can either be done by stimulation or by anatomical landmarks. We used an AD-Instruments setup and low stimulator settings (e.g., pulse width 100 μs, 1–5 mA and manual activation) in order to provoke muscle answers in the lower limbs. While the motoric roots of L4 and L5 are relatively thick and innervate the legs—stimulation results in movement of the thigh (L4) and the calf (L5)—L6 is rather thin and innervates the tail. Anatomically, L6 leaves the spinal canal after the sixth lumbar vertebra and is thin—L5 and L4 can be identified by counting cranially. Motoric roots are ventral, and sensory roots dorsal. Small pieces of 2-0 suture can be placed under the spinal nerve (Fig. 11a) in order not to lose it. We used this approach for a nerve regeneration study (Fig. 11b) where we performed a microsurgical anastomosis between the L4 and L6 ventral roots. In another project, we labeled the L4–L6 ventral roots with anterograde neural tracers, each in a different color. Therefore, we took three small sterile rubber tubes of appr. 3 mm length, opened the top of each tube, carefully inserted a nerve root in each tube, closed both sides with Vaseline® and filled the remaining space with neural tracer (Fig. 12). This ensured selective tracing of each nerve root with a specific tracing color.

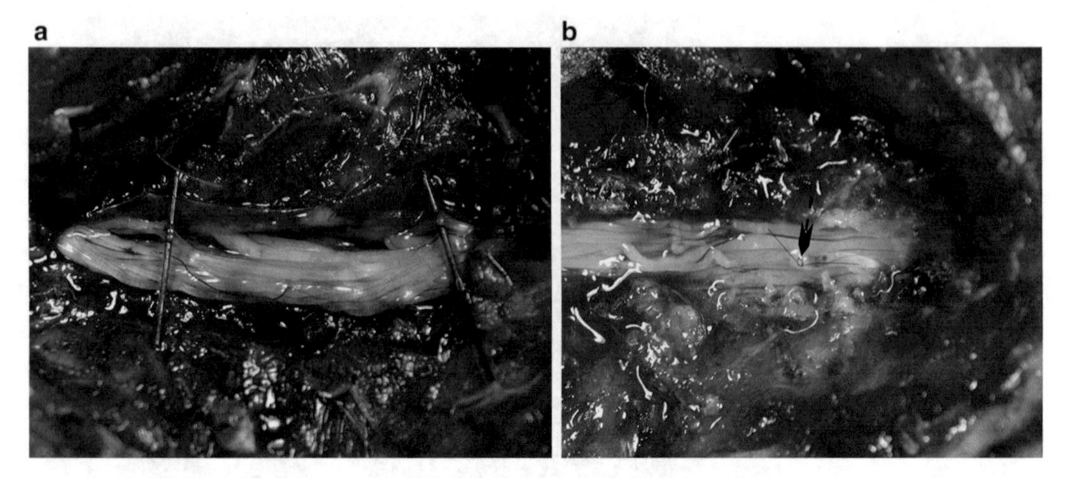

Fig. 11 (**a**) The ventral nerve roots L4 und L6 are identified and marked with tiny pieces of 2-0 suture and, in this study of nerve regeneration, an anastomosis from L4 to L6 is sutured (*arrow* in **b**)

Fig. 12 In this picture, the ventral roots of L4–L6 are put in a small rubber tube each. Both open sides of the tubes are sealed with Vaseline and a neurotracer is filled in the remaining cavity

3.2.6 Wound Closure

After surgery, a small piece of subcutaneous connective tissue is prepared and sutured to the dura; the cranial and caudal part is fixed to connective tissue of the facet joints (Fig. 13). Often, after intraspinal preparation, a CSF-tight dura closure cannot be achieved without compromising the spinal nerves. In this case the dura edges were left open and the tight suture of the muscular fascia prevented dura leaks instead. The paravertebral muscles are adapted with 5-0 Vicryl suture; tight 5-0 Vicryl subcutaneous suture closes the skin (Fig. 14). Skin glue (e.g., Dermabond®) seals the wound.

Fig. 13 Dural closure is difficult to achieve, since the suture can damage the nerve roots. Therefore, subcutaneous fascia is used to cover the spinal canal. It is either attached directly to the dura (*left side*) or to the lateral muscular wall (*right side*)

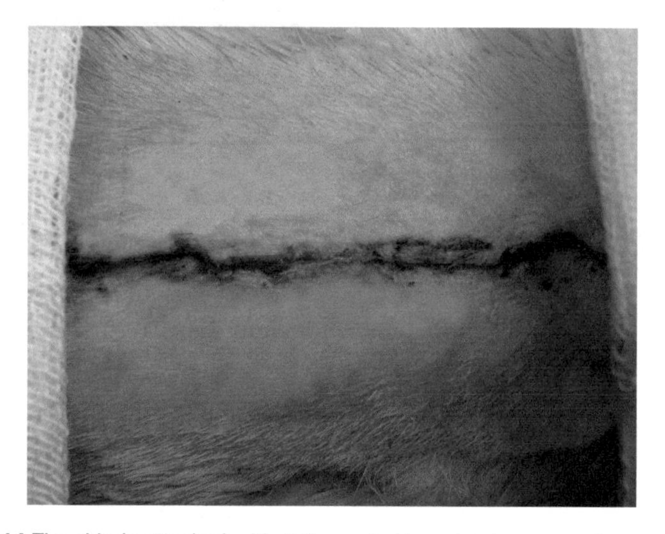

Fig. 14 The skin is attached with 5-0 resorbable, subcutaneous suture as rats tend to remove any single-knot, cutaneous sutures. If deemed necessary, skin glue is used

3.2.7 Postoperative Care

After surgery, sufficient warming via heat lamp is essential until the animal wakes up. Anesthesia can be shortened by administering Antisedan® (atipamezole, apply same volume as of medetomidine). For 3 days after surgery, the animals receive the oral antibiotic Borgal® (trimethoprim/sulfadoxine, 15 mg/kg body weight) and subcutaneous injection of Carprofen (1 mg/kg body weight q24h) as analgesic.

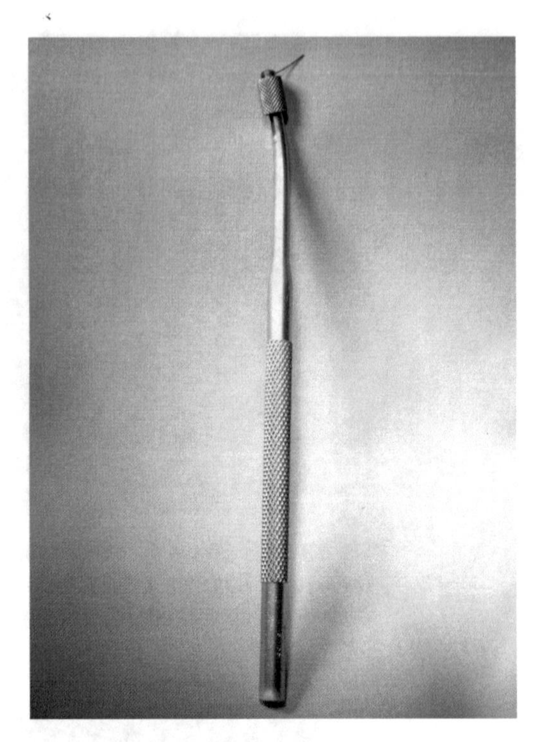

Fig. 15 A 2-0 or 3-0 suture attached to a dental instrument or a forceps is a safe way to dissect nerves

4 Notes

There are several risks that the surgeon should be aware of:

- As in humans, it is easy to be misled and operate on the wrong spinal level. Moving the spinal processes of L6 and S1 against each other helps identifying the lumbosacral junction. Orientation is even more critical in the thoracic region.
- When the paravertebral fascia is opened from caudal to cranial, the scalpel can be misled by the paravertebral tendons and cut the fascia too lateral. It is much easier to cut from cranial in a caudal direction.
- Resect subcutaneous tissue before using the drill—otherwise it might get entangled.
- Sufficient irrigation is essential during the drilling procedure. Otherwise the rat may die from overheat.
- At the end of laminectomy, a small bone layer should be left and resected with micro instruments. It is not advisable to try

to open the spinal canal only with the drill as the dura is really thin and may be injured.

- As soon as neural structures can be touched—especially nerve roots from L4 and L5, which innervate the legs—the rat should be positioned with the legs hanging downwards. If the legs touch the table, their sudden movement may cause the rat to jump up and the instruments to injure the nerves.
- Bleeding from bones can be hard to control—bone wax is really helpful.
- During intradural surgery, provide sufficient irrigation to the nerves—otherwise they may dry out.
- Depending on the type auf anesthesia and the intention of surgery, muscle relaxants can be used in order to avoid motoric responses during manipulation of the nerve roots. In some electrophysiologic settings it is, however, not possible to apply relaxants as this medication might interfere with the investigation protocol.
- Subcutaneous injection of 1 ml/h/100 g BW saline is essential to prevent renal failure.
- Single-knot non-resorbable sutures for wound closure are critical as rats tend to bite everything off they can. Tight subcutaneous suture and skin glue (e.g., Dermabond®) seem to be safer.
- Nieto et al. proposed a titanium mesh graft for reconstruction of the spinal canal in a thoracic laminectomy model [1]. We did not use this mesh as firstly we operated on the lumbar segment where only peripheral nerves are present without the risk of myelopathy. Secondly, the lumbar spine shows a higher mobility compared to the thoracic spine, resulting in an increased risk of damage to neural structure in case of dislocation of the titanium mesh.

References

1. de Medinaceli L, Wyatt RJ (1993) A method for shortening of the rat spine and its neurologic consequences. J Neural Transplant Plast 4(1):39–52
2. Kwon BK, Oxland TR, Tetzlaff W (2002) Animal models used in spinal cord regeneration research. Spine 27(14):1504–1510
3. Ofluoglu EA et al (2007) Implant-related infection model in rat spine. Arch Orthop Trauma Surg 127(5):391–396
4. Kobayashi S, Yoshizawa H, Yamada S (2004) Pathology of lumbar nerve root compression: Part 2. Morphological and immunohistochemical changes of dorsal root ganglion. J Orthop Res 22(1):180–188
5. Hoang TX, Pikov V, Havton LA (2006) Functional reinnervation of the rat lower urinary tract after cauda equina injury and repair. J Neurosci 26(34):8672–8679
6. Iwamoto H et al (1995) Production of chronic compression of the cauda equina in rats for use in studies of lumbar spinal canal stenosis. Spine 20(24):2750–2757
7. Iwamoto H et al (1997) Lumbar spinal canal stenosis examined electrophysiologically in a

rat model of chronic cauda equina compression. Spine 22(22):2636–2640

8. Takenobu Y et al (2001) Model of neuropathic intermittent claudication in the rat: methodology and application. J Neurosci Methods 104(2):191–198
9. Xiao CG, Godec CJ (1994) A possible new reflex pathway for micturition after spinal cord injury. Paraplegia 32(5):300–307
10. Brookes ZL, Brown NJ, Reilly CS (2000) Intravenous anaesthesia and the rat microcirculation: the dorsal microcirculatory chamber. Br J Anaesth 85(6):901–903
11. Nieto JH et al (2005) Titanium mesh implantation—a method to stabilize the spine and protect the spinal cord following a multilevel laminectomy in the adult rat. J Neurosci Methods 147(1):1–7

Chapter 10

Stereotaxic Injection into the Rat Spinal Cord

Charla C. Engels and Piotr Walczak

Abstract

Spinal cord is a frequent target for injection of various therapeutic agents including stem cells, growth factors, small molecules, or genetic constructs. Due to its fragility and specific anatomical localization injection has to be performed with great deal of care to minimize injury and assure precision of targeting. This chapter describes two approaches for gaining access to the spinal cord facilitating safe and efficient intraspinal injection.

Key words Stereotaxy, Spinal cord, Rat, Neurosurgery

1 Introduction

The estimated global incidence of spinal cord injury (SCI) is 15–40 cases per million. The spinal cord has minimal regenerative capacity, and, with an average age of 38 years at the time of injury onset, spinal cord injury is rightfully considered one of the most debilitating conditions, with massive social and economical consequences for both the individual and society [1–3].

Demyelination, axonal damage, and scar formation contribute to a loss of motor and sensory function after SCI. Substantial neurological disability with autonomic dysfunction is also associated with SCI, originating from sudden or sustained trauma or progressive neurodegeneration after injury [1, 2, 4].

The central nervous system (CNS) has an extremely limited intrinsic regeneration capability [3, 4]; thus, new strategies aimed at neuroprotection and/or enhancing regenerative potential are highly desirable. Stereotaxic injection into the spinal cord parenchyma is a commonly used technique often used for the delivery of stem cells or other therapeutic agents [2, 4, 5].

Electronic supplementary material: The online version of this chapter (doi:10.1007/978-1-4939-3730-1_10) contains supplementary material, which is available to authorized users.

Miroslaw Janowski (ed.), *Experimental Neurosurgery in Animal Models*, Neuromethods, vol. 116,
DOI 10.1007/978-1-4939-3730-1_10, © Springer Science+Business Media New York 2016

This chapter details two straightforward approaches for gaining access to the spinal cord and delineates a technique for precise stereotaxic injection into the spinal cord parenchyma.

2 Materials

The described protocol is suitable for adult (6–8 weeks old) female or male rats, weighing between 200 and 300 g.

2.1 Anesthesia

1. Veterinary-grade isoflurane (Aerrane®; Baxter).
2. Induction chamber (Cat# 941444; Vetequip).
3. Mobile animal anesthesia system (Cat# 901807; Vetequip).
4. Rodent anesthesia circuit mask with hose (Cat# 723026; Harvard Apparatus).
5. Fluovac Waste Gas Control (Cat# 50206; Stoelting).

2.2 Animal Prep

1. Small animal clipper (Cat# 726114; Harvard Apparatus).
2. Nair® Hair Removal Cream.
3. Betadine Surgical Scrub (Cat# 19027132; Fisher Sci.).
4. Infrared heating lamp (optional).
5. Vetericyn Animal Ophthalmic Gel (optional).

2.3 Stereotaxic and Surgical Instruments

1. Surgical microscope (Leica M320 F12).
2. Stereotaxic frame (Leica; Cat# 39463501).
3. Spinal cord surgery adaptor (Stoelting Cat# 51695): The adaptor should be customized by slightly bending (~20°) the tips, as shown in Fig. 1.
4. Motorized stereotaxic injector (Stoelting; Cat# 53311).
5. Micro drill (Roboz; Cat# RS6300).
6. Microinjection syringe 10 μl (Hamilton Cat# 7635-01).
7. Hamilton needle, 31 G, 1 in. long (Hamilton Cat# 78035).
8. Cotton swabs (Fisher Sci. Cat# 191301518).
9. Adson forceps (FST; Cat# 11027-12).
10. Dumont forceps (FST; Cat# 11251-35).
11. Scalpel handle (FST; Cat# 10004-13).
12. Scalpel blades (FST; Cat# 100020-00).
13. Fine, straight scissors (FST; Cat# 14058-11).
14. Sutures, Silk 4.0 (Ethicon; Cat# 683G).

2.4 Postoperative Period

1. Ketofen (Cat# 2193; Pfizer).
2. Injectable sodium chloride 0.9 % (Hospira Inc.; Cat# 4094888).

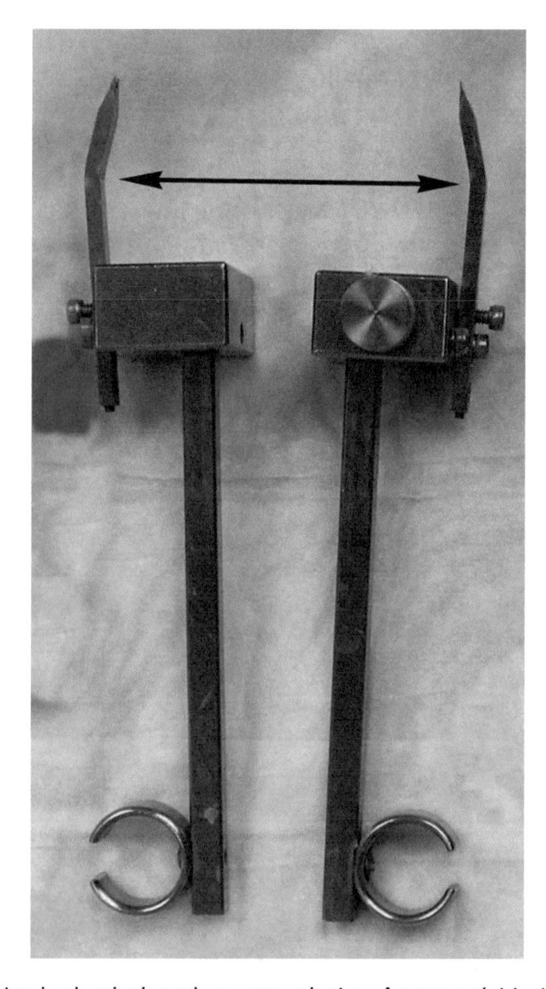

Fig. 1 Customized spinal cord surgery adaptor. *Arrows point* to the position where the pins are slightly bent downward

3 Methods

3.1 Anesthesia

Place the animal in the induction chamber and adjust the isoflurane concentration on the vaporizer to 3.0 %. As soon as the animal becomes drowsy and the respiratory rate drops to about 60/min, place the animal on the table for prepping. The table should be equipped with an anesthesia circuit mask connected to a hose for continuous anesthesia with 2 % isoflurane.

3.2 Animal Prep

1. Trim the fur over the back of the animal using a clipper.
2. Apply depilation cream, wait for 2–3 min, wipe the cream with gauze, and then wash with saline. Note: Avoid longer exposure to depilation cream as it will cause skin damage.
3. Clean the surgical area with 70 % ethanol.
4. Apply Betadine scrub.

5. Move the animal onto the stereotaxic device with the head facing the operator (Fig. 2a).

3.3 Surgery

Selection of the vertebral level is project specific. The exact level can be conveniently calculated beginning with either palpation of the final floating ribs and identification of their origin at T13, or by palpating the most prominent spinal process between the scapulae corresponding to T2. To prevent hypothermia, the rat should be placed on a bed of paper towels, thus preventing the animal from

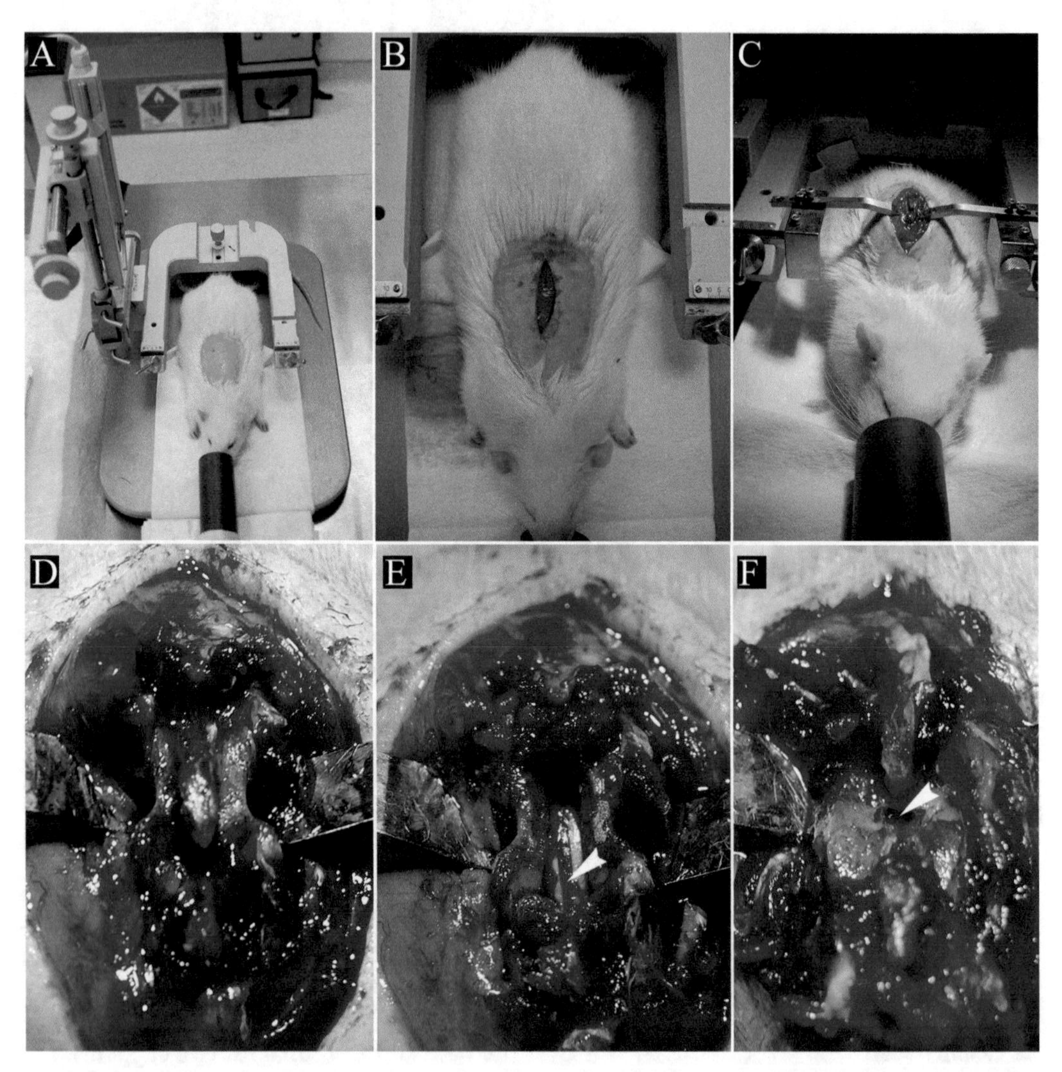

Fig. 2 Gaining access to the spinal cord. (**a**) The overall surgical setup, including the operating table with a stereotaxic frame, and an animal with the attached inhalation anesthesia line. (**b**) Placement of skin incision. (**c**) Placement of stereotaxic spinal adaptors. (**d**) Vertebral segment prepared for laminectomy procedure by removal of overlying muscles and tendons. (**e**) Vertebral segment after laminectomy with visible spinal cord. *Arrowhead points* to the posterior medial vein. (**f**) Access to the spinal cord through the intervertebral space. The spinal cord is visible within the small opening (*arrowhead*)

being in direct contact with the steel frame of the stereotaxic device. To maintain constant body temperature during longer surgeries, a heat lamp can also be used. To prevent corneal abrasion during surgery, ophthalmic gel can be applied after anesthesia induction.

3.4 Laminectomy

1. Using a scalpel, make a 2–3 cm long skin incision in the midline (Fig. 2b).
2. Clear muscles and tendons on the back and sides of two vertebral segments using scissors, scalpel, and cotton swabs. Bilateral, cranial-to-caudal incisions are made with a scalpel in the paravertebral muscles very near to the spinous processes. Bluntly separate the paravertebral muscles from the vertebrae using cotton swabs. Cotton swabs are preferred over forceps due to the advantage of absorbent properties. Some degree of bleeding cannot be avoided at this point, but can be reduced by washing the area with cold saline.
3. Fix spinal adaptors on the stereotaxic device.
4. Using Adson forceps, grasp the sides of vertebral segment just beneath the transverse processes and lift the animal to the level of the spinal adaptor pins.
5. Place the vertebra between both pins and tighten such that the pins hold the animal partly suspended (Fig. 2c).
6. Clear the lamina of the vertebrae thoroughly using cotton swabs, forceps, and a scalpel to distinctly visualize the bone edges (Fig. 2d).
7. Use a micro-drill to cut through the bone along both sides of the spinous process. This procedure is preferably performed using a surgical microscope, such as a Leica M320 F12 or the equivalent. When the lamina becomes loose, grasp the spinous process with the forceps and gently remove the lamina (Fig 3d). The spinal cord with its dorsal vein (arrowhead) in the midline should become visible. If the exposed spinal cord area is too small for adequate manipulation, the laminectomy can be extended using the micro-drill described above.

3.5 Access Through the Intervertebral Space

The initial steps for this procedure are identical to those of the laminectomy up to **step 6** of Subheading 3.4. The only difference here is that two neighboring vertebrae have to be carefully cleared from tendons and muscles.

1. Using forceps and scissors, carefully cut and remove the spinous process of the vertebra proximal to the targeted intervertebral space.
2. Make an opening in the ligaments between the arches of two vertebrae using fine forceps, e.g., Dumont 11251-35 (FST).

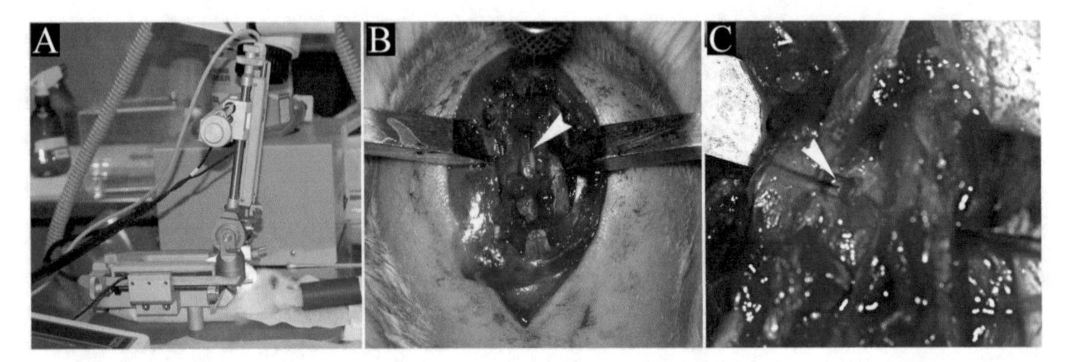

Fig. 3 Stereotaxic injection procedure. (**a**) Stereotaxic device with the angle adjusted for the perpendicular position of the needle with regard to the spinal cord. (**b**) Vertebral segment after laminectomy with the needle in place for an intraspinal infusion. *Arrowhead points* to the needle tip. (**c**) Intervertebral space with removed tendons opening access to the spinal cord. Needle tip is placed to the left of the posterior medial vein (*arrowhead*)

The spinal cord should be visible a few millimeters below the bone level (Fig. 2f). If there is no sufficient space between the vertebrae to clearly see the spinal cord, a spinal adaptor should be repositioned to achieve extreme kyphosis.

3.6 Stereotaxic Injection

1. Load the Hamilton syringe with the injection fluid. For thick, viscous solutions, such as cell suspensions, it is preferable to load from the back of the syringe. Remove the plunger, loosen the screw holding the needle (one rotation or so is enough), pull out the needle just enough to break the seal, and use a 100 μl pipette with attached plastic tip to slowly load the solution into the syringe. Insert the plunger and tighten the screw. Less viscous solutions can be loaded through aspiration via the needle. For both methods, it is important to avoid trapping air in the syringe barrel.
2. Place the Hamilton syringe with the attached needle onto the motorized injector and tighten well.
3. Set parameters on the motorized injector, including desired speed and injection volume.
4. Using a surgical microscope, manipulate the stereotaxic device to place the needle over the spinal segment to be injected.
5. Adjust the angle of the stereotaxic arm, so the needle is perpendicular to the surface of the spinal cord (Fig. 3a).
6. Identify the injection site based on the stereotaxic coordinates. The posterior medial vein is usually a convenient landmark for identifying the midline; however, in some cases, the vein has a tortuous course. An alternative landmark is the tip of the adjacent spinous process. The lateral coordinate can be adjusted as required by the application; however, the medial vein should

be avoided, as shown in Fig. 3b for a laminectomy approach or in Fig. 3c for an intervertebral approach. Puncturing this vessel causes serious bleeding and may significantly complicate the procedure.

7. Immediately prior to inserting the needle into the spinal cord, expel some fluid using the motorized injector, making sure that the system is working well and eliminating any dead space.
8. The dura mater is a very rigid membrane and cutting through it with the needle may cause some difficulty. Forcing the needle through by applying excessive pressure may damage the spinal cord; thus, it is preferable to drive the needle up and down, gradually weakening the dura and finally cutting through. Supplementary Video 1 material demonstrates how to perform this step.
9. Once the dura is penetrated, retract the needle, allowing the spinal cord to relax, and then slowly bring the needle down until it touches the surface of the spinal cord. At that point, reset the dorsal-ventral coordinate and advance the needle further into the tissue until the desired depth is reached.
10. Initialize the infusion on the motorized injector. Note that, due to the small size of the spinal cord, the injection volume should not exceed 2 μl and the injection speed should be less than 0.5 μl/min. After completing the infusion, wait for 2 min to allow for a reduction of pressure and to minimize backflow of the injected solution.
11. Slowly retract the needle, remove the rat from the stereotaxic frame, suture the muscle layer, and then suture the skin with an Ethicon 4.0 suture. Apply Betadine solution over the surgical area.

3.7 Postoperative Care

Postoperative analgesia with Ketofen, 5 mg/kg or the equivalent, should be initiated 30 min before the end of inhalation anesthesia to prevent unnecessary suffering.

Supplementary Video 1 Injection needle is brought into the field of view of the surgical microscope and moved sideways to avoid injury to the dorsal vain resulting in significant bleeding. The needle is moved up and down applying only moderate pressure to the spinal cord. This gradually weakens and cuts through the dura matter and allows for safe insertion of the needle into the spinal cord parenchyma.

References

1. Fehlings MG, Vawda R (2011) Cellular treatments for spinal cord injury: the time is right for clinical trials. Neurotherapeutics 8(4):704–720
2. Keirstead HS, Nistor G, Bernal G, Totoiu M, Cloutier F, Sharp K et al (2005) Human embryonic stem cell-derived oligodendrocyte progenitor cell transplants remyelinate and restore locomotion after spinal cord injury. J Neurosci 25(19):4694–4705
3. Bhanot Y, Rao S, Ghosh D, Balaraju S, Radhika CR, Satish Kumar KV (2011) Autologous mesenchymal stem cells in chronic spinal cord injury. Br J Neurosurg 25(4):516–522

4. Park SI, Lim JY, Jeong CH, Kim SM, Jun JA, Jeun SS et al (2012) Human umbilical cord blood-derived mesenchymal stem cell therapy promotes functional recovery of contused rat spinal cord through enhancement of endogenous cell proliferation and oligogenesis. J Biomed Biotechnol 2012:362473
5. Cummings BJ, Uchida N, Tamaki SJ, Salazar DL, Hooshmand M, Summers R et al (2005) Human neural stem cells differentiate and promote locomotor recovery in spinal cord-injured mice. Proc Natl Acad Sci U S A 102(39): 14069–14074

Chapter 11

Surgical Access to Cisterna Magna Using Concorde-Like Position for Cell Transplantation in Mice and CNS Dissection within Intact Dura for Evaluation of Cell Distribution

Miroslaw Janowski

Abstract

The CSF is increasingly considered as an attractive gateway to the central nervous system (CNS). It is warranted by the direct delivery of therapeutic agents beyond the blood-brain barrier (BBB) and widespread access to the large areas of the brain and the spinal cord. In small animals access to CSF is not trivial. The cisterna magna is the largest CSF fluid compartment; thus it was selected as a target. Here, I describe the surgical procedure for efficient and reproducible access and injection of therapeutic agents such as stem cells to cisterna magna. Due to hydromechanics, the method is distinct from previously described techniques for CSF withdrawal. Finally, I describe the method for CNS dissection within intact dura for evaluation of cell distribution.

Key words Stereotaxy, Spine, Cisterna magna, Concorde position, Spinal cord, Mouse, Neurosurgery

1 Introduction

The CSF is increasingly considered as an attractive gateway to the central nervous system (CNS). It is warranted by the direct delivery beyond the blood-brain barrier (BBB) and widespread access to the large areas of the brain and the spinal cord. There are three major routes of access to CSF: intraventricular, suboccipital, and lumbar. Intraventricular route enables most proximal access with regard to CSF circulation, but at a cost of invasiveness requiring break of CNS continuity. The lumbar route is executed by the percutaneous puncture at the level of *cauda equina*, thus excluding CNS injury. Being the least invasive it is most frequently route to CSF employed in humans and large animals. In small animals it is more complex, and practically not available for mice except of few people having extremely long training [1]. While the mice are of specific interest due to abundance of transgenic models, the

Miroslaw Janowski (ed.), *Experimental Neurosurgery in Animal Models*, Neuromethods, vol. 116,
DOI 10.1007/978-1-4939-3730-1_11, © Springer Science+Business Media New York 2016

detailed description of access to CSF in this species is highly desirable. The limited lumbar space is another constraint if larger volumes are expected to be withdrawn or injected. Thus in mice suboccipital route seems to be optimal, especially for the purpose of administration of solutions or suspensions to CSF. While the feasibility of percutaneous approach was indicated [2], due to the tiny size of cisterna magna in mice and close proximity to vital structures of brain stem, that approach is considered to be dangerous. I have developed surgical approach to cisterna magna in mice, which additionally enables efficient delivery of solutions and/or suspensions, with supreme safety due to direct visual control via operating microscopy. That approach is technically demanding; however it was reproduced by others [3], and that study supported the approval of clinical trial for the treatment of multiple sclerosis. The presented procedure overcomes the reversed leakage of solutions/suspensions, the major limitation of this route which previously forced researchers to shift from intracisternal to intraventricular route of cell delivery [4]. Here, I describe a technique of intracisternal cell application in mice, which enabled me to avoid significant cell leakage and to achieve a widespread subarachnoid distribution of transplanted cells [5].

2 Materials

2.1 Animals

The procedure was developed using B6SJL mice (Jackson Laboratories).

2.2 Anesthetics

Atropine, ketamine, and xylazine.

2.3 Cells

Human bone marrow stromal cells were used for transplantation [6]. Prior to transplantation, the cells were labeled with 1 μg/ml Hoechst 33342 (Sigma, St. Louis, MO) and 1 μg/ml 5-(and-6)-carboxyfluorescein diacetate and succinimidyl ester (CFSE; Molecular Probes, Eugene, OR) for 30 min at 37 °C, washed twice with PBS, and further incubated with fresh medium for 24 h in order to minimize efflux of non-binding dye. Prior to transplantation the cells were counted and suspended in PBS at a density of 10,000/μl.

2.4 Equipment

(a) Stereotaxic apparatus (ASI Instruments, Heidelberg, Germany).

(b) Micromanipulator arm (ASI Instruments, Heidelberg, Germany).

(c) Operating microscope (OPMI Pico, Zeiss).

(d) Ultra Micro Pump (WPI).

2.5 Tools

(a) Gauze roller of diameter 10–20 mm, depending on the size of mouse.

(b) Scalpel.

(c) Adson forceps (FST).

(d) Micro-scissors (FST).

(e) Dumont forceps × 2 (FST).

(f) Syringe RN, 20 μl (Hamilton).

(g) Needle RN, Gauge 30, Point Style 4 (Hamilton).

(h) Staples (3 M, Neuss, Germany).

3 Methods

3.1 Animal Positioning

Following pre-anesthesia with atropine (0.04 mg/kg s.c.) the animals were anesthetized with ketamine (100 mg/kg i.p.) and xylazine (16 mg/kg i.p.). Stereotaxic apparatus was used to fix the head of the animals. For more convenient placing, the posterior part of the apparatus was lifted to form a 30° angle with the table surface. After fixing the head in earbars, we put a gauze roller with its axis in an anterior-posterior dimension under the mouse. The size of the roller was adjusted so that the line connecting the most prominent parts of the cranium and the spine formed a 15° angle with the horizontal line (Fig. 1). The tooth bar was then used to press the head down on the nasion, so that the line determining the facial surface formed a 15° angle with the vertical line (Fig. 1). The lines together formed a 90° angle. In that position, the cisterna magna was nearly at the highest point of the mouse body.

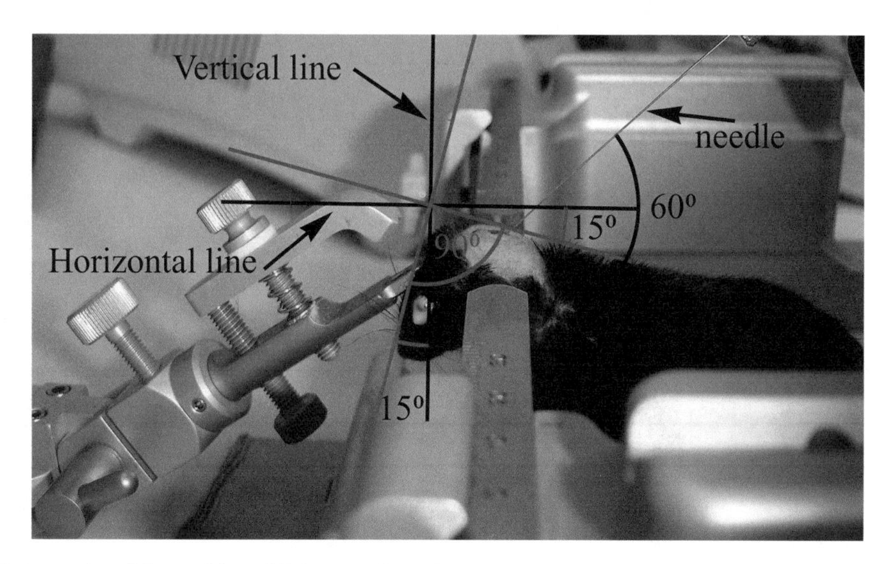

Fig. 1 Photographs of the position of the mouse head in the stereotactic frame according to the "concorde-like position" after insertion of the transplantation needle

3.2 Surgical Procedure

After appropriate positioning of animal a midline skin incision was made using scalpel and Adson forceps from the superior nuchal line to the level of C3 vertebrae. Then a strict midline blunt dissection of suboccipital muscles was performed under the operating microscope using fine Dumont forceps. As the atlanto-occipital membrane was exposed, 10 μl of cells (10,000/μl) were taken into a syringe. The syringe was then placed in the pump, which in turn was mounted on a micromanipulator arm. Subsequently the needle attached to the syringe was positioned over the midline of the atlanto-occipital membrane (where its anterior-posterior dimension is the longest) to form an angle of 60° with the horizontal line. The latter line corresponds to the line directly leading towards the cerebello-medullary fissure, which is visible under the operating microscope through the membrane. The membrane was touched by the needle tip midway between the occipital bone and the posterior arc of atlas. With a quick movement of the dial of the micromanipulator, the membrane was pierced and the needle was stopped exactly in half of the needle tip slope. It allowed the excess outflow of CSF through the needle tip (Fig. 2) giving the space for the cell suspension to be injected without excessive increase of CSF pressure. The needle was then moved forward to position the whole needle tip slope inside the cisterna magna. Under the operating microscope, it was possible to control the needle ending inside the cisterna magna to protect the brain stem from injury.

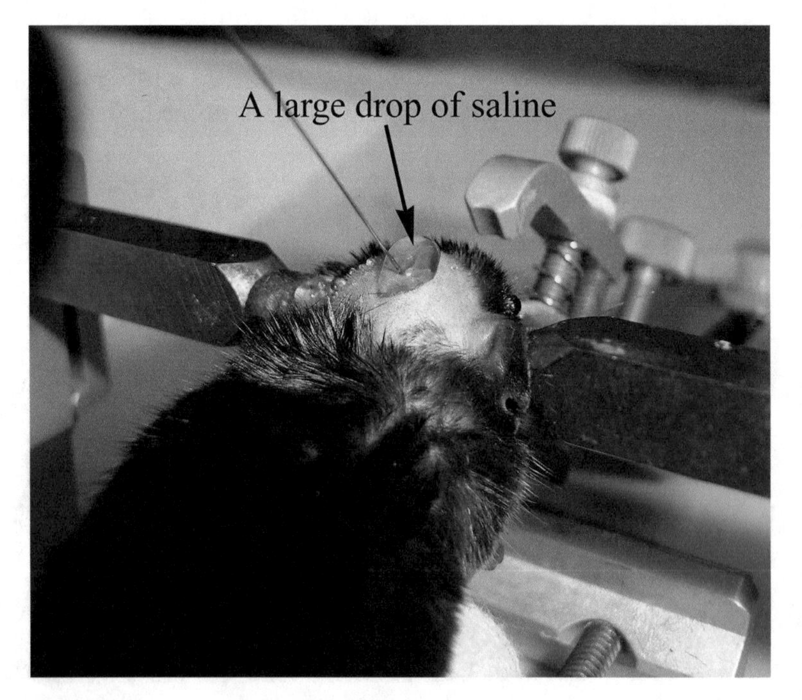

Fig. 2 Placing a drop of saline solution on *top* of the surgical field in order to avoid an outflow of the hcell suspension through the needle tract

Next, a large drop of saline (approximately 100 μl) was placed on top of the surgical field to get a convex meniscus over the wound (Fig. 3). Cells were injected over a period of 10 min. During the injection it was indispensable to observe the animal breathing movements, and as in one case of relenting and subsequent stopping of breathing, the needle had to be partially withdrawn immediately until the animal had recovered. Following injection of the

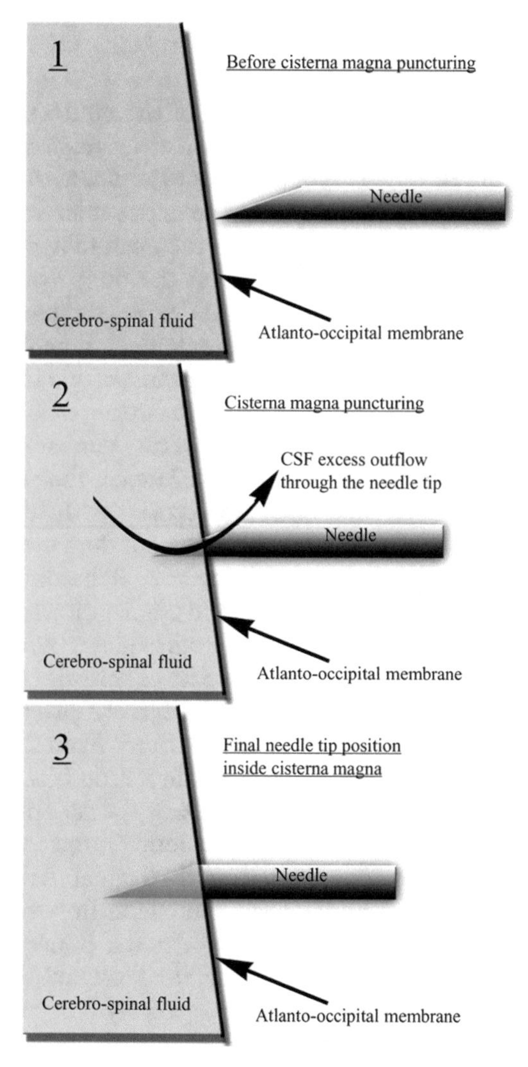

Fig. 3 Schematic representation of the piercing and injection procedure of the atlanto-occipital membrane. Scheme 1 represents the situation before penetrating the atlanto-occipital membrane. The first step (scheme 2) includes the piercing of the membrane. The needle was stopped exactly in half of the needle tip slope to allow the outflow of CSF through the needle tip to provide space for the cell suspension to be injected without excessive increase of CSF pressure. The needle was then moved forward to position the whole needle tip slope inside the cisterna magna (scheme 3)

cells the needle was kept in place for ten more minutes before withdrawal. The wound was closed with staples (3 M, Neuss, Germany).

The herein established transplantation method was subsequently used to transplant more than 150 transgenic ALS mice with various stem cell populations observing no surgery-related complications applying this technique [6]. Four mice died during the first night after surgery. This happened only once during the transplantation period, on two consecutive days. No other animals died prematurely following surgery; hence we consider the applied surgical method as being safe.

3.3 CNS Dissection Within Intact Dura for Evaluation of Cell Distribution

The dissection of the entire CNS within the intact dura mater is pivotal for precise evaluation of transplanted cell distribution. Therefore, following standard perfusion with 2% paraformaldehyde (PFA) the entire mice were postfixed in 2% PFA overnight. The CNS was then carefully prepared under the operating microscope to dissect the dura from the bone. Initially the limbs and internal organs were removed leaving skull and spine intact. Following muscle detachment, the bone dissection was initiated under operating microscope in lumbar region.

It begun with cutting of fascia between the posterior and transverse processes of two lumbar vertebral bodies with a forceps under the microscope. Hence, the posterior process of the upper vertebral body was broken with the forceps, exposing the spinal cord with its whitely lucent dura mater, followed by the removal of both transverse processes. Subsequently, all processes were removed in both caudal and cranial direction using this technique, but leaving the vertebral bodies for stabilizing the spinal cord. After removing the posterior part of the atlas exposing the atlanto-occipital membrane with the subjacent cisterna magna, the occipital and parietal bones were removed from the caudal direction after carefully breaking them into little bits. Next, the skull base was prepared from caudal paying special attention to the dura, which is tightly attached to the bones in this region. After exposing the whole skull base including the cranial nerves, the lateral, frontal, and parietal bones were removed. Importantly, one had to avoid any level motion to prevent the bones from boring into the brain parenchyma. Finally, the vertebral bodies were removed and the *bulbus olfactorius* was prepared by removing the nasal crest, which till then had served as anchorage point for the finger to lock the CNS into position. Finally, the remaining processes around the spinal cord were removed. The so-prepared CNS with intact dura mater was cryopreserved in 30% sucrose for 24 h, frozen in isopentane at −56 °C, and stored at −80 °C.

3.4 The Evaluation of Cell Delivery to Cisterna Magna

Prior to analysis of cell distribution the dissected CNS was cut and thaw-mounted on a cryostat (Leica CM3050 S) at 40 μm thick slices. The slices were evaluated for the presence of the pre-labeled cells under an Axiovert 135 inverse fluorescence microscope (Carl Zeiss, Göttingen, Germany). Estimation of cell numbers within the subdural space was done by counting Hoechst-labeled cell nuclei on every third slice throughout the CNS, i.e., ranging from the bulbus olfactorius back to the lumbar spinal cord. Exact counting was restrained by the often very tight clusters of cells found within subdural space prohibiting single-cell detection. Where feasible, single-cell nuclei were counted; else estimates based on the area occupied by the cell clusters compared to the area of single cells were evaluated.

The hMSCs within the intracranial fluid space were usually found as clusters of cells of various sizes depending on the compartment volume they were found in (Fig. 4). However, the cell density of every cluster was comparable and accounted for approximately 5000–10,000 cells per cluster. The highest number of cells was found in the cerebello-pontine angle and the ambient and basal cisterns. It ranged from 15,000 to 20,000 cells distributed over both sides of every mentioned space. Approximately 10,000 cells were found in the spinal cord (90% in the cervical and 10% in the thoracic region), and an average of 5000 cells in the IVth ventricle, prepontine, premedullar, and great cisterns, respectively, as well as on both sides of the optic chiasm and the olfactory nerve cisterns (Fig. 4). In both quadrigeminal and third ventricle the total number of cells did not exceed 1000. No cells were found in the lateral ventricles and on brain convexity. In total, we found approximately 80% of transplanted cells in all compartments described above. The variation between individual mice was minimal ($n=5$), though the indication of an average value and a standard error would be misleading, because we did only approximate cell counts due to the high density of the cell clusters prohibiting single-cell counting.

4 Notes

1. It was important that before the introduction of the needle to the *cisterna magna*, the atlanto-occipital membrane should be strained in order to facilitate its puncture. It could be achieved by slight increase of the inclination of the animal head under microscopic control.
2. To avoid cell leakage, which is mostly governed by the gravitational forces, it is critical to position cisterna magna as the highest point of the mouse body.
3. Due to the position of *cisterna magna* as the highest part of body the operative wound was lying in the horizontal plain enabling us to place and maintain a large drop of saline within the wound during the injection procedure.

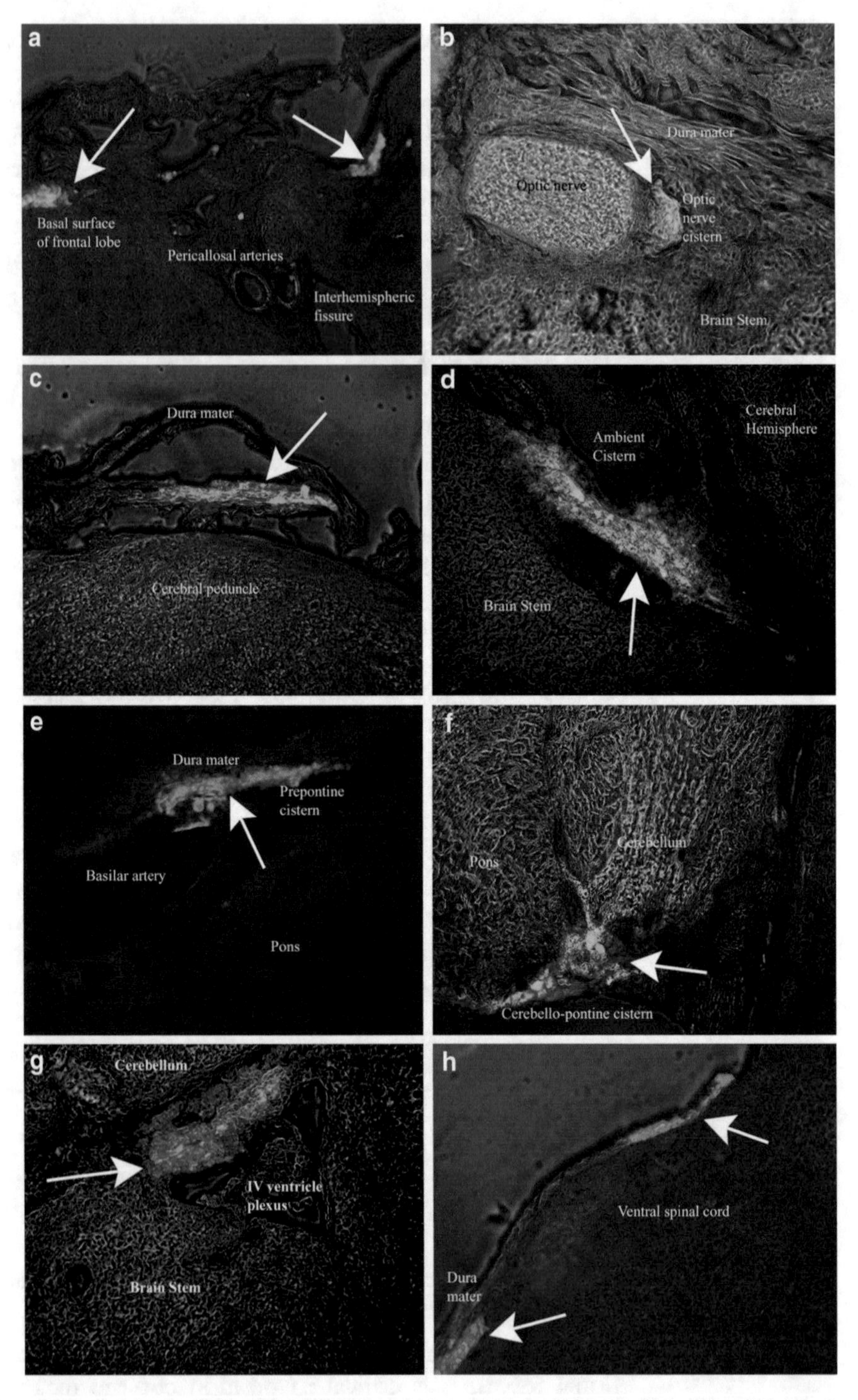

Fig. 4 Representative microphotographs of different sites of the subarachnoidal space of brain (**a**, olfactory nerve cistern; **b**, optic chiasm cistern; **c**, basal cistern; **d**, ambient cistern; **e**, prepontine cistern; **f**, cerebello-pontine cistern; **g**, IV ventricle) and spinal cord (**h**, ventral surface of cervical spinal cord) showing the distribution of hMSCs 24 h after transplantation into the cisterna magna using the "concorde-like position" method. *Arrows* indicate the transplanted cells stained with Hoechst 33342 (*blue*) and CFSE (*green*). Scale bar: 200 m

4. In contrast to other forms of intrathecal or intraparenchymal cell transplantation, the use of the concorde-like position causes cell sedimentation from the injection site towards both the cranial and caudal direction. A significant number of cells could be retrieved at the skull base and in the fourth ventricle (which are located against the CSF stream) indicating that the cell distribution depends not only on gravitation, but also on the pressure of the administrated fluid. The CSF circulation seems to be insufficient to carry the large cell clusters during the first day after transplantation.
5. Although the final cell distribution might depend on the properties of the transplanted cell type and the lesion of the brain tissue, the application of the "concorde-like position" for cell injection into the cisterna magna, as introduced in the present study, uniformly distributes the transplanted cells within the subarachnoidal space of the brain and spinal cord. This widespread cell distribution is a prerequisite for potential regenerative treatment strategies in neurodegenerative diseases without a distinct destruction of large CNS areas, which again promote cell homing towards the damaged areas. Particularly for chronic neurodegenerative diseases with a disseminated pattern of cell loss like ALS or AD, we recommend the "concorde-like position" for intrathecal cell transplantation.

References

1. Vulchanova L, Schuster DJ, Belur LR, Riedl MS, Podetz-Pedersen KM, Kitto KF, Wilcox GL, McIvor RS, Fairbanks CA (2010) Differential adeno-associated virus mediated gene transfer to sensory neurons following intrathecal delivery by direct lumbar puncture. Mol Pain 6:31
2. Lee IO, Son JK, Lim ES, Kim YS (2011) Pharmacology of intracisternal or intrathecal glycine, muscimol, and baclofen in strychnine-induced thermal hyperalgesia of mice. J Korean Med Sci 26:1371–1377
3. Harris VK, Yan QJ, Vyshkina T, Sahabi S, Liu X, Sadiq SA (2012) Clinical and pathological effects of intrathecal injection of mesenchymal stem cell-derived neural progenitors in an experimental model of multiple sclerosis. J Neurol Sci 313:167–177
4. Ohta M, Suzuki Y, Noda T, Ejiri Y, Dezawa M, Kataoka K, Chou H, Ishikawa N, Matsumoto N, Iwashita Y, Mizuta E, Kuno S, Ide C (2004) Bone marrow stromal cells infused into the cerebrospinal fluid promote functional recovery of the injured rat spinal cord with reduced cavity formation. Exp Neurol 187:266–278
5. Janowski M, Kuzma-Kozakiewicz M, Binder D, Habisch HJ, Habich A, Lukomska B, Domanska-Janik K, Ludolph AC, Storch A (2008) Neurotransplantation in mice: the concorde-like position ensures minimal cell leakage and widespread distribution of cells transplanted into the cisterna magna. Neurosci Lett 430:169–174
6. Habisch HJ, Janowski M, Binder D, Kuzma-Kozakiewicz M, Widmann A, Habich A, Schwalenstocker B, Hermann A, Brenner R, Lukomska B, Domanska-Janik K, Ludolph AC, Storch A (2007) Intrathecal application of neuroectodermally converted stem cells into a mouse model of ALS: limited intraparenchymal migration and survival narrows therapeutic effects. J Neural Transm 114:1395–1406

Chapter 12

Animal Models for Experimental Neurosurgery of Peripheral and Cranial Nerves

Joachim Oertel, Christoph A. Tschan, and Doerther Keiner

Abstract

Common experimental models for investigation of cranial and peripheral nerve function after trauma include sciatic nerve crush injuries and direct cutting of cochlear or facial nerves. Partial nerve transection, spinal nerve ligation, and chronic constriction injury are applied in neuropathic pain studies. Although these models are well established due to their potential to create reliable and reproducible results, an experimental setup for studying incomplete nerve lesions which resemble the intraoperative surgical condition was missing for years.

In neurosurgery, manipulation of peripheral or cranial nerves—such as in surgical procedures in the cerebellar-pontine angle or at the skull base—may lead to severe functional loss despite morphologically intact nerves. In the past years, different therapeutic agents for regeneration of the functional recovery have been investigated intensely. The authors' group has developed animal models to investigate the therapeutic potential of various substances in incomplete nerve injuries. In these models, the severity of the nerve lesion with distinct functional loss and recovery depends on the preset jet pressures.

Key words Waterjet dissection, Cranial nerve, Sciatic nerve, Surgical technique, Animal model, Nerve regeneration, Traumatic injury

1 Introduction

Today, several animal models for analysis of lesion and regeneration processes after peripheral and—to less extent—cranial nerve lesions are established. Partial nerve injury is commonly performed for analysis of neuropathic pain mechanisms or for the assessment of new drugs [1]. The most frequently employed models include partial nerve transection and chronic constriction injury [1–3]. These models are limited to the examination of non-traumatic lesions.

For investigation of traumatic nerve injury, nerve crush [4–8] or disruptive nerve injuries requiring primary suture or nerve grafting are well established [4, 9–12]. Sciatic nerve crush is the most commonly studied nerve injury model and has been used to test

Miroslaw Janowski (ed.), *Experimental Neurosurgery in Animal Models*, Neuromethods, vol. 116,
DOI 10.1007/978-1-4939-3730-1_12, © Springer Science+Business Media New York 2016

numerous neuroregenerative modalities [5–8]. Today, only few studies have been published which investigate the functional recovery of cranial nerves after traumatic injury [13, 14].

The functional recovery after the trauma is monitored by motor and sensory function tests and electrophysiologic evaluation [12, 15, 16]. For evaluation of functional recovery of the sciatic nerve, walking track analysis and calculation of the sciatic function index (SFI; 12) or power production of the hind legs such as extensor postural thrust [16] is performed. For analysis of the morphological recovery, histological or histomorphometric evaluation and electron microscopy of the nerves and innervated muscles are commonly performed [1, 5].

All animal models provide reliable and reproducible results, but none of the models evaluates the functional recovery of an incomplete nerve trauma with a macro-morphologically intact structure as it is often seen after neurosurgical procedures at the cranial nerves VII and VIII.

In general, neurosurgical procedures with involvement of the facial and the vestibulocochlear nerve exhibit a high risk of iatrogenic traumatic lesion resulting in functional loss of hearing and facial nerve palsy. In surgical procedures of pathologies located at the cerebellar-pontine angle such as meningiomas or acoustic neurinomas, the risk of complete functional loss of both cranial nerves is high even if nerves remain structurally intact. Due to the amelioration of microsurgical techniques and the latest development in intraoperative monitoring, the proportion of patients with preserved facial and vestibulocochlear function has been improved in the past years [17–21]. But the proportion of patients with preoperative good function of the facial and the vestibulocochlear nerve and significant deterioration after surgical procedure is still at 20% (facial nerve palsy; 18) and at 50%, respectively (hearing loss; 19).

The authors' group has developed an animal model of traumatic nerve injury that allows the investigation of functional loss of different severity. The application of the waterjet dissection device with direct contact of the jet and the nerves allows an incomplete or complete functional loss depending on the preset jet pressure.

Waterjet dissection represents a surgical technique that combines highly precise parenchymal dissection with preservation of even small vessels without thermal damage to the surrounding tissue. Basically, a water jet is pushed through a small nozzle in various pressures. Since the early 1980s, waterjet dissection has been investigated as a new technique in different surgical disciplines having its beginning in liver [22]. Nowadays it is generally accepted in this surgical discipline [23–27]. In the following years, waterjet dissection has been investigated in further surgical disciplines such as laparoscopic cholecystectomy [28], kidney [29–31], vascular [32], or ophthalmologic [33, 34] and dermatologic surgery [35].

In 1997, the research group of the authors started to apply a newly developed waterjet dissection instrument (Helix Hydro Jet, Erbe Elektromedizin, Tuebingen, Germany) in the field of neurosurgery. The instrument has been approved for clinical applications in Europe and the USA since 1996. First experimental cadaver studies in the pig and in vivo studies in the rabbit demonstrated successfully safe and accurate dissection of brain parenchyma with preservation of blood vessels [36–39]. Thus, the device has been applied clinically. Having started with meningiomas and metastases [40, 41], the operative spectrum has constantly increased to various intracranial pathologies [42–45]. Today, more than 200 surgical procedures have been performed including gliomas of all WHO grades, metastases, meningiomas, vascular tumors, and epilepsy surgery [45]. In general, the waterjet application enables precise brain parenchyma dissection in epilepsy surgery as well as tumor-brain parenchyma dissection under preservation of even small vessels and the pia/arachnoidea mater with pressures below 10 bars [43, 45]. Tumor debulking in soft and highly vascularized tumors is possible [44]. The waterjet dissection device can be applied with high safety; compared with the current literature, a higher incidence of postoperative worsening or an increased risk of deep infection or tumor spreading is not observed [45].

Since 2005, the possible use of the waterjet dissection device to protect cranial and peripheral nerves is under evaluation [46, 47].

It was shown that with jet pressures up to 6 bar the structure and function of rats' vestibulocochlear nerves remain intact. With a jet pressure of 8 bar, an incomplete functional lesion was produced with complete recovery of the auditory brainstem responses (ABRs) after 12 weeks with complete recovery of ABRs in 60 % of the animals. Jet pressures of 10 bar resulted in complete functional and—although the nerve's continuity was preserved—severe structural lesion. No recovery was found after 12 weeks [46].

In a second in vivo model, the technique was adapted to traumatic lesions of peripheral nerves [47]. In this model, a distinct and reproducible lesion of different severity was set on sciatic nerves of rats. The sciatic nerves showed complete functional recovery after 1 week if waterjet pressures up to 40 bar were used. Histologically, the nerve's structure remained intact. After 12 weeks no signs of direct nerve injury and nerve regeneration are detected. At pressures of 40–80 bar, anatomically and functionally evidence for nerve injury is found. At 40 and 50 bar, direct nerve injury and clinical signs of nerve damage are detected, but the deficits dissolve completely after 1 week. With waterjet pressures from 60 to 70 bar, clinical and electrophysiologic regeneration took up to 12 weeks to resolve. Distinct histomorphological characteristics occurred; pathological "mini" fascicles were detected. Even with a pressure set at 80 bar, the nerve's continuity was preserved, but clinical and electrophysiological regeneration was incomplete after 12 weeks.

To our knowledge, these are the first models that allow the examination of a certain therapeutic modality and its potential effect on the regeneration process in cranial and peripheral nerves after injuries of a defined force with functional deterioration and structural continuity.

2 Materials

2.1 Animal Model: Rat, In Vivo

Nerve type:

1. Sciatic nerve
2. Vestibulocochlear nerve

Adult male Sprague-Dawley rats (300–400 g for sciatic nerve dissection, respectively, 300–450 g for vestibulocochlear dissection) were used as a mammalian model for the in vivo sciatic nerve experiments. The animals were housed 1–2 per cage with a 12-h light/dark cycle and had free access to rat chow and water.

All experiments were conducted in accordance with approved protocols by the Institutional Animal Care and Use Committee and the German State Committee of Laboratory Animal Research.

2.2 Description of the Waterjet Instrument

For all experiments on the vestibulocochlear nerves, the Helix Hydro-Jet (Erbe Elektromedizin, Tübingen, Germany) was used. In 2007 it was followed by its successor Erbejet®2 (Erbe Elektromedizin, Tübingen, Germany) because of additional device features [48, 49].

For all experiments on the sciatic nerves, the Erbejet®2 by Erbe Elektromedizin Company (Tuebingen, Germany) was used. The waterjet is generated via a medium converter with an electronically controlled mechanical system (double-piston pump) with a pressure ranging from 1 to 80 bar. The medium converter is connected to a pencil-like handpiece consisting of a narrow nozzle with a diameter of 120 μm, and a surrounding suction tube. The generated water jet is a non-rotated thin laminar liquid jet (Fig. 1). Sterile 0.9 % isotonic saline is emitted as separating medium with a volume flow of 1–55 ml/min. The suction pressure can be selected from −100 to −800 mbar with a maximum suction capacity of 25 l/min. The pressure and the suction can be manually adjusted by preselection. Depending on the surgical procedure, several different settings can be selected. During surgery, the waterjet application and pressure are adjusted within the preset range by a foot pedal. The system has been approved by the regulatory authorities for surgical use in humans in Germany and the USA.

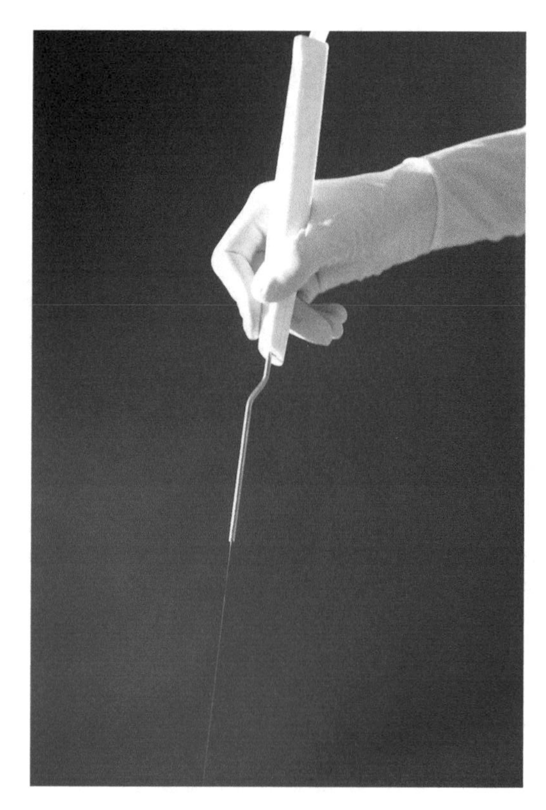

Fig. 1 Handpiece of the Erbejet®[2]. The jet nozzle diameter of the neurosurgical standard applicator is 120 μm, which creates a thin laminar liquid jet

2.3 Electrophysiological Device/Equipment

1. We use a portable Viking® and a portable Medelec™ Synergy (Version 12.2) N-EP—EMG/EP monitoring system with the high-frequency filter regulated for 5 kHz and the low-frequency filter regulated for 3 Hz.
2. For the monitoring of the sciatic nerves, subdermal paired needle electrodes (Ambu A/S, Ballerup, Denmark) were used for recording electrodes and for stimulation.
3. For ABR recording (Fig. 2a), subdermal-paired needle electrodes (Ambu A/S, Ballerup, Denmark) were used as recording electrodes (Fig. 2b). For click stimulation, in-ear headphones were adapted to the small size of the external rat's auditory canal (Fig. 2c).

2.4 Surgical Equipment

2.4.1 Sciatic Nerve Dissection

1. Standard surgical microscope (Zeiss, Germany) with photo/video documentation.
2. A no. 15 scalpel blade.
3. Standard microsurgical instruments.
4. A wound expander.
5. Vicryl 5-0 sutures (Ethicon, Germany) for muscle fascia.

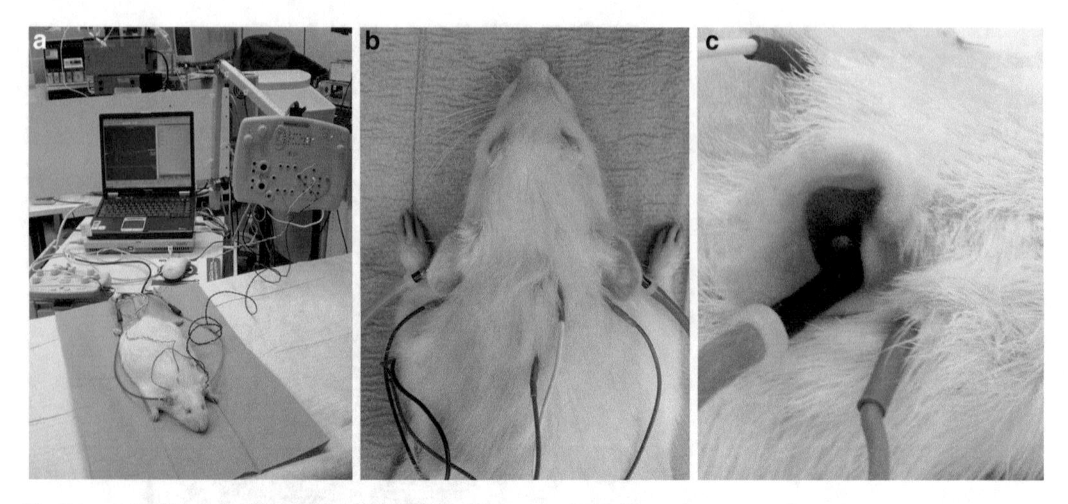

Fig. 2 (**a–c**) ABR recording by a portable Medelec™ Synergy (Version 12.2) monitoring system (**a**). Subcutaneous needle electrodes are placed over the left and right posterior convexity, vertex, and neck (**b**). Click stimuli of 80 db are conducted through tubal earphones, inserted into the rat external auditory canal (**c**)

6. Ethibond 4-0 sutures (Ethicon, Germany) for skin.
7. Chloralhydrate solution.
8. Tramadol (Tramal®).

2.4.2 Vestibulocochlear Nerve Dissection

1. Standard surgical microscope (Zeiss, Germany) with photo/video documentation.
2. A no. 15 scalpel blade.
3. Standard microsurgical instruments.
4. A wound expander.
5. A diamond drill (1–2 mm).
6. A bipolar forceps (Aesculap, Germany).
7. Tabotamp (Ethicon, Germany).
8. Wound batting (Merocel®, Medtronic, Germany).
9. Vicryl 5-0 sutures (Ethicon, Germany) for muscle fascia.
10. Ethibond 4-0 sutures (Ethicon, Germany) for skin.
11. Ketamine (Ketanest®).
12. Xylazine.
13. Tramadol (Tramal®).

3 Methods

3.1 Waterjet Dissection on Peripheral Nerves

3.1.1 Surgical Procedure

1. Anesthetize the animals with chloralhydrate solution ip at a dose of 36 mg/kg body weight before surgery and perform analgesia with tramadol ip at a dose of 50 mg/kg body weight (the sedation is to be maintained for the duration of each individual experiment).
2. For exact positioning of the needle electrodes, both hindlimbs have to be shaved and disinfected.
3. Perform a posterior-laterally skin incision parallel to the right femur with the no. 15 scalpel and open the muscle fascia of the gluteus muscles (Fig. 3a, b).
4. Expose the sciatic nerve carefully at midthigh level with the aid of a wound expander (Fig. 3c, d).
5. Mobilize the nerve carefully under microscopic view with a microforceps and microscissors from its surrounding muscle fascia until it is exposed from the sciatic notch exit to the part, where the nerve divides into its specific motor branches (Fig. 3e, f). Care has to be taken to avoid excessive spreading of the tissue to prevent nerve damage due to tension.
6. Apply the water jet to the right sciatic nerve. Ideally, the rat is placed on a computer-controlled linear device that moves with continuous speed and enables to fixate the handpiece of the

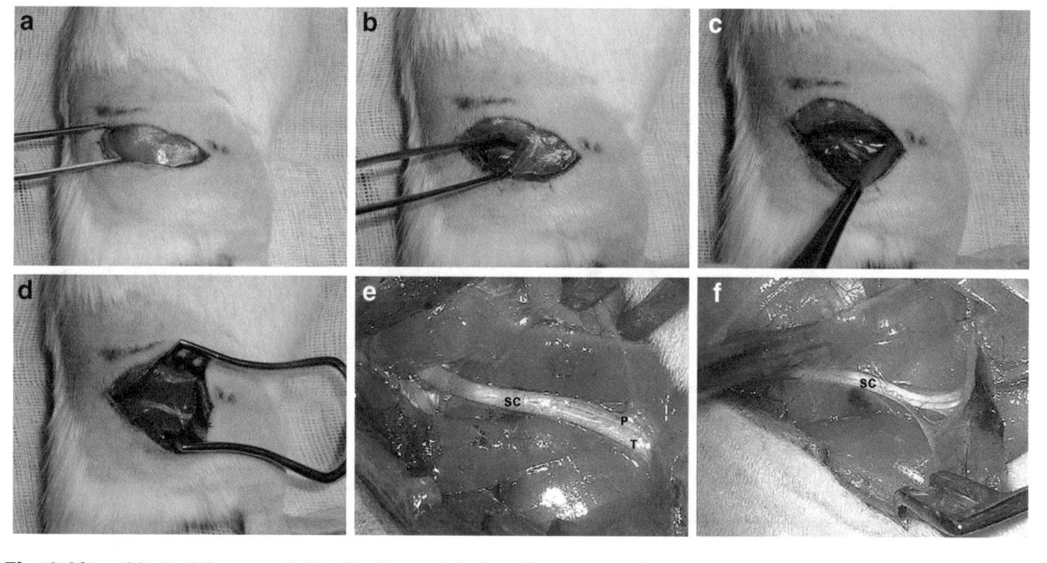

Fig. 3 After skin incision parallel to the femur (**a**), the gluteus muscles are incised (**b**) and the sciatic nerve is identified (**c**). The cut is enlarged (**d**) and the nerve is dissected carefully from its surrounding tissue under microscopic view (**e**, **f**) leaving the epineurium intact. Finally, the nerve is exposed from the sciatic notch exit to the part, where the specific motor branches are divided. *SC* sciatic nerve, *P* peroneal fascicle, *T* tibial fascicle

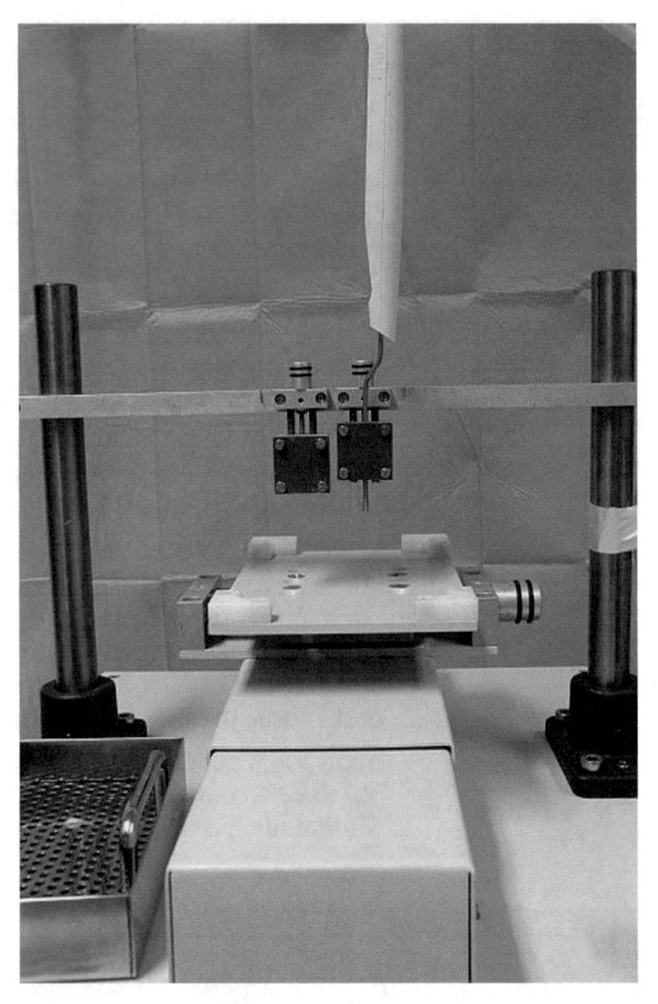

Fig. 4 Computer-controlled linear device for reliable adjustment of the cutting distance from the nozzle tip to the nerve surface to obtain accurate and comparable results

waterjet dissector. For this, we use a specially developed computer-controlled linear device that was employed for accurate and comparable results (Erbe Elektromedizin GmbH, Tübingen, Germany; Software Servomanager 6.4.1., Parker Automation). The device allows a standard adjustment (we use a 2 mm cutting distance from the nozzle's tip to the nerve surface and of a cutting speed of 1 cm/s; Fig. 4).

7. The muscle fascia is closed with 5-0 vicryl single stitches followed by intracutaneous skin closure preventing wound gnawing.
8. Postoperatively, oral analgetics (tramadol 50 mg/kg body weight) are administered po for 1 week.

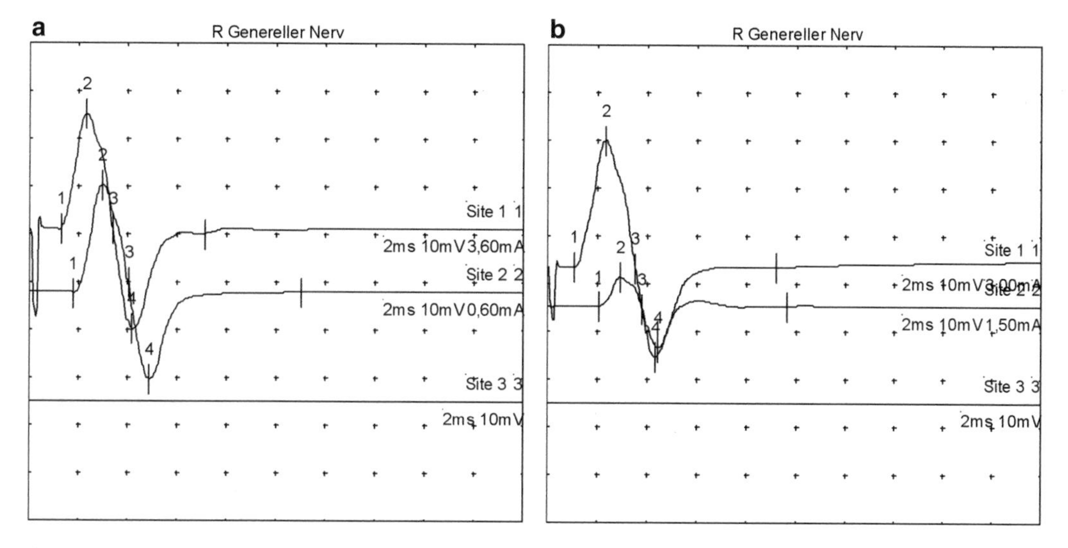

Fig. 5 (**a**, **b**) Perioperative measurement of the distal motor latency and the compound muscle action potential amplitude for calculation of the motor nerve conduction velocity before (**a**) and after (**b**) waterjet lesion

3.1.2 Electrophysiological Examination

1. The distal motor latency and the compound muscle action potential amplitude are measured and the motor nerve conduction velocity is calculated (Fig. 5a, b). All measurements are taken for the proximal and the distal site of the sciatic nerve lesion to determine the neurophysiologic decline proximal of the lesion.
2. Electrophysiological measurements are performed before surgery, after nerve exposure to verify the nerve's electrophysiological integrity, directly after the wajerjet-induced nerve lesion and at the end of the surgical procedure after wound closure.
3. If necessary, nerves are moistened with 0.9% saline solution to avoid desiccation during the examination.
4. The pre- and postoperative electrophysiologic measurements are performed on both sciatic nerves.
5. The recording (different) electrode is placed under the skin above the tibialis anterior muscle and its reference (indifference) electrode is placed 1.0–1.5 cm distally above the tendon of the tibialis anterior muscle. For stimulation, the cathode is placed at the popliteal fossa for distal stimulation and at the sciatic notch proximal to the site of the lesion for proximal stimulation; the anode is placed in the paraspinal muscles. The ground electrode is placed in the tail.
6. We apply a 20 ms supramaximal stimulus to generate an action potential, but care is taken to keep the stimulation intensity to less than 7 mA.
7. We carried out serial follow-up neurophysiologic examinations at day 1, and at 1 and 12 weeks. For examination, the animals have to be anesthetized (see Sect. 3.1.1).

3.2 Waterjet Dissection on Cranial Nerves

3.2.1 Surgical Procedure

1. Anesthetize the animals with 10% ketamine ip at a dose of 1 ml/kg body weight and 2% xylazine ip at a dose of 0.5 ml/kg body weight before surgery (the sedation is to be maintained under sedation for the duration of each individual experiment).
2. The skin of the neck has to be shaved and disinfected.
3. The head of the rats is fixed in anteflexion and prone position in a stereotactic frame system.
4. In general, the right side (or the best ABR-responding side) is chosen for surgery.
5. Perform a straight skin incision with a no. 15 scalpel parallel to the midline, and then incise the neck muscles. The suboccipital cranium is exposed with the aid of a wound expander.
6. Under microscopic view, perform a lateral suboccipital craniectomy with the aid of a diamond drill of 1–2 mm caliber (Fig. 6a). The transversal and sigmoid sinus has to be exposed on the edge of the craniectomy.
7. Open the dura mater and drape it in the direction of the sigmoid sinus. The cerebellum is retracted upward and to the middle (Fig. 6b). Use wound batting, if necessary.
8. To get more space, open the cerebellopontine cistern to let out the cerebrospinal fluid.
9. Expose the vestibulocochlear nerve carefully below the flocculus on its course from the brainstem to the internal auditory canal (Fig. 6c). During this step of nerve exposure changes in ABR recording are observed in some cases. If this is the case, the complete recovery of the ABR has to be awaited.
10. Apply the water jet directly to the vestibulocochlear nerves' surface without the use of the nozzle-integrated suction.

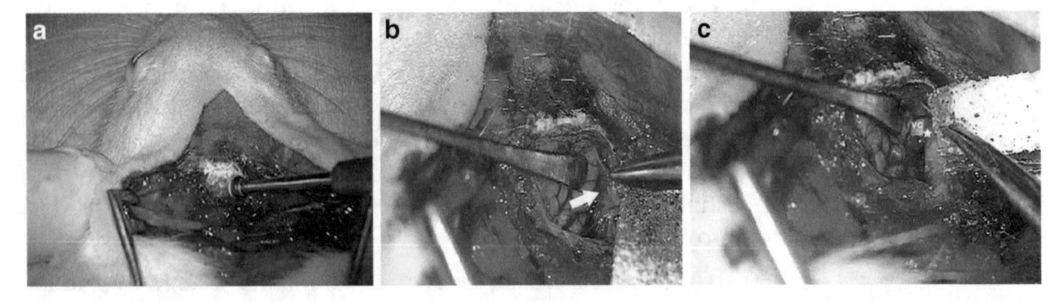

Fig. 6 (**a–c**) Lateral suboccipital craniectomy by a diamond drill (**a**). The dura mater is opened and draped in the direction of the sigmoid sinus (*arrow*, **b**). The cerebellum is retracted to the midline. Below the flocculus the vestibulocochlear nerve is exposed on its course from the brainstem to the internal auditory canal (*asterisk*, **c**). *VC* vestibulocochlear nerve

11. The dura mater and the skull are closed with autologous fascia and fibrin gel foam. The muscles and the muscle fascia have to be tightly sutured by 4-0 ligature. Close the skin with 3-0 ligature single stitches.
12. Observe your animals closely until awake. Oral analgetics (tramadol 50 mg/kg body weight or novaminsulfon 20–50 mg/kg body weight) have to be administered postoperatively for 1 week. Animals presenting with neurological complications have to be immediately sacrificed.

3.2.2 Electrophysiological Examination

1. The ABRs are recorded preoperatively, intraoperatively before and directly after lesion and after wound closure.
2. The pre- and postoperative recording has to be performed bilaterally. Before surgery, select the best responding site.
3. Subcutaneous needle electrodes are placed over the left and right posterior convexity, vertex, and neck. Click stimuli of 80 db are conducted through tubal earphones that are inserted into the rat external auditory channel.
4. We carried out serial follow-up examinations bilaterally at day 1, week 1, week 2, and week 6 on anesthetized animals (see Sect. 3.2.1).

4 Notes

1. For peripheral nerve surgery, it is important to examine animals of the same weight in the whole experimental trial, because a direct proportionality between the size of a rat and the number of sciatic nerve fibers has been shown [50].
2. To obtain comparable results it is of importance to take care of the same cutting distance and the same cutting speed as it was received in the authors' experimental groups by using the computer-controlled linear device (see Sect. 3.1). There is a tendency of an increasing lesion with aggravated neurological deficits in case of a larger distance between the nozzle tip and the nerve.
3. Heating lamps and/or homeothermic blanket control units should be used during the surgical procedure and the neurophysiologic examinations to maintain a body temperature approximately at 37 °C. Lower temperatures can lead to a false (positive) pathological distal motor latency or a broadened muscle action potential.
4. During the surgical procedure, iatrogenic excessive spreading of sciatic nerve surrounding tissue should be avoided to prevent nerve damage due to tension. Prior to the nerve lesion, the epineurial covering should be left intact to maintain the nerve integrity.

References

1. Lindenlaub T, Sommer C (2004) Partial sciatic nerve transection. In: Luo ZD (ed) Methods in molecular medicine, vol 99, Pain research: methods and protocols. Humana, Totowa, NJ, pp 47–53
2. Bennet GJ, Xie YK (1988) A peripheral monoeuropathy in rat that produces disorders of pain sensation like those seen in man. Pain 33:87–107
3. Ma W, Bisby MA (2000) Calcitonin gene-related peptide, substance P and protein gene product 9.5 immunoreactive axonal fibers in the rat footpad skin following partial sciatic nerve injuries. J Neurocytol 29(4):249–262
4. Magill CK, Tong AMS, Kawamura D et al (2007) Reinnervation of the tibialis anterior following sciatic nerve crush injury: a confocal microscopic study in transgenic mice. Exp Neurol 207(1):64–74
5. Pan HC, Chen CJ, Cheng FC et al (2009) Combination of G-CSF administration and human amniotic fluid mesenchymal stem cell transplantation promotes peripheral nerve regeneration. Neurochem Res 34(3):518–527
6. Yan JG, Matloub HS, Yan Y et al (2010) The correlation between calcium absorption and electrophysiologcal recovery in crushed rat peripheral nerves. Microsurgery 30(2): 138–145
7. Dinh P, Hazal A, Palispis W et al (2009) Functional assessment after sciatic nerve injury in a rat model. Microsurgery 29(8):644–649
8. Kalender AM, Dogan A, Bakan V et al (2009) Effect of Zofenopril on regeneration of sciatic nerve crush injury in a rat model. J Brachial Plex Peripher Nerve Inj 4:6
9. Huang JH, Cullen DK, Browne KD et al (2009) Long-term survival and integration of transplanted engineered nervous tissue constructs promotes peripheral nerve regeneration. Tissue Eng Part A 15(7):1677–1685
10. Dodla MC, Bellamkonda RV (2008) Differences between the effect of anisotropic and isotropic laminin and nerve growth factor presenting scaffolds on nerve regeneration across long peripheral nerve gaps. Biomaterials 29(1):33–46
11. Brunelli G, Spano P, Barlati S et al (2005) Glutamatergic reinnervation through peripheral nerve graft dictates assembly of glutamatergic synapses at rat skeletal muscle. Proc Natl Acad Sci U S A 102(24):8752–8757
12. Bain JR, Mackinnon SE, Hunter DA (1989) Functional evaluation of complete sciatic, peroneal, and posterior tibial nerve lesions in the rat. Plast Reconstr Surg 83:129–136
13. Sekiya T, Tanaka M, Shimamura N et al (2001) Macrophage invasion into injured cochlear nerve and its modification by methylprednisolone. Brain Res 905(1–2):152–160
14. Costa HJZR, da Silva CF, Korn GP et al (2006) Posttraumatic facial nerve regeneration in rabbits. Rev Bras Otorrinolaringol 72(6): 786–793
15. Martins RS, Siqueira MG, da Silva CF et al (2005) Electrophysiologic assessment of regeneration in rat sciatic nerve repair using suture, fibrin glue or a combination of both techniques. Arq Neuropsiquiatr 63(3-A): 601–604
16. Koka R, Hadlock TA (2000) Quantification of functional recovery following rat sciatic nerve transection. Exp Neurol 168:192–195
17. Seol HJ, Kim CH, Park CK et al (2006) Optimal extent of resection in vestibular schwannoma surgery: relationship to recurrence and facial nerve preservation. Neurol Med Chir (Tokyo) 46:176–181
18. Samii M, Gerganov V, Samii A (2006) Improved preservating of hearing and facial nerve function in vestibular schwannoma surgery via the retrosigmoid approach in a series of 200 patients. J Neurosurg 105(4):527–535
19. Samii M, Matthies C, Tatagiba M (1997) Management of vestibular schwannomas (acoustic neuromas): auditory and facial nerve function after resection of 120 vestibular schwannomas in patients with neurofibromatosis 2. Neurosurgery 40(4):696–705
20. Samii M, Matthies C (1997) Management of 1000 vestibular schwannomas (acoustic neuromas): the facial nerve-preservation and restitution of function. Neurosurgery 40(4): 684–694
21. Isaacson B, Kileny PR, El-Kashlan H, Gadre AK (2003) Intraoperative monitoring and facial nerve outcomes after vestibular schwannoma resection. Otol Neurotol 24(5): 812–817
22. Papachristou DN, Barters R (1982) Resection of the liver with a water jet. Br J Surg 69:93–94
23. Baer HU, Stain SC, Guastella T et al (1993) Hepatic resection using a water jet dissector. HPB Surg 6:189–198
24. Hata Y, Sasaki F, Takahashi H et al (1994) Liver resection in children, using a water-jet. J Pediatr Surg 29:648–650

25. Izumi R, Yabushita K, Shimizu K et al (1993) Hepatic resection using a water jet dissector. Surg Today 23:31–35
26. Rau HG, Schauer R, Pickelmann S et al (2001) Dissektionstechniken in der Leberchirurgie. Chirurg 72:105–112
27. Rau HG, Duessel AP, Wurzbacher S (2008) The use of water-jet dissection in open and laparoscopic liver resection. HPB (Oxford) 10:275–280
28. Shekarriz H, Shekarriz B, Kujath P et al (2003) Hydro-Jet-assisted laparoscopic cholecystectomy: a prospective randomized clinical study. Surgery 133:635–640
29. Hubert J, Mourey E, Suty JM et al (1996) Water-jet dissection in renal surgery: experimental study of a new device in the pig. Urol Res 24:355–359
30. Penchev RD, Losanoff JE, Kjossev KT (1999) Reconstructive renal surgery using a water jet. J Urol 162:772–774
31. Moinzadeh A, Hasan SM et al (2005) Water jet assisted laparoscopic partial nephrectomy without hilar clamping in the calf model. J Urol 174:317–321
32. Kobayashi M, Sawada S, Tanigawa N et al (1995) Water jet angioplasty—an experimental study. Acta Radiol 36:453–456
33. Lipshitz I, Bass R, Loewenstein A (1996) Cutting the cornea with a waterjet keratome. J Refract Surg 12:184–186
34. Gordon E, Parolini B, Abelson M (1998) Principles and microscopic confirmation of surface quality of two new waterjet-based microkeratomes. J Refract Surg 14:338–345
35. Siegert R, Danter J, Jurk V et al (1998) Dermal microvasculature and tissue selective thinning techniques (ultrasound and water-jet) of short-time expanded skin in dogs. Eur Arch Otorhinolaryngol 255:325–330
36. Oertel J, Gaab MR, Knapp A et al (2003) Water jet dissection in neurosurgery: experimental results in the porcine cadaveric brain. Neurosurgery 52:153–159
37. Oertel J, Gaab MR, Pillich D-T et al (2004) Comparison of waterjet dissection and ultrasonic aspiration: an in vivo study in the rabbit brain. J Neurosurg 100:498–504
38. Oertel J, Gaab MR, Schiller T et al (2004) Towards waterjet dissection in neurosurgery: experimental in-vivo results with two different nozzle types. Acta Neurochir (Wien) 146: 713–720
39. Piek J, Wille C, Warzok R et al (1998) Waterjet dissection of the brain: experimental and first clinical results. Technical note. J Neurosurg 89:861–864
40. Oertel J, Gaab MR, Piek J (2003) Waterjet resection of brain metastases—first clinical results with 10 patients. Eur J Surg Oncol 29:407–414
41. Oertel J, Gaab MR, Warzok R et al (2003) Waterjet dissection in the brain: review of the experimental and clinical data with special reference to meningioma surgery. Neurosurg Rev 26:168–174
42. Piek J, Oertel J, Gaab MR (2002) Waterjet dissection in neurosurgical procedures: clinical results in 35 patients. J Neurosurg 96: 690–696
43. Oertel J, Gaab MR, Runge U et al (2005) Waterjet dissection versus ultrasonic aspiration in epilepsy surgery. Neurosurgery 56: 142–146
44. Oertel J, Wagner W, Gaab MR et al (2004) Waterjet dissection of gliomas—experience with 51 procedures. Minim Invasive Neurosurg 47:154–159
45. Keiner D, Gaab MR, Backhaus V, Piek J, Oertel J (2010) Water jet dissection in neurosurgery: an update after 208 procedures with special reference to surgical technique and complications. Neurosurgery 67:342–354
46. Tschan CA, Gaab MR, Krauss JKK, Oertel J (2009) Waterjet dissection of the vestibulocochlear nerve: an experimental study. J Neurosurg 110(4):656–661
47. Tschan CA, Keiner D, Müller HD, Schwabe K, Gaab MR, Sommer C, Oertel J (2010) Waterjet dissection of peripheral nerves: an experimental study of the peripheral nerves of rats. Neurosurgery 67(368):376
48. Tschan CA, Tschan K, Krauss JK, Oertel J (2009) First experimental results with a new waterjet dissector: Erbejet 2. Acta Neurochir (Wien) 151(11):1473–1482
49. Tschan CA, Tschan K, Krauss JK, Oertel J (2010) New applicator improves waterjet dissection quality. Br J Neurosurg 24(6): 641–647
50. Schmalbruch H (1986) Fibre composition of the rat sciatic nerve. Anat Rec 215(1):71–81

Chapter 13

Surgery of the Brain and Spinal Cord in a Porcine Model

Jan Regelsberger

Abstract

Working time restrictions and economic pressure hinder surgical specialties to implement an adequate and structured training program. Alternative training forms seem to be requested in which a most realistic setup is required imitating the daily routine. An in vivo swine model was evaluated for its practical use in training neurosurgical residents in the past years (Regelsberger et al. Cen Eur Neurosurg 72:192–195, 2010 [2]). Surgical procedures included craniotomy, dura opening, brain surgery with sulcal preparation, and excision of an artificial tumor as well as laminectomy or other dorsal approaches to the spine with exposure of the dural sack and nerve roots. Microscopy and bleeding management were an integrated part of training and were found to be very useful supplements for young neurosurgeons. Our experiences with these unique in vivo training model are outlined and its advantages and pitfalls described.

Key words Neurosurgical training, Neuroanatomy of the swine, Craniotomy in porcine model, Spine surgery in pigs, Wet-lab training

1 Introduction

European working time directive (EWTD) may be one of the main reasons at present why training programs are still judged to be inadequate and insufficient [3]. Shifting and alternating duties are in contrast with predictable schedules in the operating rooms and structured education plans [4]. The high level of medical supply, pressure of time, and costs may be further strong factors hindering residents to get skilful surgeons [5]. Therefore adjuvant education forms seem to be necessary. While hands-on cadaver courses have spread across the countries we were starting with an in vivo swine model in cranial and spinal neurosurgery in 2005.

2 Training Model

A swine model was chosen for a neurosurgical training course and found to be an adequate in vivo object as cranial vault, brain, and spinal anatomy are easy to mediate allowing young neurosurgeons

Miroslaw Janowski (ed.), *Experimental Neurosurgery in Animal Models*, Neuromethods, vol. 116,
DOI 10.1007/978-1-4939-3730-1_13, © Springer Science+Business Media New York 2016

to focus quickly on their neurosurgical skills. Ideal conditions were found at the European Surgical Institute (Norderstedt, Johnson&Johnson, Germany) with a veterinary backup and full anesthetic equipment where practical experiences in the laboratory could be exclusively focused on the learning aims [1, 2]. Microscopes (Möller-Wedel, Wedel, Germany), bipolar coagulation, suction, and microsurgical instruments (Codman, Johnson&Johnson) were provided to install an almost realistic copy of the daily situation in the operating room (OR).

2.1 Neuroanatomy and Brain Surgery in a Porcine Model

The central nervous system (CNS) of all vertebrates is topologically equivalent and bilaterally, rostrocaudally orientated. This is not different in swine with a developmental architecture of a prosencephalon including diencephalon and telencephalon, brain stem with mesencephalon and rhombencephalon, completing with the spinal cord caudally. The CNS is covered by the meninges and the brain is protected by the skull.

Main differences, compared to human anatomy, can be seen in the skull and cranial base architecture. Looking from above or anteriorly, a plane forehead and vertex are ending in a high, sloping crest posteriorly where the neck muscles are assessing (Fig. 1a, b). The vertex plane is limited laterally by the parietal bones (occipital) and orbitae (frontal) restricting the craniotomy in the biparietal width. Midline sagittal suture is identified easily covering the sagittal sinus whereas coronal and lambdoid sutures will not be seen in a conventional approach. The bone itself will appear soft and less mineralized, especially in young pigs with a thickness of about 1 cm or more at the lateral edges of the vertex plane.

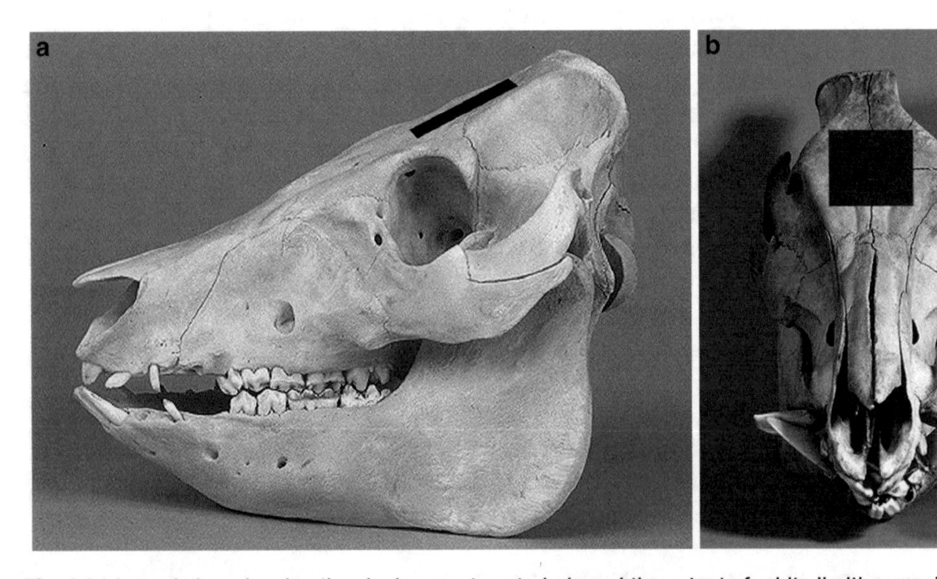

Fig. 1 (**a**) Lateral view showing the sloping crest posteriorly and the extent of orbita limiting craniotomy anteriorly. The extent of craniotomy is marked *black*. (**b**) Anterior view from above with marking of the craniotomy

The cerebral hemispheres are divided into lobes and main sulci comparable to human anatomy. Topographic allocations, cranial nerves, their origin at the brain stem, and their pathways are very similar as well (Figs. 2 and 3). The olfactory bulb is somewhat bigger but will not be seen following craniotomy. Ventricles containing CSF are small and do not allow an endoscopic access in an appropriate way for learning reasons. Compared with humans, arterial blood supply of the brain in pigs is different as a plexus of very small vessel discharge into the internal carotid artery. This circumstance and Willis cross perfusion circumvent cerebral infarction models. Anterior, middle, and posterior cerebral arteries, latter one fusing with the basilar artery, are found as well as venous drainages including sinuses and jugular veins are similar again to cerebral blood supply in humans.

Following a midline incision starting at the coronal suture and ending at the sloping crest posteriorly, unilateral and bilateral approaches to the brain can be performed. Attention has to be driven to the limited access of about 3.5 cm width and 4 cm length (anterior-posterior) in bilateral approaches, a thin dura which may be easily injured during craniotomy and venous bleedings out of the bone or violated sinus (Fig. 4). Frontal sinuses are large and will be opened by trepanation. Drills are needed in some cases to uncover the dura but craniotomy is performed in the ordinary way. Orbital roof and thickness of bone at the lateral edge may restrict craniotomy. Bone wax and other hemostyptics should be prepared in advance as well as bone punches may facilitate a more safe approach for beginners, especially in bilateral exposures crossing the sagittal sinus. The approach should be extended by drills and

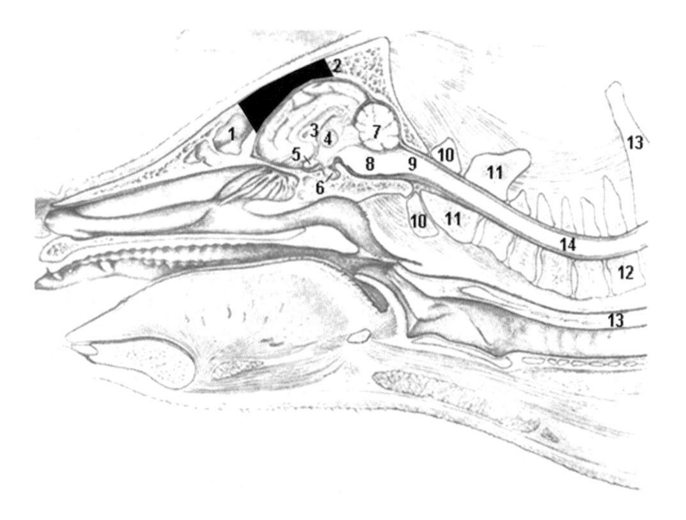

Fig. 2 Neuroanatomy of the pig with topographic landmarks and anterior-posterior extent of craniotomy. *1* frontal sinuses, *2* skull, *3* hemisphere, *4* interhemispheric commissura, *5* chiasm, *6* pituitary, *7* cerebellum, *8* brain stem, *9* medulla, *10* atlas, *11* axis, *12* vertebral column, *13* anterior spinal processes, *14* spinal cord

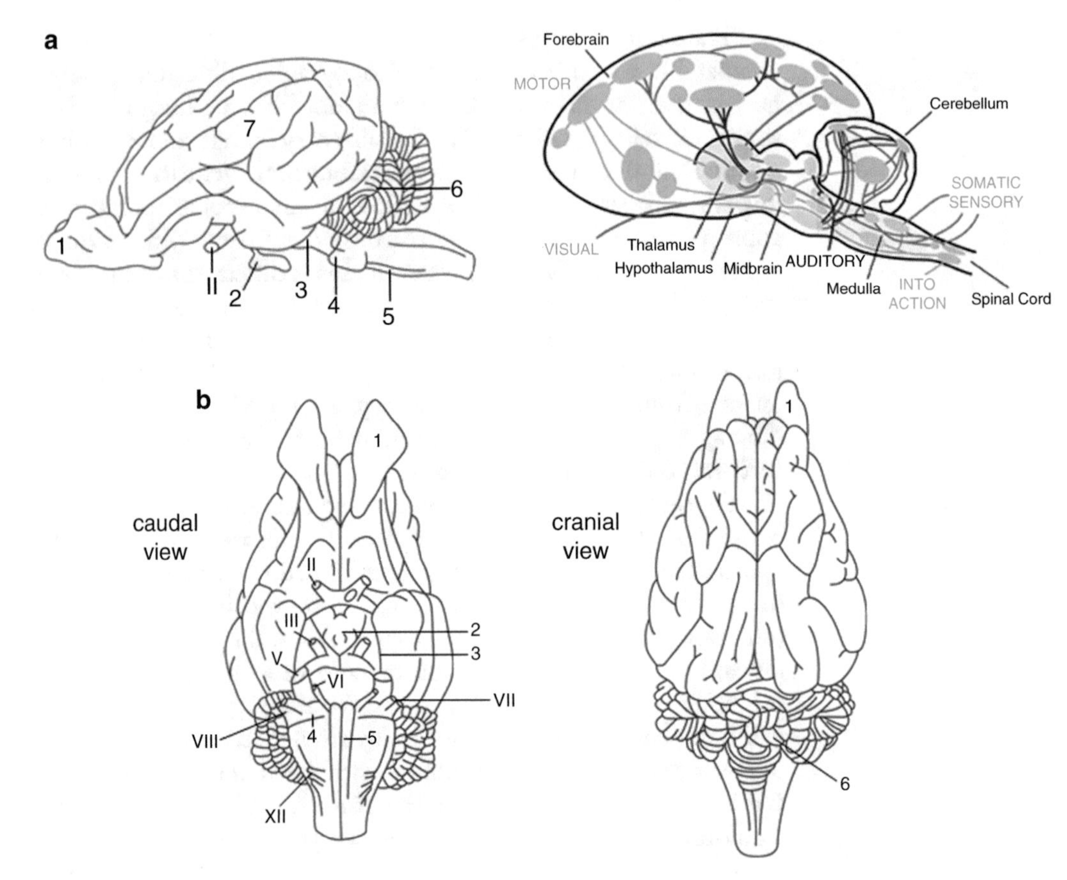

Fig. 3 (**a**) Cranial nerves and functional areas of the porcine brain. (**b**) Cranial nerves in the caudal and cranial view

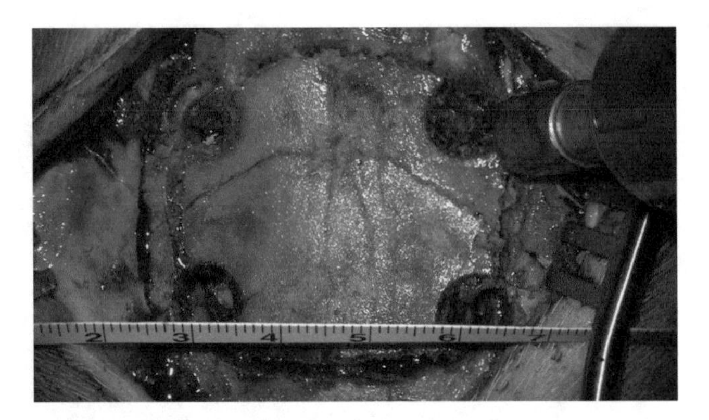

Fig. 4 Burr holes and craniotomy with size of about 3.5 × 3.5 cm

bone punches to the lateral, frontal, and occipital limits allowing a maximum view on the cortical surface later.

Dura is lifted by one suture laterally followed by incision and opening in a crescent manner leaving the sagittal sinus untouched (Fig. 5). Arachnoid granulations may bleed and should be preserved, otherwise compressed by hemostyptics or transected under microscopic magnification. Careful microscopic dissection should be made on bridging veins draining into the sagittal sinus (Fig. 6). Further preparation may include arachnoid opening, sulcal dissection, and resection of a gyrus or an artificially inserted tumor (Fig. 7). Colored fibrin glue (dura sealants, amount of 0.5–1 ml) is

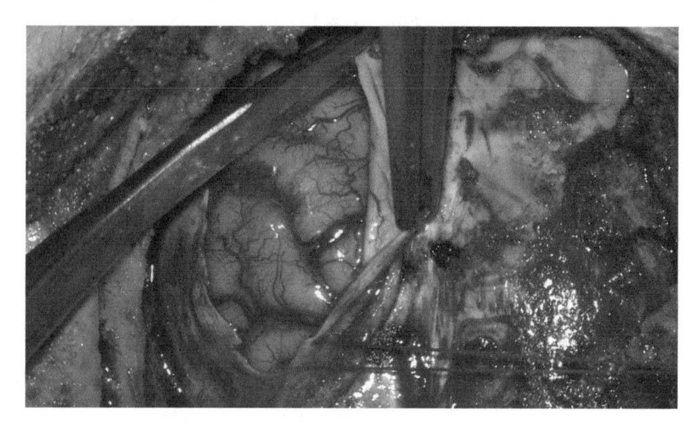

Fig. 5 Dural opening preserving the sagittal sinus in the midline

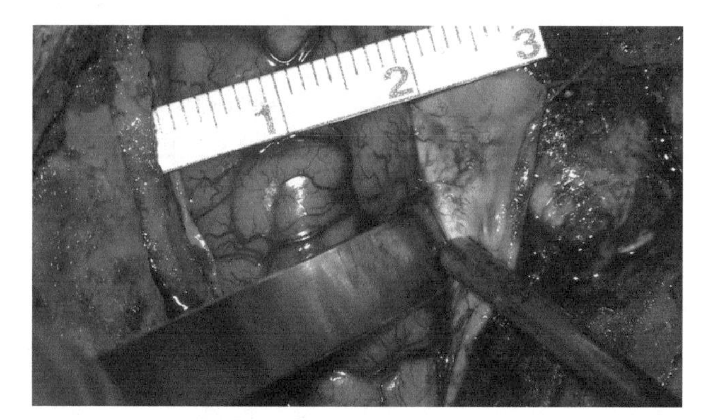

Fig. 6 Bridging veins are mobilized to get full exposure to the interhemispheric region

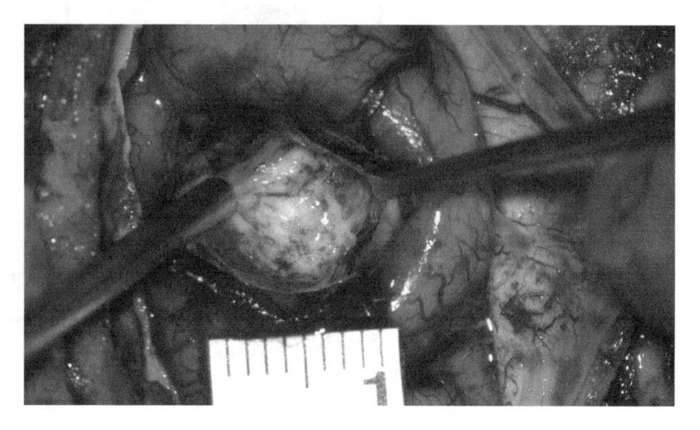

Fig. 7 Sulcal preparation is followed by incision of the white matter. Resection by suctioning is comparable to the real situation of glioma resection

injected via a needle transcortically into the brain. The cranial base including cranial nerves and basal arteries is only reached by an extended bone resection and/or resection of the frontal brain. Subtemporal, fronto-temporal, or occipital approaches are not comparable to surgery in humans as bone and soft tissue have to be prepared extensively in which orbital rims on the one hand and the entire muscles of the neck have to be dissected.

2.2 Anatomy of the Spinal Cord and Surgery on Nerve Roots and Intradural Lesions in a Porcine Model

Spine surgery in swine is limited to the microscopic and minimal invasive techniques predominantly. Even porcine vertebrae possess similar ligamentous structure and facet joint orientation; they are smaller, have anterior processes, and are less mineralized. Screws or more complex instrumentations will less likely find a sufficient stay in our experience; therefore human cadaver models may be favored in these special issues.

The vertebral column in pigs protects the spinal cord by neural arches composed of lamina with transverse and articular processes, just like in humans. Therefore porcine spine is an ideal training model for dorsal or dorsolateral approaches to intraspinal epidural including exposure of the nerve roots or intradural extra- or intramedullary lesions.

Midline incision comprises the extent of three to four spinal processes in minimum. Paraspinal muscles are of distinctive strength and have to be removed from the midline to the facet joints laterally (Fig. 8). Laminectomy is simply performed by bone punches and rongeurs or by bone saws allowing to mediate the techniques of laminoplasty and/or laminotomy. Epidural and intradural lesions are reached if all bleedings from the bone and the epidural venous plexus are stopped in the conventional manner by bone wax, drilling without rinsing, bipolar coagulation, hemostyptics, and compression. Comparable to the cranial approaches spinal dura is thin and may be easily injured (Fig. 9). The dura is opened

Fig. 8 Midline, bilateral exposure to the spinal cord with the possibilities of interlaminar approach, transforaminal approach, hemilaminectomy, laminectomy, and unilateral undercutting procedures

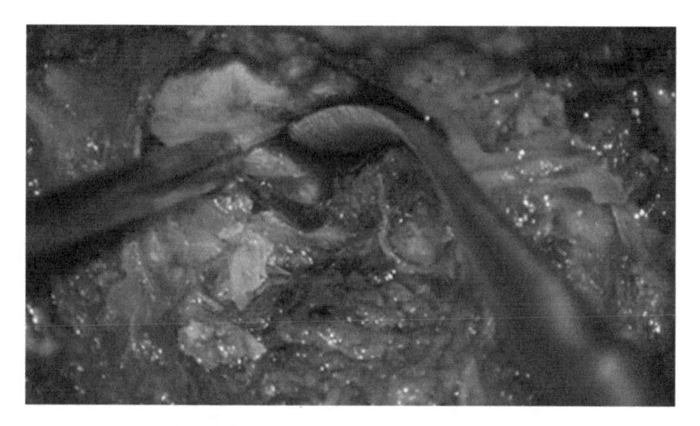

Fig. 9 Interlaminar window with access to the nerve root (here dural opening)

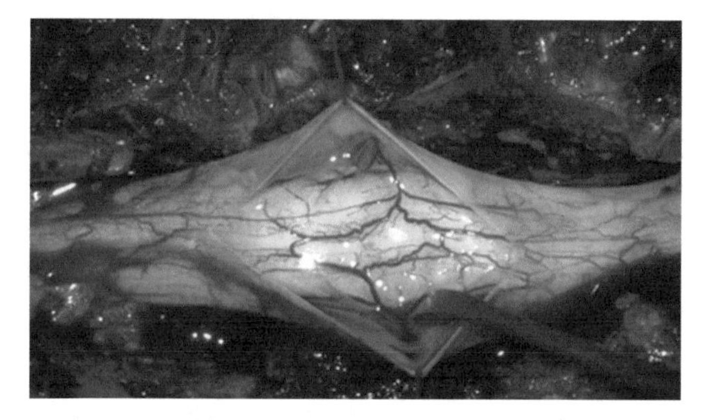

Fig. 10 Laminectomy requires careful bleeding management, especially epidurally. Dural opening allows further preparation on extramedullar lesions following the nerve roots or the dural sheet as well as intramedullar lesions

under microscopic magnification and fixed by sutures keeping it open (Fig. 10). Myelotomy is done in the midline following incision of the arachnoid membrane. While nerve fibers and filum terminale are found in the cauda equina of the lumbar spine, the myelon impresses more pulpy than in humans.

Transmuscular, tube-assisted approaches to the interlaminar space and ordinary dorsal approaches to the intervertebral space may be as well performed. Following these steps disc surgery may be imitated by flavectomy and/or extended bone removal of the laminar edges. Once again, bleedings of the epidural venous plexus have to be carefully managed before nerve roots are followed into the osseous foramen.

Microscopic dissection and bleeding management of a pulsating brain and spinal cord with the daily used OR equipment are particular challenges which can be trained in this setup with calmness and patience. Failure is not life threatening for the patient (swine). Nevertheless it will end up in an unclear, blood-filled cavity or damaged brain. Suctioning while bleeding, handling the microscope, changing the instruments, and decision making in the in vivo model is as complex as in daily practice but their individual stress factors aggravate the difficulty significantly. Learning is mediated by the successful or disappointing in vivo dissection and depends on individual experiences.

Team approach can be appointed by the realistic technical setup in which assistance has to take over the nursing job, has to anticipate the next surgical step, and is asked to exchange his experiences with the performing surgeon to achieve a most blood-dry and clear resection cavity. In our point of view the in vivo model presents as an ideal opportunity for microsurgical training including even social skills which are required in a competitive and achievement-orientated field today.

3 Notes

3.1 Brain Surgery in a Porcine Model

- Midline incision large enough to expose coronal sutures, orbital rims, and sloping crest posteriorly.
- Bilateral parietal burr holes and craniotomy of about 3 by 4 cm, less mineralized bone may require drills, to avoid bleeding from a lacerated sagittal sinus bone punches are more safe crossing the midline.
- Dura is thin and easy to violate; lift it by one suture before incision is made; leave the sagittal sinus untouched.
- Ideal training model for learning microneurosurgery, focused on surgery of the cortex and/or resection of an artificially inserted brain tumor.
- Endoscopic approaches of the ventricles are limited by the small and narrow size.
- Vascular procedures exposing the basal arteries require brain resection and are again limited by the small diameter of arteries and veins.
- Access for researchers may be the large and safe exposure of brain.

3.2 Spine Surgery in a Porcine Model

- Predominantly used for microscopic and minimal invasive procedures as instrumentations will less likely find a sufficient stay.
- Midline approach with interlaminar access to nerve roots, laminotomy, laminoplasty, or simple laminectomy to expose the spinal canal.
- Epidural venous plexus of major concern in bleeding management.
- Dura is thin and may easily be lacerated.
- Ideal training model for interlaminar approaches to nerve roots, extra- and intradural, extra- and intramedullary lesions.

References

1. Regelsberger J, Heese O, Horn P, Kirsch M, Eicker S, Sabel M, Westphal M (2010) Training microneurosurgery—four years experiences with an in vivo model. Cen Eur Neurosurg 72:192–195
2. Regelsberger J, Eicker S, Siasios I, Hänggi D, Kirsch M, Horn P, Winkler P, Signoretti S, Fountas K, Dufour H, Barcia JA, Sakowitz O, westermaier T, Sabel M, Heese O (2015) In vivo porcinetraining model for cranial neurosurgery. Neurosurgical review 38(1):157–63
3. Brennum J (2000) European neurosurgical education—the next generation. Acta Neurochir (Wien) 142:1081–1087
4. Mazotti LA, Vidyarthi AR, Wachter RM, Auerbach AD, Katz PP (2009) Impact of duty-hour restriction on resident inpatient teaching. J Hosp Med 4:476–480
5. Reulen HJ, Hide RA, Bettag M, Bodosi M, Cunha ESM (2009) A report on neurosurgical workforce in the countries of the EU and associated states. Task Force "Workforce Planning" UEMS Section of Neurosurgery. Acta Neurochir (Wien) 151:715–721

Chapter 14

Real-Time Convection Delivery of Therapeutics to the Primate Brain

Dali Yin, Massimo S. Fiandaca, John Forsayeth, and Krystof S. Bankiewicz

Abstract

Convection-enhanced delivery (CED) has been developed as a drug delivery strategy and represents a powerful methodology for targeted therapy in the brain. Our group has extensively studied and refined this approach for distributing various agents, including small molecules, macromolecules, viral particles, nanoparticles, and liposomal drugs into the brain parenchyma by means of a procedure called real-time convection-enhanced delivery (RCD). We also defined infusion parameters referred to as "red," "blue," and "green" zones for cannula placements that result in poor, suboptimal, and optimal volumes of distribution, respectively, in the target area of brain of nonhuman primates (NHP). We have defined the scale differences between NHP brains and those of humans. Furthermore, we applied the ClearPoint® system to the RCD procedure, which allows RCD to be carried out with a high level of precision, predictability, and safety. This approach may improve the success rate for clinical trials involving intracerebral drug delivery by direct infusion. These innovations may have important implications in ensuring effective delivery of therapeutics into brain targets utilizing NHP stereotactic coordinates translated via stereotactic MRI localization procedures in humans. These delivery innovations should be considered when localized therapeutic delivery, such as gene transfer or protein administration, is being translated into clinical treatments. In this chapter, we review recently developed methods that ensure controlled distribution of therapeutic agents in the brain.

Key words Real-time convection-enhanced delivery, RGB zones, ClearPoint system, Nonhuman primates

1 Introduction

Convection-enhanced delivery (CED) was an interstitial central nervous system (CNS) delivery technique [1] that circumvents the blood–brain barrier in delivering therapeutics into the CNS. Traditional local delivery of most therapeutic agents into the brain has relied on diffusion, which depends on a concentration gradient. The rate of diffusion is inversely proportional to the size of the therapeutic and is usually slow with respect to tissue clearance. Thus, diffusion results in a nonhomogeneous distribution of most

Miroslaw Janowski (ed.), *Experimental Neurosurgery in Animal Models*, Neuromethods, vol. 116,
DOI 10.1007/978-1-4939-3730-1_14, © Springer Science+Business Media New York 2016

delivered agents restricted to a few millimeters from the source. In contrast, CED uses a fluid pressure gradient established at the tip of an infusion catheter and bulk flow to propagate substances within the extracellular fluid space [1]. CED allows the extracellularly infused material to further propagate via the perivascular spaces and the rhythmic contractions of blood vessels acting as an efficient motive force for spreading of the infusate [2]. As a result, a higher concentration of drug was distributed more evenly over a larger area of the targeted structure than what would be seen with a simple injection. CED has been developed as a drug delivery strategy and represents a powerful methodology for targeted therapy in the fields of neurodegenerative diseases, such as Parkinson's disease [3, 4], and neuro-oncology [5, 6]. Laboratory investigations with CED cover a broad field of application, such as the delivery of small molecules [7, 8], macromolecules [1], viral particles [9], magnetic nanoparticles [10], and liposomes [11].

One of the more recent advances in CED was real-time convective delivery (RCD), which uses MRI to visualize the CED process during infusion with the aid of gadolinium-loaded liposomes (GDL) that co-distribute with the infused therapeutic to the non-human primate (NHP) target [12, 13]. Our RCD methodology has evolved through extensive modeling in NHP over the years. Visualizing infusions in real time allows active feedback on (1) cannula placement, (2) physical and anatomic diffusion parameters, and (3) control of drug delivery for optimizing gene transfer, thereby reducing the potential for adverse effects. Initially described by Oldfield and colleagues who employed albumin-linked surrogate tracers [14], our current technique of RCD employs interventional MRI (iMRI) to monitor the distribution of therapeutic agents that are co-infused with gadolinium-based tracers [15]. Our initial work with GDL [16, 17] has progressed to the co-infusion of free gadoteridol for predicting the distribution of protein [18] and AAV2 vectors [15, 17, 19]. A similar strategy was used to co-infuse therapeutic agent with Gd-diethylenetriamine pentaacetic acid (Gd-DTPA) in a clinical study in two patients with intrinsic brainstem lesions at the National Institutes of Health [20]. RCD allowed us to monitor infusion of liposomal drugs into brain tumors [21] and viral gene therapy vectors into parenchyma [17]. Visualizing infused drug distribution was necessary to ensure accurate delivery of therapeutic agents into target sites while minimizing exposure of healthy tissue. Moreover, because infusions could be visualized, we were able to define quantitative relationships between infusate volume (Vi) and subsequent volume of distribution (Vd) for both white and gray matter [12]. This method has given us the ability to directly monitor the local delivery of therapeutic agents and has improved the efficacy of CED in animals.

During RCD, the Vd for a given agent depends on the structural properties of the tissue being convected, such as hydraulic conductivity, vascular volume fraction, and extracellular fluid fraction

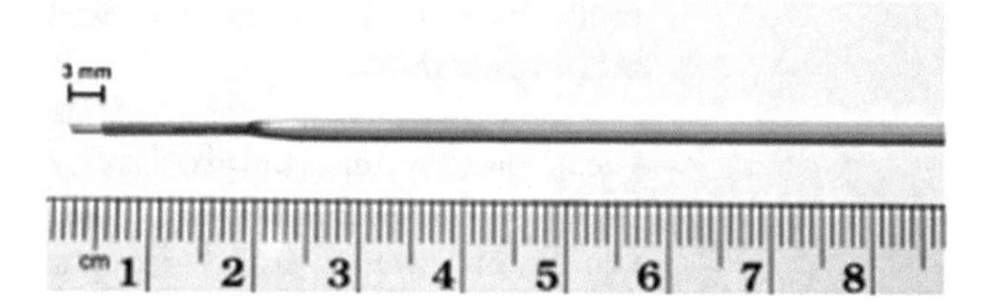

Fig. 1 Step-design cannula. The length of each infusion cannula was measured to ensure that the distal tip extended 3 mm beyond the length of the respective guide. This created a stepped design at the tip of the cannula to maximize fluid distribution during RCD procedures and minimize reflux along the cannula tract. We refer to this transition from fused silica tip to a fused silica sheath as the "step"

[22]. It also depends on the technical parameters of the infusion procedure such as cannula design, cannula placement, infusion volume, and rate of infusion [23–25], with the overall aim of improving delivery efficiency while attempting to limit the spread of the therapeutic into regions outside the target. Development of the optimal cannula type for effective CED delivery in the brain has also been critical. We examined several types of cannulae with respect to size and design, and concluded that a stepped design (Fig. 1) with a fused silica tip provided us with the most consistently robust brain delivery [26–28]. The stepped cannula dramatically reduces reflux along the infusion device by restricting initial backflow of fluid flow beyond the step. Furthermore, in our experience, a key component of successful CED was the site of cannula placement within the targeted area, such as putamen, thalamus, and brain stem. The distance from cannula step to its entry point in the target region was found to be critical for optimal distribution of therapeutics. We have developed the concept of RGB (red, green, and blue zones) zones for cannula placement during RCD. We defined these zones on the basis of containment of infusate within the target region. Within each region so far investigated, we were able to define a subset of cannula locations associated with complete containment within the target (green), substantial containment (blue), or poor containment (red). Infusate escape into nearby ventricles or white matter tracts was driven by proximity to these structures. We defined three-dimensional RGB zones in the putamen [29], thalamus, and brain stem [30] of NHP. To obtain the most effective distribution of the infused therapeutic within the intended target, it was essential to understand the optimal site of placement of the step and tip of the infusion cannula within that target, preferably within the green zone [29, 30]. Such optimal placement will reduce distribution into surrounding white matter tracts that serve as leakage points [31]. Such consideration allows more precise delivery of the therapeutic to the target structure(s) but lessens the risk of inadvertent spread into surrounding brain regions. These factors may also explain some of the reported failures of CED in both NHP studies and human clinical

trials, which may have been related to suboptimal targeting of the infusion cannula.

In order to further optimize the RCD technology, the ClearPoint® system (Surgivision Inc, Irvine, CA) has been adopted by us to translate targeting from the NHP brain into humans. ClearPoint was a novel integrated hardware (skull-mounted SmartFrame device)/software platform for RCD that provides prospective stereotactic guidance for the cannula placement and performance of RCD. This platform was based on the concept of prospective stereotaxy, the alignment of a skull-mounted trajectory guide within an MRI system [32], already used in clinical studies to perform brain biopsies [33, 34] and placement of DBS leads [34–36]. In anticipation of upcoming gene therapy clinical trials, we adapted this "off-the-shelf" device to RCD of therapeutics via a customized infusion cannula. The targeting accuracy of this delivery system and the performance of the infusion cannula were validated in NHP. The ClearPoint system allows RCD to be performed with a high level of precision, safety, and predictability. This technique should increase the utility of RCD for expanding the scope of drug delivery studies. Clinical application of this guidance platform was likely to improve the success of clinical trials employing intracerebral drug delivery. Although continued refinements in RCD may be expected, our work over the past decade has resulted in a new paradigm for direct parenchymal delivery that may find increasing application in the treatment of currently intractable diseases like Alzheimer's and Parkinson's disease, brain tumors, and other movement disorders. Here we describe in detail our methods for optimal RCD.

2 Materials

2.1 Experimental Subjects

Normal rhesus macaques and cynomolgus monkeys (aged from 8 to 18 years; mean age = 11.9 years, weight = 4–9.4 kg) were the subjects in our study. Experimentation was performed according to the National Institutes of Health guidelines and to protocols approved by the Institutional Animal Care and Use Committee at the University of California San Francisco (San Francisco, CA). Adult monkeys were individually housed in stainless steel cages. Each animal room was maintained on a 12-h light/dark cycle and room temperature ranged between 64 and 84 °F. Prior to assignment to the study, all animals underwent at least a 31-day quarantine period mandated by the Centers for Disease Control and Prevention (Atlanta, GA).

2.2 Liposome Preparation

1. 1,2-Dioleoyl-sn-glycero-3-phosphocholine (DOPC).
2. Cholesterol.
3. 1,2-Distear-oyl-sn-glycero-3-[methoxy(polyethylene glycol)-2000] (PEG-DSG).

4. Chloroform/methanol (90:10, v/v).
5. Ethanol.
6. 0.5 M Gadoteridol (10-(2-hydroxy-propyl)-1,4,7,10-tetraaza-cyclododecane-1,4,7-triacetic acid).
7. Double-stacked polycarbonate membranes with a pore size of 100 nm.
8. Sephadex G-75 size-exclusion column.
9. HEPES-buffered saline (pH 6.5, 5 mM HEPES, 135 mM NaCl, pH adjusted with NaOH).
10. Rhodamine.
11. 20 mM Sulforhodamine B in pH 6.5 HEPES-buffered saline.

2.3 Magnetic Resonance Imaging

1. Periosteal elevator (Fine Scientific Tools, Foster City, CA, USA), rongeur calipers (Fine Scientific Tools, Foster City, CA, USA), gelfoam (Baxter, Deerfield, IL, USA), dental acrylic, gauze, syringes (5 and 50 mL), latex gloves, stopwatch timer.
2. Reflux-resistant infusion cannula.
3. Teflon tubing for secondary and loading lines (1.57 mm outer diameter, 0.76 mm inner diameter; Upchurch Scientific, West Berlin, NJ, USA).
4. Plastic cannula guide ports.
5. Gadoteridol (ProHance®; Bracco Diagnostics Inc., Monroe Township, NJ, USA).
6. Skull-mounted aiming device (SmartFrame®, MRI Interventions Inc., Memphis, TN, USA) and software (ClearPoint®, MRI Interventions Inc., Memphis, TN, USA).
7. Sterile hardware: Plastic screws, pens, rulers, screwdriver, dummy catheter (Upchurch Scientific, West Berlin, NJ, USA), large animal MRI-compatible stereotactic frame (Kopf Instruments, Tujunga, CA, USA), 3500 Medfusion pump (Strategic Applications Inc., Lake Villa, IL, USA), Tefzel ferrule connectors and Luer-Lock adapters (Upchurch Scientific, West Berlin, NJ, USA), impaction drill (3.5 mm round drill bit; Stryker, Portage, MI, USA).
8. 1.5-T MRI scanner (Signa LX; GE Medical Systems, Waukesha, WI, USA), 5-in. circular surface MRI coil (MR Instruments Inc., Hopkins, MN, USA).
9. OsiriX® software (v5.5.2; Pixmeo, Bernex, Switzerland).

2.4 Infusion Procedure

1. Ketamine and xylazine.
2. Isoflurane.
3. MRI-compatible stereotactic frame.
4. MRI-compatible guide cannula.

5. Dental acrylic.
6. Stylet screw.
7. Step-design cannula.
8. Loading line (containing GDL or free gadoteridol).
9. Infusion line with oil and another infusion line with trypan blue solution.
10. Syringe 1 ml filled with 1 % trypan blue solution.
11. Stereotactic holder.
12. Micro-infusion pump.

2.5 ClearPoint System

1. SmartFrame®.
2. Infusion cannula.
3. Software system.

3 Methods

3.1 Liposome Preparation

Separate liposomes were prepared for detection either by MRI or by histology. Liposomes containing the MRI contrast agent were composed of 1,2-dioleoyl-sn-glycero-3-phosphocholine (DOPC)/cholesterol/1,2-distearoyl-sn-glycero-3-[methoxy(polyethylene glycol)-2000] (PEG-DSG) with a molar ratio of 3:2:0.3. DOPC was purchased from Avanti Polar Lipids (Alabaster, AL), PEG-DSG from NOF Corporation (Tokyo, Japan), and cholesterol from Calbiochem (San Diego, CA). The lipids were dissolved in chloroform/methanol (90:10, v/v), and the solvent was removed by rotary evaporation, resulting in a thin lipid film. The lipid film was dissolved in ethanol and heated to 60 °C. A commercial US Pharmacopeia solution of 0.5 M gadoteridol (10-(2-hydroxy-propyl)-1,4,7,10-tetraazacyclododecane-1,4,7-triacetic acid) (Prohance; Bracco Diagnostics, Princeton, NJ) was heated to 60 °C and injected rapidly into the ethanol/lipid solution. Unilamellar liposomes were formed by extrusion (Lipex; Northern Lipids, Vancouver, Canada) by 15 passes through double-stacked polycarbonate membranes (Whatman Nucleopore, Clifton, NJ) with a pore size of 100 nm, resulting in a liposome diameter of 24–124 nm as determined by quasi-elastic light scattering (N4Plus particle size analyzer; Beckman Coulter, Fullerton, LA). Unencapsulated gadoteridol was removed with a Sephadex G-75 (Sigma, St. Louis, MO) size-exclusion column eluted with HEPES-buffered saline (5 mM HEPES, pH 6.5, 135 mM NaCl). Liposomes loaded with rhodamine for histological studies were formulated with the same lipid composition and preparation method as the gadoteridol-containing liposomes, except that the lipids were hydrated directly with 20 mM sulforhodamine B (Sigma) in pH 6.5 HEPES-buffered saline by six successive cycles of rapid freezing and thawing rather than by ethanol injection. The sulforhodamine B liposomes had a diameter of 90 ± 30 nm (used alone for histological analysis) or 115 ± 40.1 nm (used for co-infusion with GDL in the MRI-monitoring study). For the preparation of liposomes containing a DiI-DS fluorescent probe, 1,1′-dioctadecyl-3,3,3′,3′-tetramethylindocarbocyanine-5,5′ disulfonic acid (DiIC18(3)-DS) (DiI-DS; Molecular Probes, Eugene, OR) was added to the lipid solution at a concentration of 0.2 mol% of the total lipid. DiI-DS liposomes had a diameter of 110 ± 40 nm. For further detail regarding formulation of liposomes loaded with chemotherapeutic agents, refer to Krauze et al. [37].

During our studies it became apparent that, even when using the same method (CED), the volumes of distribution for different compounds were inconsistent. For simple infusates, CED distribution was significantly increased if the infusate was more hydrophilic or had weaker tissue affinity [38]. Encapsulation of tissue-affinitive molecules by liposomes significantly increased their tissue distribution. However, it appeared that liposomal surface properties (cationic versus neutral liposome, surface charge, and percentage of PEGylation) affected parenchymal volume of distribution. We found that PEGylation, which provided steric stabilization and reduced surface charge, yielded the greatest Vd when compared with volume of infusion (Vi) of all other liposomal formulations [38].

3.2 Quantification of Liposome-Entrapped Gadoteridol by Magnetic Resonance Imaging

The concentration of gadoteridol entrapped in the liposomes was determined from nuclear MR relaxivity measurements. The relationship between the change in the intrinsic relaxation rate imposed by a paramagnetic agent (ΔR), also known as "T1 shortening," and the concentration of the agent is defined by the equation $\Delta R = \mathrm{r1}[\mathrm{agent}]$, in which r1 = relaxivity of the paramagnetic agent and $\Delta R = (1/\mathrm{T1observed} - 1/\mathrm{T1intrinsic})$. The relaxivity of gadoteridol had been empirically derived previously on the same system, a 2-T Brucker Omega scanner (Brucker Medical, Karlsruhe, Germany), and had a value of 4.07 $\mathrm{mM}^{-1}\ \mathrm{s}^{-1}$. The concentration of the encapsulated Gadoteridol was then calculated with the following equation: $[\mathrm{Gadoteridol}] = [(1/\mathrm{T1wGado}) - (1/\mathrm{T1w/oGado})]/4.07$.

3.3 Infusion Catheter Design

We developed a new stepped design cannula for CED that effectively prevents reflux. Cannula design has been one of the most neglected features of brain delivery protocols. Reflux was defined as the phenomenon of the movement of infusate back up the outside of the cannula rather than into the tissue. Although earlier studies showed that smaller cannula diameters permit better delivery, the crucial problem of reflux was either not assessed or not measurable. In our early studies, we confirmed that smaller cannula diameters allowed faster delivery rates but the smallest available cannulae were associated with increasing reflux when the rate of infusion exceeded 0.5 μl/min [26], clearly a significant problem when infusing large volumes. Recently, we have been able to increase the infusion rate to 5 μl/min without reflux by means of an innovative stepped cannula [26], which dramatically reduces reflux along the infusion device by restricting initial backflow of fluid flow beyond the step. The early metal cannula has been replaced by one made of silica that also features sharp transitions in outer diameter that prevent reflux (Fig. 1) [4]. The larger diameter of the stem of the cannula had an outer and inner diameter of 0.53 mm and 0.45 mm, respectively. The outer and inner diameters of the tip segment were 0.43 mm and 0.32 mm, respectively. The length of each infusion cannula was measured to ensure that the distal tip extended 3 mm beyond the length of the respective guide. This created a stepped design at the tip of the cannula to maximize fluid distribution during CED procedures and minimize reflux along the cannula tract. In the text, we refer to this transition from fused silica tip to a fused silica sheath as the "step," and all positioning data are derived from the position of this step because of its unambiguous visibility on MRI. Pushing flow rates significantly above 5 μl/min, however, can induce reflux even with this catheter, as we have shown in canine and primate studies [31]. Robust reflux-free delivery and distribution of liposomes was achieved with the stepped design cannula in both rats and nonhuman primates. This stepped design cannula may allow reflux-free distribution and shorten the duration of infusion in future clinical applications of CED in humans.

3.4 Infusion Procedure

Primates received a baseline MRI before surgery to visualize anatomical landmarks and to generate stereotactic coordinates of the proposed target infusion sites for each animal. NHP underwent neurosurgical procedures to position the MRI-compatible guide cannula over the target. Each customized guide cannula was cut to a specified length, stereotactically guided to its target through a burr hole created in the skull, and secured to the skull by dental acrylic. The tops of the guide cannula assemblies were capped with stylet screws for simple access during the infusion procedure. Animals recovered for at least 2 weeks before initiation of infusion procedures. Animals were anesthetized with isoflurane (Aerrane; Ohmeda Pharmaceutical Products Division, Liberty Corner, NJ) during real-time MRI acquisition. Each animal's head was placed in an MRI-compatible stereotactic frame, and a baseline MRI was performed. Vital signs, such as heart rate and PO_2, were monitored throughout the procedure. Briefly, the infusion system (Fig. 2) consisted of a reflux-resistant fused silica cannula that was connected to a loading line (containing GDL or free gadoteridol), an infusion line with oil, and another infusion line with trypan blue solution. A 1 ml syringe (filled with trypan blue solution) mounted onto a micro-infusion pump (BeeHive; Bioanalytical System, West Lafayette, IN) regulated the flow of fluid through the system. Based on MRI coordinates, the cannula was manually guided to

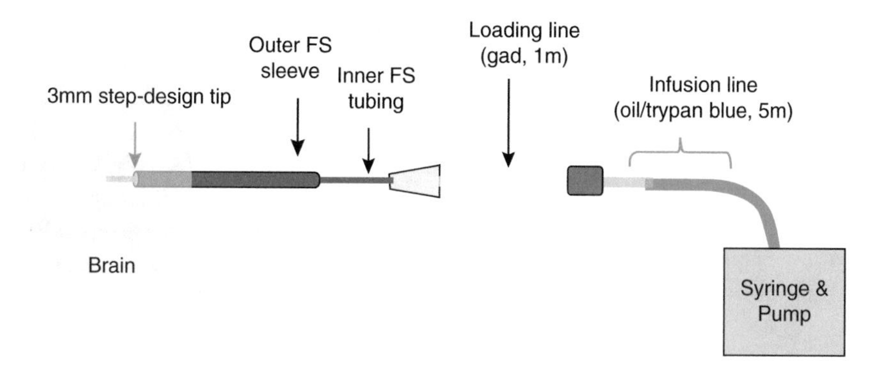

Fig. 2 Schematic of infusion system. *FS* fused silica

the targeted region of the brain through the previously placed guide cannula. After securing placement of the infusion cannula, the CED was initiated with real-time MRI data simultaneously acquired (RCD). We used the same infusion parameters for every NHP infused throughout the study. Infusion rates were as follows: 0.1 μl/min was applied when lowering cannula to targeted area and increased at 10-min intervals to 0.2, 0.5, 0.8, 1.0, and 2.0 μl/min. After turning off delivery pumps at the end of each infusion session, the catheter remained in place for another 5 min to allow the built-up tissue pressure to abate. Animals received up to three infusion procedures into the same anatomic location. Each animal had at least a 4-week interval between each infusion procedure.

3.5 Magnetic Resonance Image

NHP were sedated with a mixture of ketamine (Ketaset, 7 mg/kg, IM) and xylazine (Rompun, 3 mg/kg, IM). After sedation, each animal was placed in an MRI-compatible stereotactic frame. The ear-bar and eye-bar measurements were recorded, and an intravenous line was established. MRI data were then obtained, after which animals were allowed to recover under close observation until able to right themselves in their home cages. MR images of brain in NHP were acquired on a 1.5-T Siemens Magnetom Avanto (Siemens AG, Munich, Germany). Three-dimensional rapid gradient echo (MP-RAGE) images were obtained with repetition time (TR) = 2110 ms, echo time (TE) = 3.6 ms, flip angle = 15°, number of excitations (NEX) = 1 (repeated three times), matrix = 240 × 240, field of view (FOV) = 240 × 240 × 240, and slice thickness = 1 mm. These parameters resulted in a 1 mm^3 voxel volume. The scan time was dependent on the number of slices needed to cover the extent of infusion and ranges from 9 min 44 s to 11 min 53 s.

MR images were obtained from each RCD and used to measure distance from cannula step to corpus callosum (CC), internal capsule (IC) and external capsule (EC) for infusion into the putamen, and from the cannula step to the midline, to cannula entry point in the target region, and to the lateral border of the target regions for infusion into thalamus or brainstem of NHP. The measurements were made on an Apple Macintosh G4 computer with OsiriX® Medical Image Software (v2.5.1). OsiriX software reads all data specifications from DICOM (digital imaging and communications in medicine) formatted MR images obtained via local picture archiving and communication system (PACS). For each image, the default window and level settings were used throughout the study; that is, there was no attempt to alter or manipulate settings from one experiment to another. The distances from cannula step to each structure mentioned above were manually defined and then calculated by the software. All the distances were measured in the same manner on MRI sections. These data were used to define RGB zones in the putamen, thalamus, and brain stem.

MR images were also used for volumetric quantification of distribution of gadoteridol. The Vd of gadoteridol in the brain of each subject was quantified on an Apple Macintosh G4 computer. Region of interest (ROI) derived in the target and white matter tract (WMT) were manually defined, and software then calculated the area from each MR image and established the volume of the ROI based on area defined multiplied by slice thickness (PACS volume). The boundaries of each distribution were defined in the same manner in the series of MRI sections. The defined ROI volumes allowed for 3D image reconstruction with BrainLAB software (BrainLAB, Heimstetten, Germany).

3.6 Coordinates for Green Zone in the Putamen, Thalamus and Brainstem of Three-Dimensional Brain Space in NHP

The *X*, *Y*, and *Z* axial values of cannula step location in green zone were determined with 2D orthogonal MR images generated by OsiriX software, where MR images were projected in all three dimensions (axial, coronal, and sagittal). We used midpoint of the anterior commissure–posterior commissure (AC–PC) line as zero point (0,0,0) of three-dimensional (3D) brain space. Briefly, AC–PC line was drawn on mid-sagittal plane of MRI, and the midpoint of AC–PC line was determined. The horizontal and vertical plane through the midpoint of AC–PC line was then obtained, and they could be shown on all the three plans simultaneously. The *X*, *Y*, and *Z* axial values of cannula step were then obtained by measurements of distance from cannula step to midline on coronal MRI plane (*X* value), distance anterior (or posterior) to the midpoint of AC–PC line of the coronal MRI plane (*Y* value) and the distance above (or below) axial plane incorporating the AC–PC line on MRI (*Z* value). All the distances were measured (in millimeters) in the same manner on MRI sections for each case.

The results obtained were used to determine a set of 3D stereotactic coordinates that define an optimal site for infusions into putamen, thalamus, and brainstem in NHP. Based on the coordinate calculations for the cannula step by MRI, the target for green zone in the putamen were $x = 11.85 \pm 0.56$ mm lateral (X coordinate), $y = 7.36 \pm 0.49$ mm anterior to the AC–PC midpoint (Y coordinate), and $z = 3.62 \pm 0.40$ mm superior to the AC–PC axial plane (Z coordinate). The mean coordinates for placing the step in the thalamic green zone were $x = 6.9 \pm 0.7$ mm lateral (range 4.1–10.2), $y = 1.2 \pm 0.2$ mm posterior (range 0.4–1.9 mm), and $z = 3.1 \pm 0.4$ mm superior (range 1.4–4.7 mm). The mean coordinates for green zone in the brainstem were $x = 2.3 \pm 0.2$ mm lateral (range, 1.6–3.5 mm), $y = 4.0 \pm 0.5$ posterior (range, 2.6–6.0 mm), and $z = 8.9 \pm 1.0$ mm inferior (range, 4.8–11.9). We think that cannula placement and definition of optimal (green zone) stereotactic coordinates have important implications in ensuring effective delivery of therapeutics into the target utilizing routine stereotactic MRI localization procedures.

3.7 Cannula Placement Guidelines

Optimal results in the direct delivery of therapeutics into primate brain depend on reproducible distribution throughout the target region. In our recent studies, we retrospectively analyzed MRI of RCD infusions into the putamen, thalamus and brainstem of NHP, and defined infusion parameters referred to as "red," "blue," and "green" zones (RGB zones) for cannula placements that result in poor, suboptimal, and optimal volumes of distribution, respectively. The most robust data was achieved in putamen, and the reason for this is that problematic structures (ventricles, corpus callosum) surround this region. So it was relatively easy to define RGB zones in this setting (Fig. 3a). In contrast, thalamus and brain stem, much larger structures in any case, really do not present this kind of challenge. Accordingly, infusions in thalamus defined G and B zones but not R (Fig. 3b). In brain stem, we only identified coordinates that gave excellent containment of infusate (Fig. 3c). Clearly, each new target region will impose its own anatomical constraints and optimization of RCD will require empirical determinations to some extent. However, the three regions we have investigated suggest the following rules of thumb. When infusate emanates from the tip of stepped cannulae, the infusate forms an

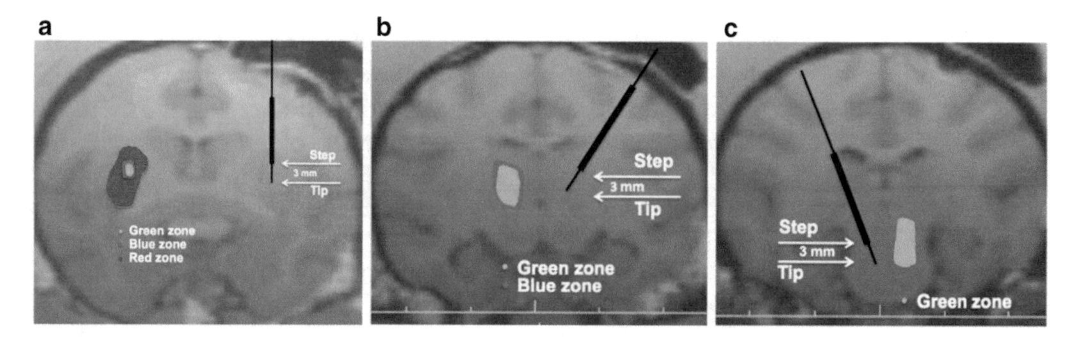

Fig. 3 RGB zones for step placement outlined in the putamen (**a**), thalamus (**b**), and brain stem (**c**) of NHP

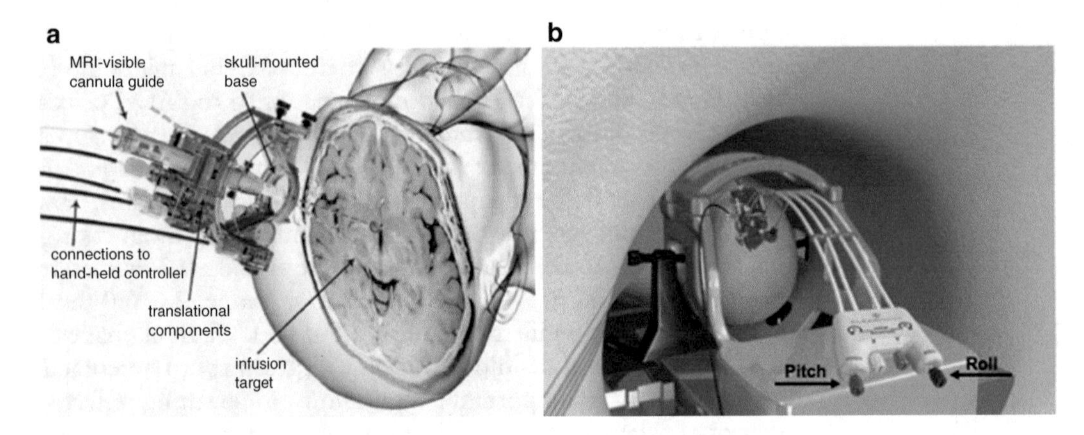

Fig. 4 (**a**) Schematic showing the basic components of the SmartFrame. (**b**) The fluid stem was aligned to the target trajectory via both "pitch and roll" axes and an X–Y translational stage

ovoid pattern with the cannula as the vertical axis. The upper dimension of the ovoid extends upwards about somewhat less than the length of the step-tip. Thus, a 3 mm step-tip will generate a little less than 3 mm backflow. In the smaller rat striatum, we adjusted the cannula tip to 1 mm and placed the step approximately 1–2 mm from the corpus callosum in order to place the cannula tip nicely within the striatum while maintaining a clear separation of the leading edge of the backflow from the entry point [28]. This rule should be followed for the design of cannulae in smaller structures. With respect to peri-ventricular zones, we found in putamen that the cannula should be placed at least 3 mm from external and internal capsules. In general then, a cannula trajectory in the monkey that can maintain a distance of 3 mm or more from sensitive structures seems to be a good place to start. In humans, of course, these distances are correspondingly enlarged. The size of the striatum in humans is about fivefold that of the Rhesus monkey [39], and consideration of such target volume differences is an important factor in clinical planning.

3.8 ClearPoint System

The ClearPoint® system consists of the SmartFrame® (Fig. 4a), an infusion cannula, and a software system that communicates with both the MRI console and the operating neurosurgeon in the MRI suite. The ClearPoint software allows registration of the AC and PC from an initial MRI scan, selection of a target for cannula tip placement in AC-PC space, and planning of the cannula trajectory. Although the entry point was relatively fixed in the NHP due to use of the adapter plug, in the clinical system the entry point was modifiable in the pre-craniotomy planning stage as the trajectory was adjusted. The SmartFrame houses an MRI-visible (gadolinium-impregnated) fluid stem and integrated fiducials that are detected by the software. The fluid stem, which also serves as the infusion cannula guide, was aligned to the target trajectory via both "pitch and roll" axes and an X–Y translational stage (Fig. 4b). This was accomplished with an attached hand controller resting at the opening of the MRI bore, according to directions generated by the software in response to serial T1 MRI sequences, until the fluid stem alignment matches the chosen target trajectory.

3.9 RCD with ClearPoint System

3.9.1 Surgical Procedure

Infusions were performed in a research magnet shared between human and NHP use. Due to institutional regulations prohibiting procedures that expose animal blood products in such an area, the surgical placement and removal of the skull mounted aiming device (SmartFrame) did not occur in the iMRI suite as would occur in patients. Two weeks prior to infusion, NHP underwent stereotactic placement of skull-mounted, MRI-compatible, threaded plastic adapter plugs (12 mm diameter × 14 mm height) (Fig. 5a) for later attachment of the SmartFrame. After performing bilateral craniectomies, one plug was secured to the skull over each hemisphere with dental acrylic. After placement of the adapter plugs, animals recovered for at least 2 weeks before initiation of iMRI infusion procedures.

3.9.2 Trajectory Planning and Cannula Insertion

On the day of infusion, NHP were sedated with ketamine (Ketaset, 7 mg/kg, intramuscular) and xylazine (Rompun, 3 mg/kg, intramuscular), intubated, and placed on inhaled isoflurane (1–3%). The plug adapter was prepared sterile and the NHP was placed supine in an MRI-compatible stereotactic frame. The SmartFrame

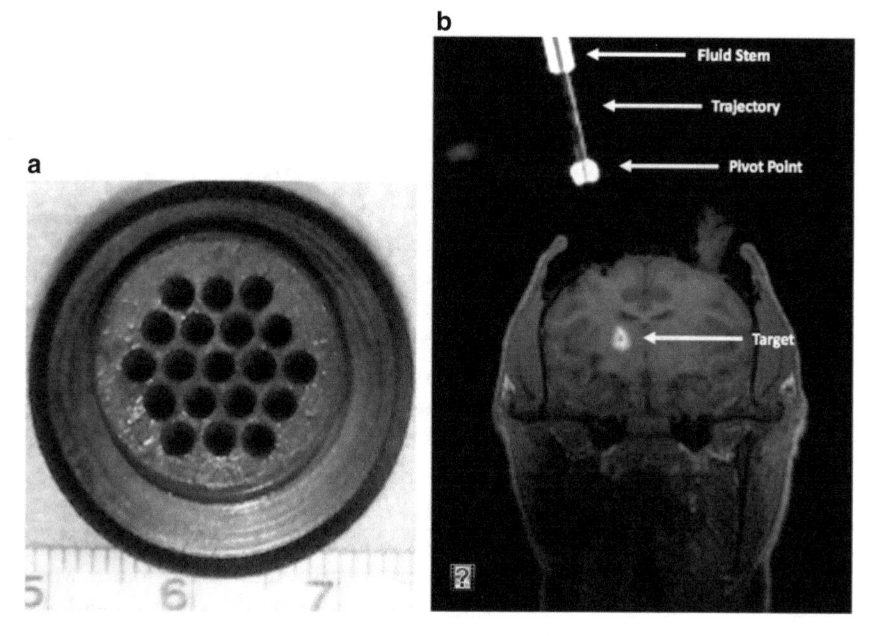

Fig. 5 (**a**) Plastic adapter plug. (**b**) Sagittal screenshot of target trajectory alignment. The T1 MRI-visible fluid stem, which holds the infusion cannula, has been aligned by translating the SmartFrame around a fixed pivot point so that the trajectory meets the target

was attached by screwing the base onto the adapter plug over one hemisphere. The NHP was moved into the bore and a controller was attached to the SmartFrame by inserting guide wires into each of four adjustment knobs. This controller allows the surgeon to manually "dial in" distance changes to align the cannula to the desired trajectory in four planes (pitch, roll, anterior-posterior, medial-lateral) (Fig. 4b) as instructed by the ClearPoint software.

First, a high-resolution anatomical MR scan was acquired for target identification and surgical planning. Specific details of our MRI scanning may be found in Fiandaca et al. [40]. The scan was a 9-min 3D Magnetization Prepared Rapid Gradient Echo (MPRAGE). The MPRAGE images were then transferred to the ClearPoint system, where the target for cannula tip placement was selected. Next, rapid scans were obtained that allowed the ClearPoint software to detect the position and orientation of the SmartFrame fluid stem. First, a 6-s 2D turbo-spin echo (TSE) was acquired through the distal fluid stem in an orientation perpendicular to the desired trajectory. The software used this image to compare the current SmartFrame trajectory to the target trajectory in order to calculate an expected error for tip placement and generate instructions to adjust SmartFrame alignment via the pitch and roll. After these adjustments were made, the scan was re-acquired to measure the new expected error and this process was repeated as necessary.

When the expected error fell below 1.0 mm, the pitch and roll axes on the SmartFrame were locked and a 26-s 2D TSE scan was acquired along the sagittal and coronal planes of the guide stem for fine adjustment of the SmartFrame X–Y stage. Seven slices at 1 mm isotropic resolution were acquired over a 180×240 mm FOV with a TE of 22 ms, a TR of 500 ms, two repetitions, an echo train length of 7 and a bandwidth of 250 Hz/pixel. The ClearPoint software used these images to generate instructions for fine adjustment of the trajectory, achieved by dialing in distance changes on the SmartFrame X–Y stage. This process was repeated until the software reported an expected error of less than 0.5 mm that typically required no more than two iterations. The infusion system included a customized, ceramic, fused silica reflux-resistant cannula developed in accord with previously reported principles developed in our laboratory [16, 26]. For infusions, the cannula was connected to a loading line containing 1 mM gadoteridol, and flow was regulated with 1 ml syringe filled with trypan blue, mounted onto a MRI-compatible infusion pump (Harvard Bioscience Company). With the aiming device aligned in its final position, the software reported the expected distance from the target to the top of the guide stem, and this distance was measured from the cannula tip and marked on the cannula with a sterile ink marker. A depth-stop was then secured at the marked location and the measured insertion distance was verified. The infusion pump was started at 1 μl/min, and after visualizing fluid flow from the cannula tip when held at the height of the bore, the cannula was inserted through the SmartFrame guide stem and into the brain. When the depth-stop touched the top of the guide stem, it was secured with a locking screw.

3.9.3 Infusion and Imaging

After cannula insertion, repeated multi-planar Fast Low Angle Shot (FLASH) images were obtained every 5 min throughout the duration of the infusion. The FLASH images were acquired at an in-plane resolution of $0.7 \times 0.7 \times 1$ mm with 128 slices over the 180 mm FOV at a TE of 4.49 ms, a TR of 17 ms with two repetitions and a bandwidth of 160 Hz/pixel. The first scan was acquired with a 4° flip angle to produce a proton-density-weighted image for visualization of the cannula tip. All subsequent scans were acquired with a 40° flip angle to increase the T1 weighting and highlight the signal enhancement from gadoteridol in the infusate.

Upon visualization of gadoteridol infusion at the cannula tip, the infusion rates were increased from an initial rate of 1 μl/min in a ramping fashion, 0.5 μl/min every 5 min, to reach a maximum of 3 μl/min. The interface between trypan blue and gadoteridol within the loading line was also marked at the start and finish of the infusion in order to verify that the infused volume matched that reported by the pump. Each NHP first received a small volume infusion in the thalamic area (16–25 μl) to allow calculation of targeting error, followed by a larger volume infusion (187–230 μl) into the ipsilateral thalamus. In general, the total time under general anesthesia for NHP that received four sequential infusions was approximately 6 h.

3.9.4 Imaging Data Analysis

Images obtained during RCD were transferred to the ClearPoint system for analysis of targeting error. With the target position hidden from view, the location of the cannula tip was manually selected in the ClearPoint console by identifying the center of the gadoteridol signal in the lower one third of the infusion volume on the first scan demonstrating convection following cannula insertion (Fig. 5b). The software then automatically reported the vector distance between the target site and the actual position of the cannula tip. The average target error for all infusions was later calculated and the 95 % confidence interval was determined. Spearman's rank-order correlation was used as a non-parametric measure of the statistical dependence between depth to target and target error.

4 Results of RCD with Application of ClearPoint System

Based on the results of RCD, the ClearPoint system appears to be highly accurate. Satisfactory cannula placement was achieved on the first attempt (without the need for repositioning) in all cases (Fig. 6) (Richardson et al., Stereotactic and Functional Neurosurgery in Press). The ClearPoint system automatically calculated each targeting error, defined as the three-dimensional distance between the

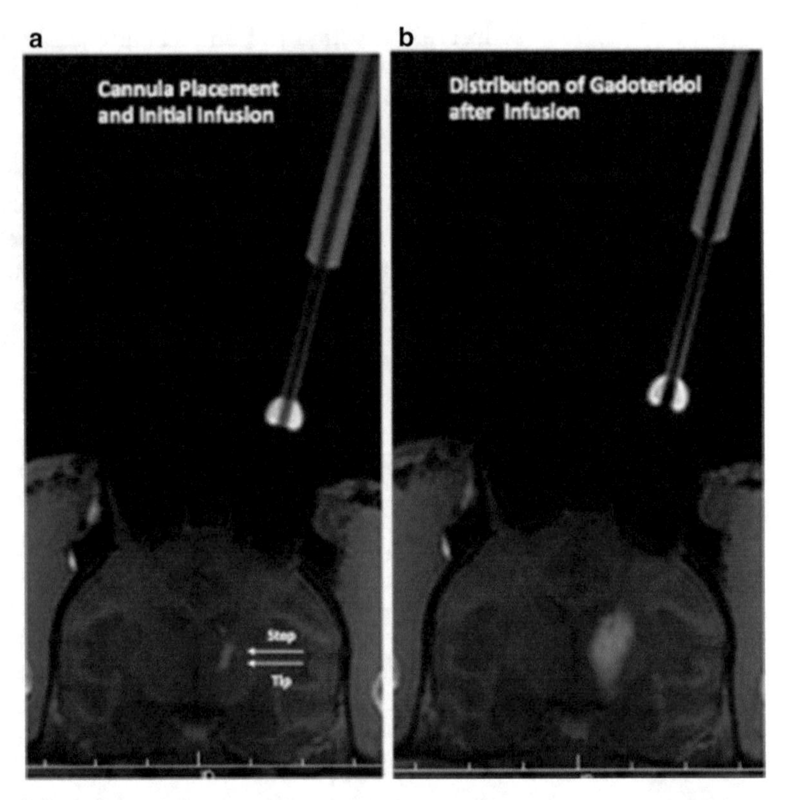

Fig. 6 Cannula placement and initial infusion in the thalamus with application of ClearPoint system are shown in panels **a** for green zone. Panel **b** shows distribution of gadoteridol in the thalamus after infusion into green zone. Note that infusion into green zone (**b**) resulted in tracer distribution in thalamus only

expected cannula tip location and the actual location measured on post-insertion imaging. The average targeting error for all targets ($n = 11$) was 0.8 mm (95 % CI = 0.14 mm). We demonstrated that this system could place two infusions in close proximity without producing reflux in the initial cannula tract during the course of the second infusion. No technical limitations were encountered in redirecting the cannula for infusing multiple targets in the same hemisphere. No infusions in any target produced occlusion, cannula reflux or leakage from adjacent tracts, and no signs of unexpected tissue damage were observed. In terms of cannula safety, no MRI-visible hemorrhages occurred during cannula placement, and no adverse events occurred during RCD. Standard postoperative care assessments indicated that no RCD-related side effects were observed over the course of these experiments (2 months).

The accuracy of ClearPoint system surpasses that of our previous experience with RCD in NHP where either a guide cannula or a multiport guide array was placed stereotactically in reference to a baseline MRI, prior to the iMRI procedure. Infusion data obtained by these methods were analyzed recently to determine the optimal zones for cannula placement within the putamen, thalamus, and brain stem [29, 30] that predict contained distribution within the target region. Images obtained during RCD in those studies showed that distribution of gadolinium tracer outside of the target structure occurred in 64% of putaminal infusions and in 43% of thalamic infusions, clearly demonstrating the need for prospective stereotaxy. In the current study, no infusions in the thalamus produced significant extra-thalamic distribution into white matter tracks or leakage into the CSF space due to poor cannula placement. In fact, the accuracy of this system guarantees insertion of the infusion cannula at preselected coordinates within a "green zone" for optimal distribution within target structures. In addition to the ability to choose the target and trajectory in real time, there are other advantages to the ClearPoint system that may explain improved performance over previous experimental studies. In comparison to the cannula used in our most recent NHP studies [29], the internal diameter of the current cannula was smaller (200 mm vs. 324 mm). Additionally, the skull-mounted SmartFrame provides a rigid housing for the cannula that restricts axial movement during brain insertion. Therefore, the ClearPoint system allows RCD to be performed with a high level of precision, predictability, and safety. This technique should increase the utility of RCD for expanding the scope of drug delivery studies. Clinical application of this platform was likely to improve the success rate for clinical trials employing cerebral drug delivery by direct infusion.

5 Future Development

The ability of this RCD platform to deliver controlled volumes of drug to any structure in the NHP brain with highly accurate localization (on the order of 1 mm), and the capacity to monitor the infusion in real time, expands the utility of the NHP brain to model human disease and development of novel therapies. The NHP brain is uniquely suited for neurosurgical investigation of therapeutic delivery, due to similarities between human and primate anatomy and physiology that cannot be closely modeled in other species [9]. We anticipate that this system will facilitate the creation of new NHP disease models due to the novel ability for precise infusion of therapeutic agents within discrete brain regions. Our recent investigations comparing linear measures in humans versus NHP will allow us to translate our NHP stereotactic (RGB) targeting data to humans [40], thereby facilitating advancement of these techniques into the clinic. In addition, we expect that ongoing studies will allow modeling of specific patterns of viral vector distribution and subsequent gene expression in structures to be targeted, such as mapping infusions in the putamen for Parkinson's disease. The evolution of this system also may also include tools that aid the neurosurgeon in planning, delivering, and anticipating the functional outcome of infusions into multiple brain locations. For instance, initiatives to incorporate the auto-segmentation of target structures and surrounding anatomy, as well as auto-segmentation of infusion volumes in real time, are under way. Eventually, analysis of retrospective and prospective infusion data in the NHP brain should allow the development of predictive algorithms that will ultimately allow the system software to forecast areas of drug distribution or transgene expression based on a selected location for cannula placement.

6 Notes

Our strategy for real-time convection delivery (RCD) in the primate brain is to minimize reflux and leakage of therapeutics infused while maximizing its distribution in the target. Attention should be paid to several points during RCD. When the infusion line is set up, it is necessary to keep all of the lines at the same level as the cannula and remove air bubble from the lines. When the cannula is lowered to a targeted area, the pump should be infusing at rate of 1.0 μl/min to prevent occlusion. Poor cannula placement or any lateral movement during insertion is often a problem during RCD. It is important to position the cannula in the right area (green zone) of the target. Real-time MRI and using the ClearPoint system would improve the accuracy of cannula placement. Infusion volume (Vi) of therapeutics in the target is also important. A small Vi would have limited distribution of therapeutics while a Vi that is too large would result in reflux and leakage. High infusion rate, such as >5 μl/min, could easily cause reflux and leakage. Therefore, all these factors need to be considered when RCD is planned.

References

1. Bobo RH et al (1994) Convection-enhanced delivery of macromolecules in the brain. Proc Natl Acad Sci U S A 91(6):2076–2080
2. Hadaczek P et al (2006) "Perivascular pump" driven by arterial pulsation is a powerful mechanism for the distribution of therapeutic molecules within the brain. Mol Ther 14(1):69–78
3. Gill SS et al (2003) Direct brain infusion of glial cell line-derived neurotrophic factor in Parkinson disease. Nat Med 9(5):589–595
4. Eberling JL et al (2008) Results from a phase I safety trial of hAADC gene therapy for Parkinson disease. Neurology 70(21):1980–1983
5. Kunwar S (2003) Convection enhanced delivery of IL13-PE38QQR for treatment of recurrent malignant glioma: presentation of interim findings from ongoing phase 1 studies. Acta Neurochir Suppl 88:105–111
6. Mardor Y et al (2001) Monitoring response to convection-enhanced taxol delivery in brain tumor patients using diffusion-weighted magnetic resonance imaging. Cancer Res 61(13):4971–4973
7. Carson BS Sr et al (2002) New approach to tumor therapy for inoperable areas of the brain: chronic intraparenchymal drug delivery. J Neurooncol 60(2):151–158

8. Lonser RR et al (1999) Convection-enhanced selective excitotoxic ablation of the neurons of the globus pallidus internus for treatment of parkinsonism in nonhuman primates. J Neurosurg 91(2):294–302
9. Richardson RM, Larson PS, Bankiewicz KS (2008) Gene and cell delivery to the degenerated striatum: status of preclinical efforts in primate models. Neurosurgery 63(4):629–644, discussion 642–644
10. Kroll RA et al (1996) Increasing volume of distribution to the brain with interstitial infusion: dose, rather than convection, might be the most important factor. Neurosurgery 38(4):746–752, discussion 752–754
11. Krauze MT et al (2006) Real-time imaging and quantification of brain delivery of liposomes. Pharm Res 23(11):2493–2504
12. Krauze MT et al (2005) Real-time visualization and characterization of liposomal delivery into the monkey brain by magnetic resonance imaging. Brain Res Brain Res Protoc 16(1-3): 20–26
13. Saito R et al (2005) Gadolinium-loaded liposomes allow for real-time magnetic resonance imaging of convection-enhanced delivery in the primate brain. Exp Neurol 196(2):381–389
14. Nguyen TT et al (2003) Convective distribution of macromolecules in the primate brain demonstrated using computerized tomography and magnetic resonance imaging. J Neurosurg 98(3):584–590
15. Richardson RM et al (2009) Future applications: gene therapy. Neurosurg Clin N Am 20(2):205–210
16. Fiandaca MS et al (2008) Image-guided convection-enhanced delivery platform in the treatment of neurological diseases. Neurotherapeutics 5(1):123–127
17. Fiandaca MS et al (2009) Real-time MR imaging of adeno-associated viral vector delivery to the primate brain. Neuroimage 47(Suppl 2):T27–T35
18. Gimenez F et al (2011) Image-guided convection-enhanced delivery of GDNF protein into monkey putamen. Neuroimage 54(Suppl 1):S189–S195
19. Su X et al (2010) Real-time MR imaging with gadoteridol predicts distribution of transgenes after convection-enhanced delivery of AAV2 vectors. Mol Ther 18(8):1490–1495
20. Lonser RR et al (2007) Real-time image-guided direct convective perfusion of intrinsic brainstem lesions. Technical note. J Neurosurg 107(1):190–197
21. Dickinson PJ et al (2008) Canine model of convection-enhanced delivery of liposomes containing CPT-11 monitored with real-time magnetic resonance imaging: laboratory investigation. J Neurosurg 108(5):989–998
22. Sykova E (2004) Diffusion properties of the brain in health and disease. Neurochem Int 45(4):453–466
23. Chen MY et al (1999) Variables affecting convection-enhanced delivery to the striatum: a systematic examination of rate of infusion, cannula size, infusate concentration, and tissue-cannula sealing time. J Neurosurg 90(2):315–320
24. Krauze MT et al (2005) Effects of the perivascular space on convection-enhanced delivery of liposomes in primate putamen. Exp Neurol 196(1):104–111
25. Szerlip NJ et al (2007) Real-time imaging of convection-enhanced delivery of viruses and virus-sized particles. J Neurosurg 107(3):560–567
26. Krauze MT et al (2005) Reflux-free cannula for convection-enhanced high-speed delivery of therapeutic agents. J Neurosurg 103(5): 923–929
27. Sanftner LM et al (2005) AAV2-mediated gene delivery to monkey putamen: evaluation of an infusion device and delivery parameters. Exp Neurol 194(2):476–483
28. Yin D, Forsayeth J, Bankiewicz KS (2010) Optimized cannula design and placement for convection-enhanced delivery in rat striatum. J Neurosci Methods 187(1):46–51
29. Yin D et al (2011) Optimal region of the putamen for image-guided convection-enhanced delivery of therapeutics in human and non-human primates. Neuroimage 54(Suppl 1):S196–S203
30. Yin D et al (2010) Cannula placement for effective convection-enhanced delivery in the nonhuman primate thalamus and brainstem: implications for clinical delivery of therapeutics. J Neurosurg 113(2):240–248
31. Varenika V et al (2008) Detection of infusate leakage in the brain using real-time imaging of convection-enhanced delivery. J Neurosurg 109(5):874–880
32. Truwit CL, Liu H (2001) Prospective stereotaxy: a novel method of trajectory alignment using real-time image guidance. J Magn Reson Imaging 13(3):452–457
33. Hall WA et al (2001) Brain biopsy sampling by using prospective stereotaxis and a trajectory guide. J Neurosurg 94(1):67–71
34. Martin AJ et al (2008) Minimally invasive precision brain access using prospective stereotaxy and a trajectory guide. J Magn Reson Imaging 27(4):737–743

35. Martin AJ et al (2005) Placement of deep brain stimulator electrodes using real-time high-field interventional magnetic resonance imaging. Magn Reson Med 54(5): 1107–1114
36. Starr PA et al (2010) Subthalamic nucleus deep brain stimulator placement using high-field interventional magnetic resonance imaging and a skull-mounted aiming device: technique and application accuracy. J Neurosurg 112(3): 479–490
37. Krauze MT et al (2007) Convection-enhanced delivery of nanoliposomal CPT-11 (irinotecan) and PEGylated liposomal doxorubicin (Doxil) in rodent intracranial brain tumor xenografts. Neuro Oncol 9(4):393–403
38. Saito R et al (2006) Convection-enhanced delivery of Ls-TPT enables an effective continuous low-dose chemotherapy against malignant glioma xenograft model. NeuroOncology 8(3):205–214
39. Yin D et al (2009) Striatal volume differences between non-human and human primates. J Neurosci Methods 176(2):200–205
40. Fiandaca MS et al (2011) Human/nonhuman primate AC-PC ratio—considerations for translational brain measurements. J Neurosci Methods 196(1):124–130

Chapter 15

Focal Cerebral Ischemia by Permanent Middle Cerebral Artery Occlusion in Sheep: Surgical Technique, Clinical Imaging, and Histopathological Results

Björn Nitzsche, Henryk Barthel, Donald Lobsien, Johannes Boltze, Vilia Zeisig, and Antje Y. Dreyer

Abstract

According to the recommendation of international expert committees, large animal stroke models are demanded for preclinical research. Based on a brief introduction to the ovine cranial anatomy, a sheep model of permanent middle cerebral artery occlusion (MCAO) will be described in this chapter. The model was particularly designed to verify several therapeutic strategies during both, acute and long-term studies, but is also feasible for development of diagnostic procedures. Further, exemplary application of imaging procedures and imaging data analyses using magnetic resonance imaging (MRI) and positron emission tomography (PET) are described. The chapter also includes recommendations for appropriate animal housing and medication.

Key words Large animal model, Sheep, Experimental neurosurgery, Craniotomy, Experimental stroke, Middle cerebral artery occlusion, MRI, PET

1 Introduction

1.1 The Role of Animal Models in Preclinical Stroke Research

Worldwide, ischemic stroke represents a major cause of death and is the most important reason for permanent disability in adulthood [1]. Thrombolysis by recombinant tissue plasminogen activator is currently the only pharmacological approved therapy for this disease, and the time window for intervention has recently been extended to 4.5 h [2]. Nevertheless, because of this still narrow time window and due to rapidly decreasing therapeutic efficacy within that time [3], the vast majority of stroke patients remain untreated or only benefits from minor therapeutic effects. Despite significant research activities and promising results in preclinical tests, not a single experimental treatment strategy was successfully translated into clinical routine so far [4]. The underlying reasons

Miroslaw Janowski (ed.), *Experimental Neurosurgery in Animal Models*, Neuromethods, vol. 116,
DOI 10.1007/978-1-4939-3730-1_15, © Springer Science+Business Media New York 2016

are considered multiple and, among others, comprise inappropriate or nonpredictive animal models in preclinical research [5].

The standard species in preclinical stroke research are rats and mice. Rodent animal models offer numerous advantages such as easy animal housing, well established methodology including the availability of genetically modified strains, and excellent tools to assess functional outcome.

However, international expert committees recommend the use of large animal models to verify results previously obtained in small animal studies using neuroprotective [6] or cell-based approaches [7].

Large animals may allow for long-term assessment of safety and efficacy up to years, whereas the life span of a common laboratory rodent is usually restricted to several months post stroke. Moreover, therapeutic efficacy can be tested in gyrencephalic brains, and influences related to brain size or bodyweight (e.g., after systemic injection of the therapeutic agent) may be assessed more thoroughly. Large animal models also permit testing in specific neuroanatomical structures, in particular white matter [6], and the use of clinical imaging techniques. Moreover, it offers the possibility to deliver stem cells of several autologous sources. Thus, large animal models are considered to be of increasing relevance for preclinical and translational stroke research.

1.2 Large Animal Models of Stroke

Focal ischemic stroke is predominantly induced by permanent or transient occlusion of the middle cerebral artery occlusion (MCAO). Existing large animal models of focal cerebral ischemia comprise rabbit [8], canine [9], feline [10], and swine [11] models. MCAO studies with rabbits have been conducted for a long time but are restricted by the relatively low weight and small size of the animals. For anatomical reasons, canine, feline and porcine stroke models mostly require enucleation to assess the MCA. This may lead to severe behavioral abnormalities restricting the use of these models for long-term safety and efficacy assessments. Nonhuman primate models, which are essential to investigate acute stroke pathophysiology, are often associated with high mortality rates thus preventing long-term observations [6]. An exception is the marmoset. However, MCAO studies using this nonhuman primate [12] did not predict the failure of the neuroprotectant NYX-059 in clinical trials [13].

An ovine model of permanent distal MCAO (*see* **Note 1**) has been developed avoiding enucleation by using a temporal transcranial approach to the MCA [14]. Ischemic lesion size and subsequent functional outcome can be controlled by varying the number of occluded cortical MCA branches. Further, the sheep model benefits from the easy availability of clinical imaging modalities. Widely available clinical scanners, including computer tomography (CT), magnetic resonance imaging (MRI), and positron emission tomography (PET) can be used to visualize the progress of the infarction as well as the effects of potential therapeutic agents.

A detailed description of the surgical methodology, used materials, application of sophisticated imaging techniques, and post mortem pathohistology is given in this chapter. We also briefly refer to options of stereotaxic or image guided surgery using this animal model (*see* **Note 2**).

2 Materials and Animals

All devices, materials, consumables, and drugs mentioned in the protocols can principally be replaced by equivalent products from other suppliers.

2.1 Animal Housing, Care, and Brief Anatomical Description of the Ovine Skull

Some general information regarding animal handling, ovine skull anatomy, and cerebral blood supply is needed to ensure adequate and reproducible experimental results. The following paragraphs provide a brief introduction to the mentioned aspects.

2.1.1 Experimental Subjects and Animal Housing

The herein neurosurgical approach for MCAO induction necessitates hornless subjects for easy accessibility of cranial structures. Merino sheep may be used preferably as many hornless strains can be found in this widely available breed. Weight (ewe: 75–85 kg; ram: 120–140 kg) and body size (height at withers: 0.8–0.9 m) of adult Merino sheep [15] allows relatively easy handling. Species appropriate housing, feeding (*see* Table 1) as well as thorough medical inspections and blood screening (*see* Table 2), medication and vaccination (*see* Tables 1 and 3) ensure a significantly reduced risk of postoperative complications and thereby enhance study quality. Frequent and early human contact facilitates familiarization and improves the handling, especially during long-term studies.

2.1.2 Ovine Skull Anatomy

The ovine skullcap is less convex as compared to humans, primates, dogs, and cats. In sheep, the oral rim is comparatively small but the species has a long pharyngeal cavity with a massive torus linguae and a long epiglottis, the latter being situated dorsal of the soft palate. The frontal bone includes widespread air-filled sinuses (that may be extend up to the origin of the horns, *see* Fig. 1). Those sinuses must not be opened due to a ·possibly fatal sinusitis/osteomyelitis.

2.1.3 The Ovine Brain

The mean weight of an adult sheep brain is 120 g. The mean horizontal circumference including the cerebellum measures about 20 cm, the mean vertical circumference is approximately 12 cm. When adjusting on BW and age, the cerebral tissue volumes are 51.5 ± 3.9 mL (GM), 35.6 ± 3.2 mL (WM), 29.7 ± 3.3 mL (CSF), and 87.1 ± 6.0 mL (total brain volume) [16]. Sulci and gyri of the

Table 1
Housing, feeding, and general health care

Food	- Water and hay ad libitum - Add mash daily (avoid sweet corn, use oat or barley) or silage - Provide mineral food supplements (salt block)
Housing	- Flock, separated by sex - Individual enclosure for 2 days postsurgical with visual contact to flock members - Space: ≥1.5 m^2 per subject - Ground litter: wooden shaving (sterilized) or straw - Ambient temperature: 5–28 °C, heating may be required during winter - No air conditioning is necessary during summer, but shaving of subject before the summer months - Outdoor housing for at least 2 h per day (above 5 °C) - Minimum light intensity 80 lx - Maximal environmental gas concentration: 3.500 ppm CO_2, 30 ppm NH_3, 5 ppm H_2S
Parasite prophylaxis	- Doramectin - Toltrazuril
Vaccination	- Should be completed before entering the experimental facility - Clostriadia, pneumococcus (e.g., Ovilis Heptavac P, Intervet, Germany): in postnatal week 3, booster vaccination in week 7, followed by annual boostering - Footrot (e.g., Footvax® Schering-Plough, Animal Health, Germany): in postnatal week 3, boostering in week 7, followed by annual boostering - Ecthyma contagiosum (e.g., Ecthybel ad us. vet., Merial, France): in postnatal week 7, followed by annual boostering - Bluetongue disease boostering (e.g., BTVPUR AlSap™ 8, Merial, France): in postnatal week 4, followed by annual boostering - Rabies (e.g., Rabisin®, Merial, France): in postnatal week 8, followed by boostering every third year
Vital parameters	- Rumen: normally three fluctuations in 2 min - Body temperature: 38.0–40.0 °C - Breathing frequency: 20–100 min^{-1} - Heart beat rate: 70–110 min^{-1} - Blood pressure: 80/55/60 mmHg (systolic/diastolic/mean)

ppm parts per million, *CO2* carbon dioxide, *NH3* ammonia, *H2S* hydrogen sulfide

neocortex, like in primates, are assembled in an individual configuration in any animal (*see* Fig. 2; *see* **Note 3**).

The pyramidal tracts in primates are of utmost importance for motor functions. Almost all pyramidal fibers cross to the contralateral hemisphere. In contrast, only about 50% of the motor fibers cross at the pyramidal decussation in sheep [17]. Therefore, a unilateral loss of central motor innervation (as seen following stroke) can partly be compensated. Consequently, even large cortical deficits caused by MCAO in sheep result in hemiparesis, but not hemiplegic conditions. Motoric dysfunction can clearly be observed and quantified, while the animal is still able to move and join a flock. This dramatically reduces poststroke mortality and complications, which represents a major advantage in particular for long-term studies and observations.

2.1.4 Cerebral Blood Supply in Sheep

The blood supply of the ovine circle of Willis (CW) originates from the so called rete mirabile epidurale rostrale. The rete is supplied by the maxillary artery [18]. This is a major difference to the anatomical situation found in humans and primates. The rete, being embedded in a venous sinus, comprises a dense network of intercommunicating, small arteries. Because of their very small diameter, these arteries do not allow intravascular approaches to the CW. Only in lambs, the CW is supplied by a carotid artery (originating from maxillary artery), which obliterates in the first postnatal months. Thus, a functional intact internal carotid artery does not exist in adult sheep.

At the level of CW, the blood supply of the ovine brain is comparable to the situation in primates and humans. Strong anterior

Table 2
Normal range of relevant blood parameters in sheep

Hematology			Blood chemistry		
Parameter	**Unit**	**Value (range)**	**Parameter**	**Unit**	**Value (range)**
Leukocytes	10^9/L	5.0–11.0	Urea	mMol/L	2.8–7.1
Erythrocytes	10^{12}/L	7.0–11.0	Creatinine	μMol/L	100–125
Hemoglobin	mMol/L	5.6–9.3	Total protein	g/L	60–78
HKT	L/L	0.27–0.40	ASAT	U/L	<65
MCV	fL	28.0–40.0	ALAT	U/L	<14
MCH	fMol	0.6–0.7	GGT	U/L	<60
MCHC	mMol/L	19.0–23.0	Bilirubin	μMol/L	<4
Platelets	10^9/L	280–650	AP	U/L	<60
Neutrophile granulocytes	10^9/L	0.1–4.8	LDH	U/L	<600
Lymphocytes	10^9/L	2.2–9.2	CK	U/L	13–230
Monocytes	10^9/L	0–2.0	Na	mMol/L	139–152
Eosinophile granulocytes	10^9/L	0–2.0	Cl	mMol/L	95–110
Basophile granulocytes	10^9/L	0–1.0	Fibrinogen	g/L	1.0–5.0

ALAT alanine amino transaminase, *AP* alkaline phosphatase, *ASAT* aspartate aminotransaminase, *CK* creatine kinase, *Cl* chloride, *GGT* gamma-glutamyltransferase, *HKT* hematocrit, *LDH* lactate dehydrogenase, *MCH* mean corpuscular hemoglobin, *MCHC* mean corpuscular hemoglobin concentration, *MCV* mean corpuscular volume, *Na* sodium

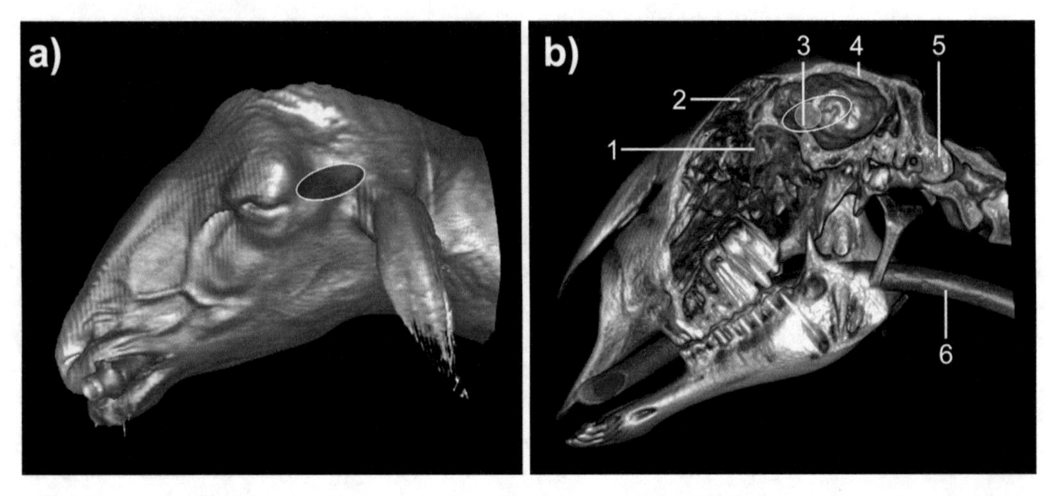

Fig. 1 Gross anatomy of the ovine head. Three-dimensional reconstruction was performed using a high resolution CT data set using OsiriX 3.8.1 [33]. (**a**) Head of a sheep: The area of surgical access is indicated by the hemitransparent, elliptic overlay. (**b**) Topography of the sheep skull: The approximate position of area of surgical access is indicated by the elliptic overlay. The burr hole is indicated by the *hemitransparent circle*. The extensive nasal sinus (*1*) and the frontal sinus (*2*) must not be opened during transcranial surgery. The Sella turcica (*3*) indicates the level of the middle cerebral artery. The roof of the compact Os parietale (*4*) is easily achievable for any kind of cranial surgery in the species. However, the MCA cannot be reached using a high parietal approach. The atlanto-occipital junction (*5*) includes a massive axis. Note reconstruction of the endotracheal tube (*6*) which is placed in the diastema between incisor and premolar teeth

cerebral arteries (ACA) are connected by a communicating branch, which is present in virtually all subjects. The middle cerebral artery (MCA) originates from the CW at level of pituitary infundibulum. In further progress, the MCA runs rostrally to the Lobus piriformis inside the lateral sulcus (M1 segment). It splits up into two or three branches (M2 segments). The posterior communicating arteries of the CW give rise to the posterior cerebral arteries (PCA), as well as to the rostral cerebellar artery before converging into the basilar artery [19]. A detailed visualization of the cerebral blood supply in sheep is given in Fig. 3 (*see* **Note 4**).

2.2 Anesthesia, Arterial, and Venous Access

2.2.1 Initial Sedation, Arterial, and Venous Access

1. Ketamine, xylazine, diazepam or midazolam, and propofol (*see* Table 3; *see* **Note 5**).
2. 50 mL 0.9 % sterile sodium chloride solution.
3. Syringes: 3 mL (1×), 5 mL (3×), 10 mL (3×), 20 mL (1×).
4. 18G needles (10×; Braun Melsungen AG, Germany; *see* **Note 6**).
5. 14G venous cannula (1×; length: 80 mm, e.g., Braunüle MT, Braun Melsungen) with in-stopper (Braun Melsungen).

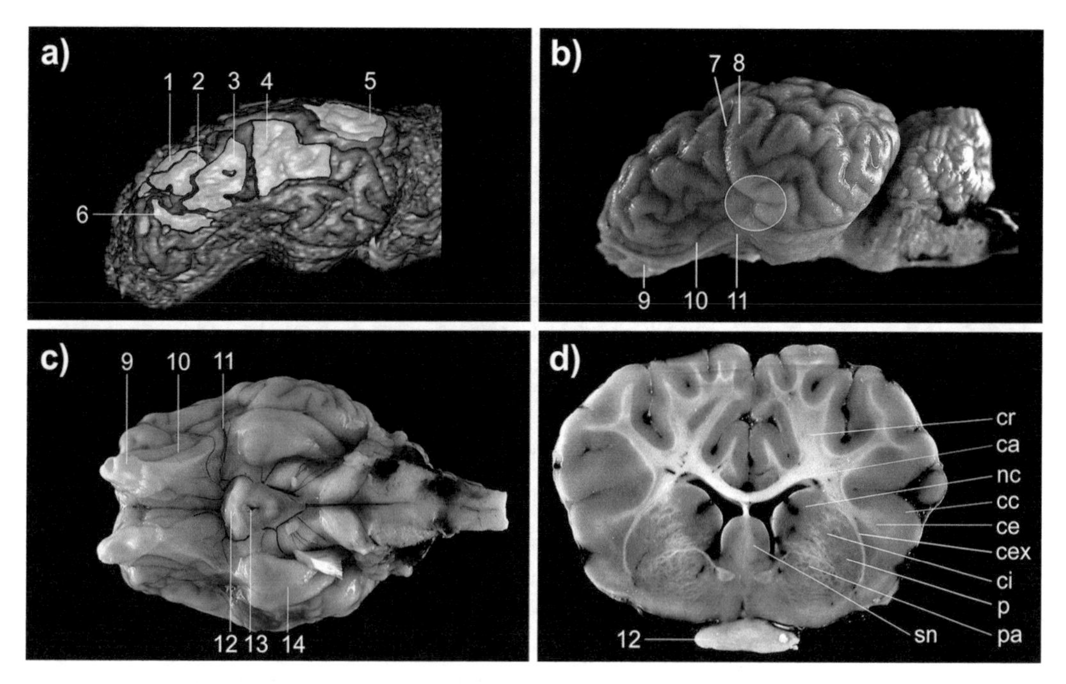

Fig. 2 Anatomy of the sheep brain and functional organization of the neocortex. (**a**) Functional areas of the sheep neocortex are highlighted in a MRI 3D reconstruction [33]. The brain in (**b**–**d**) was removed from the skull after automated perfusion with 20 L 4 % paraformaldehyde, followed by immersion fixation in 4 % paraformaldehyde for 3 days. (**b**) Lateral view of the brain: The approximate location of the drill hole is indicated by the *hemitransparent circle* overlay. (**c**) Basal view of a brain. (**d**) Coronal brain slice at level of optic chiasm. Legend: (*1*, *2*) somatosensory area I (face, lips, tongue); (*3*) somatosensory area II (face, forelimb, hindlimb); (*4*) auditory area; (*5*) visual cortex; (*6*) motor area (eye face, head, forelimb, hindlimb, tongue); (*7*) lateral sulcus; (*8*) caudal sylvian gyrus; (*9*) olfactory bulb; (*10*) lateral rhinal fissure; (*11*) middle cerebral artery; (*12*) optic chiasm; (*13*) infundibulum; (*14*) piriform lobe; (*ca*) corpus callosum; (*cc*) claustrocortex; (*ce*) extrem capsule; (*cex*) external capsule; (*ci*) internal capsule; (*cr*) corona radiata; (*nc*) caudate nucleus; (*p*) putamen; (*pa*) globus pallidus; (*sn*) septal nuclei

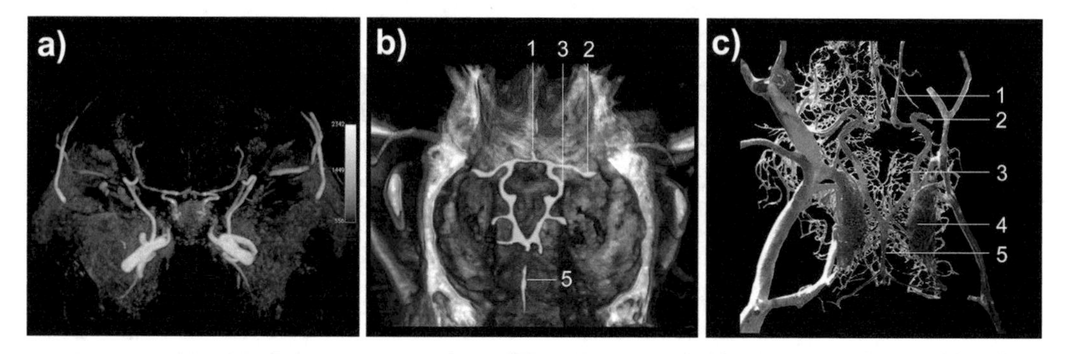

Fig. 3 Cerebral blood supply in sheep. (**a**) MRI time-of-flight (TOF) visualization of the cerebral blood supply at the skull base. (**b**) Dorsal view into the cranial cavity (3D reconstruction of a CT-angiography). The major arterial vessels were highlighted digitally. (**c**) Corrosion cast model made by intra-arterial delivery of Mallocryl M®, basal view. Legend: (*1*) anterior cerebral artery with communicating branches; (*2*) occluded middle cerebral artery (note missing ipsilateral capillaries); (*3*) caudal communicating branches of the Circle of Willis; (*4*) rete mirabile epidurale rostrale; (*5*) basal artery

6. 18G LEADER-Catheter set (1×; diameter: 1.2 mm, length: 18 cm, Vygon GmbH & Co.KG, Germany).
7. Tube guide rod (aluminum, semiflexible, diameter: 2–3 mm, length: 60 cm).
8. Laryngoscope (Heine AG, Germany) with Miller blade, size 4 (Heine Classic F.O., Germany).
9. Endotracheal tube (size: 10 or 11 mm, length: 140 mm; Pharmazeutische Handelsgesellschaft, Germany) and bandage (e.g., Idealast® C, Paul Hartmann AG).
10. Pulse oximeter (PM-60, Mindray Ltd., China) with tongue probe (Eickemeyer KG, Germany).
11. Stethoscope (Littmann®, 3M GmbH, Germany).
12. Temperature probe/clinical thermometer (Henry-Schein GmbH, Germany).
13. Animal balance (e.g., Kern EOS300 K200XL, Wägetechnik Koch GmbH, Germany).
14. Small (size: 1.6 mm e.g., Golden A fine, Averde GmbH, Germany) and large (size: 2.8 mm, e.g., Golden A medium) electric clippers (e.g., Oster Power UltraPro, Averde GmbH, Germany).

2.2.2 Inhalative Anesthesia During Surgery and Imaging Procedures

1. 2 L sterile Ringer-lactate (*see* Table 3).
2. Medical gases: oxygen and isoflurane (*see* Table 3; *see* **Note 7**).
3. Eye ointment (e.g., Corneagel®, Bausch & Lomb GmbH, Germany).
4. Respirator (e.g., Draeger Primus®, Draeger AG, Germany) with oxygen, isoflurane vapors absorber, capnography sensor, and breathing filter with heat/moisture exchanger (ClearGuard3, Intersurgical GmbH, Germany). A MR-compatible machine is required for ventilation anesthesia during MRI (*see* **Note 8**).
5. Electrocardiogram (ECG) and vital signs monitor (e.g., Infinity Gamma XXL, Draeger AG) with crocodile clips (Eickemeyer KG), noninvasive blood pressure (NIBP) cuff (size: pediatric) and rectal temperature probe (Draeger AG).
6. Automated infusion system (Infusomat FM®, Braun Melsungen AG) and tubing.
7. Heidelberger extension and 3-way-stopcock (Discofix®, Braun Melsungen AG) (Table 3).

Table 3
Medication schemes

Purpose	Recommended drug	Supplier	Route	Dose	Time of application	Duration
Parasite prophylaxis	Doramectin (e.g., Dectomax®)	Pfizer	i.m.	0.2 mg kg^{-1}	After animal arrival at experimental facility	Single injection
	Toltrazuril (e.g., Baycox 5 %®)	Bayer	per os	20 mg kg^{-1}		Single application
Antibiosis	Enrofloxacin (e.g., Baytril®)	Bayer	i.m.	5 mg kg^{-1}	7× daily post MCAO	Single injection
Antiphlogistic and analgesia	Flunixin meglumin	CP Pharma	i.m.	2.2 mg kg^{-1}	5× daily post MCAO	Single injection
	Buprenorphine (e.g., Temgesic®)	RB Pharmaceuticals	i.m.	0.01 mg kg^{-1}	Every 8 h for 2 days postsurgical	Single injection
Induction of anesthesia (surgery and imaging)	2 % xylazine hydrochloride (e.g., Xylazin®)	Ceva Sante Animal	i.v. bolus	0.1 mg kg^{-1}	Prior to anesthesia	Single injection
	Ketamine hydrochloride (e.g., Ketamin®)	Medistar	i.v. bolus	4 mg kg^{-1}	Prior to anesthesia	Single injection
	Diazepam (e.g., Faustan®) or	Temmler Pharma	i.v. bolus	0.2 mg kg^{-1}	Prior to surgery	Single injection
	Midazolam (e.g., Midazolam®)	Braun Melsungen	i.v. bolus	0.2 mg kg^{-1}	Prior to imaging	Single injection
Inhalation anesthesia	Isoflurane	CP Pharma	Ventilation	1.5–2.0 %	During surgery	Up to 5 h
	Oxygen	Linde Medical Gases		20– 40 %		

(continued)

Table 3 (continued)

Purpose	Recommended drug	Supplier	Route	Dose	Time of application	Duration
Infusion anesthesia	Midazolam	Braun Melsungen	i.v. infusion	0.1 mg kg^{-1} h^{-1}	During imaging	Up to 5 h
	Ketamine hydrochloride (e.g., Ketamin®)	Medistar	i.v. infusion	2 mg kg^{-1} h^{-1}		
	2% propofol (e.g., Propofol Lipuro®)	Braun Melsungen	i.v. infusion	6 mg kg^{-1} h^{-1}		
Hydration	Physiological saline solution	Braun Melsungen	i.v. infusion	3 mL kg^{-1} h^{-1}	During imaging	Up to 4 h
	Ringer-Lactate	Braun Melsungen	i.v. infusion	3 mL kg^{-1} h^{-1}	During surgery	
Ruminal stimulation	Propionic acid	Raiffeisen	per os	12.5 g	5× daily post MCAO	Twice a day
	Butafosfan (e.g., Catosal 10%®)	Bayer Healthcare	i.m.	0.5 mL kg^{-1}		Single injection
	Menbuton (e.g., Genabil®)	Boehringer Ingelheim	i.m.	5 mL per animal		Single injection
	Amynin	CP Pharma	i.v.	5 mL kg^{-1}		Single injection
Sacrifice	Pentobarbital (e.g., Eutha77®)	Veterinaria	i.v. Bolus	80 mg kg^{-1}	Prior to decapitation; only during anesthesia!	Single injection

i.m. intramuscular, *i.v.* intravenous, kg^{-1} ($\times$ h^{-1}): per kilogram body weight (and hour)

2.3 Surgery and Postsurgical Care

2.3.1 Surgical Approach for MCAO

1. 2 L 0.9% sterile sodium chloride solution (Braun Melsungen AG).
2. Iodine-containing solution for skin disinfection.
3. Folio drape (1 × 1 m, 2×), adhesive folio drape (0.3 × 0.3 m, 1×) and fenestrated folio drape (0.9 × 0.9 m, 1×; all Heiland VET GmbH, Germany).
4. Adhesive tape (1×, Eickemeyer KG).
5. Eye swab (minimum: 10×) and surgical pads (minimum: 30×, all Eickemeyer KG).
6. Bone wax (1 package, Braun Melsungen AG) and neurosorb® patties (2 packages, Vostra GmbH, Germany).
7. 0-0, 2-0, and 6-0 resorbable filaments (2× each, Ethicon Ltd., Germany).
8. Electrosurgery and cauterization device (ME 411, KLS Martin, Germany) with straight and bayonet-shaped neurosurgical bipolar forceps (1× each, Aesculap AG, Germany).
9. Electric power system and surgical motor (e.g., Microspeed® uni mini inclusive straight handpiece HiLan XS size II, scil animal care company GmbH, Germany) with 4 mm Rosen burr and 6 mm Barrel burr (size II, scil animal care company GmbH).
10. Standard surgical instruments including Williger raspatories (size: 2 × 2 mm (1× sharp, 1× blunt) and 1 × 4 mm), atraumatic retractor (1×), curved Wullstein retractor (1×), dura hook (Fisch dura retractor, length: 185 mm, 1×, e.g., catalog-no: FD376R, Braun Melsungen AG), surgical and anatomical forceps (3× each), atraumatic Adson-Brown forceps (1×), bayonet-shaped neurosurgical forceps (length: 240 mm, 1×), Roberts artery forceps (1×), curved and straight Metzenbaum scissor (1× each), straight and angled spring type micro scissors (length: 120 mm, 1× each), ligature scissor (1×), scalpel (size 22, 1×), Kerrison rongeurs (size: 2 and 4 mm, 1× each), Backhaus towel clamps (12–20×), standard needle holder (e.g., Mathieu Durogrip, 1×).
11. Head light system with magnifiers (e.g., 3s LED Headlight PR, Heine Optotechnik GmbH & Co.KG, Germany).

2.3.2 Postsurgical Care

1. Drugs: enrofloxacin, flunixine-meglumine, buprenorphine (*see* Table 3).
2. Aseptic wound spray (Betadona®, Mundipharma GmbH, Germany).
3. Wound cover spray (Alu spray silver®, Eskadron GmbH, Germany).
4. 5 mL syringes (2×) and 18G needles (10×, Braun Melsungen AG).

5. Adhesive tape (1×, Eickemeyer KG, Germany).
6. Elastic bandage (2×), cotton (1×) and tape (1×, Eickemeyer KG).

2.4 Imaging Procedures

2.4.1 General Materials for Imaging

1. Adhesive tape (Eickemeyer KG).
2. Folio drape (2× minimum, 1.5 × 1.5 m, Heiland GmbH) for covering.
3. Neck crest (height: 0.15–0.25 m).

2.4.2 MR Imaging

1. 1.5 T Scanner (e.g., Gyroscan Intera, Philips, Netherlands) with a flexible double loop RF-coil (Sense Flex M, Philips) or
2. 3 T Scanner (e.g., Magnetom Trio, Siemens, Erlangen, Germany) with a four channel flex coil.

2.4.3 PET Imaging

1. High-resolution clinical PET scanner (e.g., ECAT EXACT HR+; Siemens/CTI, USA).
2. NeuroShield® (Scanwell Systems, Canada).
3. $[^{15}O]H_2O$; synthesized by a catalyst-mediated reaction between $[^{15}O]O_2$ and H_2 (from PETtrace cyclotron, GE Healthcare, USA), followed by dialysis exchange in an automated system (Veenstra, The Netherlands), that performs tracer injection subsequently.
4. $[^{18}F]$Fluordesoxyglucose (FDG), synthesized by a standard nucleophilic substitution with alkaline hydrolysis.
5. Blood sampler (e.g., ALLOGG AB, Allogg Mariefred, Sweden).
6. MRI data set with individual anatomical information for coregistration with functional PET data; preferentially use T1 or T2 T2 (turbo spin echo (TSE), fluid attenuation inversion recovery (FLAIR)) sequences.

2.4.4 Imaging Data Interpretation and Analysis

1. Vendor-specific MRI postprocessing software for magnetic resonance angiography (MRA, e.g., Syngo, Siemens) and for diffusion tensor imaging (DTI; e.g., DTI studio, Center for Imaging Science, Johns Hopkins University, Baltimore, USA).
2. ImageJ image postprocessing software (National Institute of Health, USA).
3. Vendor-specific PET postprocessing software: PMOD software (version 3.0, PMOD Technologies, Ltd., Zürich, Switzerland) for combined interpretation of image data sets derived from different modalities (PET, MRI, and others). Further, the PMOD software can be used for kinetic modeling like cerebral blood flow (CBF) quantification.

2.5 Sacrifice and Post Mortem Analyses

1. Pentobarbital (*see* Table 3).
2. 3 L Phosphor-buffered sodium (PBS) solution (pH 7.4, store at 4 °C).
3. 20 L buffered paraformaldehyde (PFA, pH 7.4, store at 4 °C).
4. 5 L 30% sucrose in PBS (pH 7.4).
5. 12G steel cannula (2×, Carl-Roth AG, Germany), 100 or 500 mL syringes (2×, Braun Melsungen AG), large surgical (2×) and anatomical (2×) forceps, stout thread (4×, length: 0.25 m), knife with 180 mm blade (2×), oscillating saw (1×, e.g., HEBUmedical GmbH, Germany).
6. Roller pump (pericyclic pump) with two rollers (e.g., Cyclo II®, Carl-Roth GmbH).
7. 20 L canister and adapter for tubing (Carl-Roth GmbH).
8. Tubing (2×3 m).
9. Embedding system (e.g., HyperCenter®, Shandon, Germany).
10. Antibodies and histological staining reagents according to desired analyses. For more details and suggestions, please refer to [14].

3 Methods

3.1 Animal Housing

1. Prior to the trail, subject all animals to a familiarization period of at least 7 days (14 days for long-term studies) in the experimental facility.
2. Perform parasite prophylaxis immediately upon arrival (*see* Table 3).
3. Collect blood samples for routine hematological and general health screening prior to any trial.
4. Exclude subjects with nonphysiological values in routine hematological screening (*see* Tables 1 and 2), obvious preexisting neurofunctional deficits, or other illnesses including parasite infestation.
5. Prior to MCAO or any anesthesia, deprive subjects of food for at least 18 h (optimum: 24 h) by using common multiperforated calf muzzle (diameter 20 cm).
6. Allow *ad libitum* water access (drinking is possible with the multiperforated muzzle).
7. For venous access via the jugular veins, shave the lateral neck around the jugular sulcus on each side using electric clippers, followed by disinfection with 70% alcohol. To prepare arterial access, shave the area of the tarsus.
8. Check the weight of the animal prior to sedation for adequate drug dosages (*see* Table 3).

3.2 Anesthesia, Arterial, and Venous Access

3.2.1 Initial Sedation, Arterial, and Venous Access

1. Prepare syringes for anesthesia: 1× ketamine and xylazine mix (can be used in one syringe, so called "Hellabrunner mix"), 1× diazepam or midazolam, 3× propofol according to individual dosages (*see* Table 3), and 20 mL of 0.9% sterile sodium chloride solution.
2. Repeat disinfection of both jugular sulci with 70% alcohol.
3. Sedate the subject via slow intravenous injection of ketamine and xylazine directly into the right jugular vein. Wait until animal loses consciousness and catch it when it falls down.
4. Place the animal in a lateral position on the right side. Always maintain this position for transportation and during the surgical approach to avoid torsion of stomach. Place animal in prone position through imaging.
5. Cannulate the left jugular vein with the venous catheter, fix the catheter with skin suture and place the in-stopper (*see* **Note 7**).
6. Slowly inject diazepam (for surgery) or midazolam (for imaging) via the venous catheter.
7. Intubate the animal, using the guide rod and laryngoscope. The guide rod is useful to lift the large soft palate which often interferes direct tube insertion. In case the swallowing reflex still persists, administer a propofol bolus intravenously. Finally, inflate the blocker balloon of the endotracheal tube, and fix the tube using elastic bandage.
8. Place a naso-oesophageal reflux collector (diameter: 1 cm) below the ventral conches. Guide along the tracheotubes through the diastema between incisor and premolar tooth.
9. Continuously monitor breathing frequency using the stethoscope. Control for heart rate and oxygen saturation using the pulse oximeter and the tongue clip. Make sure the tongue clip is in an adequate position.
10. Check rectal temperature using the thermometer every 10 min.
11. For arterial access, disinfect the tarsus area and (*see* step 2 of this section) and place the LEADER-catheter in the tarsal artery using the Seldinger technique. The arterial line is used for invasive blood pressure measurements (IBP) during surgery and for blood sampling during PET. Carefully fix the catheter with skin suture and tape.
12. During any transportation of the animal (to the imaging facility or to the operation room), maintain continuous monitoring, and protect venous and arterial lines. You can use a standard stretcher for transportation.

3.2.2 Inhalative Anesthesia During Surgery and Imaging Procedures

1. Recommended concentrations for isoflurane and oxygen are given in Table 3. These recommendations may be ignored and be replaced by alternative concentrations in case the anesthetist

will find this appropriate. During surgery, in particular trepanation, do not use less than 1.5% of isoflurane.

2. Set breathing volume to 15 mL kg^{-1} bodyweight and respiration frequency to 12–18 min^{-1}. Use oxygen concentration between 20 and 40% (*see* **Note 7**).
3. Set inspiratory pressure to 18 cm H_2O.
4. Infuse the Ringer-lactate intravenously with 2–10 mL h^{-1} kg^{-1} bodyweight.
5. Use eye ointment to protect eyes from drying.
6. Monitor the peripheral pulse and oxygen saturation during MCAO and imaging. Further monitor capnography, ECG, NIBP, IBP, rectal temperature (*see* Table 1) during MCAO surgery using the mentioned monitoring systems.
7. Alternatively, maintain anesthesia by infusion method if applicable (*see* **Note 8**).

3.3 Surgery and Postsurgical Care

3.3.1 Surgical Approach for MCAO

1. Place the animal in a right lateral position.
2. Ensure deep unconsciousness by testing for the blink reflex and ensure anesthesia by repeated moderate pain stimuli (e.g., use a small needle to stab between the claws of a forelimb).
3. Apply the eye ointment in both eyes. Then attach the ear straight behind the neck using adhesive tape. Fix the head thoroughly using adhesive tape.
4. Shave the area between ear and eye and use iodine-containing solution for thorough disinfection. Wait for at least 3 min before the next step.
5. Cover the head with the folio drape surgical covers, only exposing the area between ear and eye. Also cover the body to avoid secondary contamination.
6. Remove an elliptic skin lobe at the temporal region between lateral eye rim and ear (*see* **Note 9**).
7. Carefully expose the superficial temporal artery and the corresponding vein. First occlude the artery electrosurgically using the bipolar forceps, followed by the vein (*see* **Note 10**).
8. Dissect the retroorbital fat directly behind the orbital rim to fully expose the temporal muscle.
9. Incise the temporal muscle at the temporal line. Importantly, make sure to leave a small rim of connective tissue at the bone to readapt the muscle after MCAO. Then carefully elevate the temporal muscle using a Wullstein retractor (*see* Fig. 4; *see* **Note 11**). The coronoid process of the mandibula is covered by the tissue of the temporal muscle (*see* **Note 12**), and must not be exposed. Thus proceed with step 10.

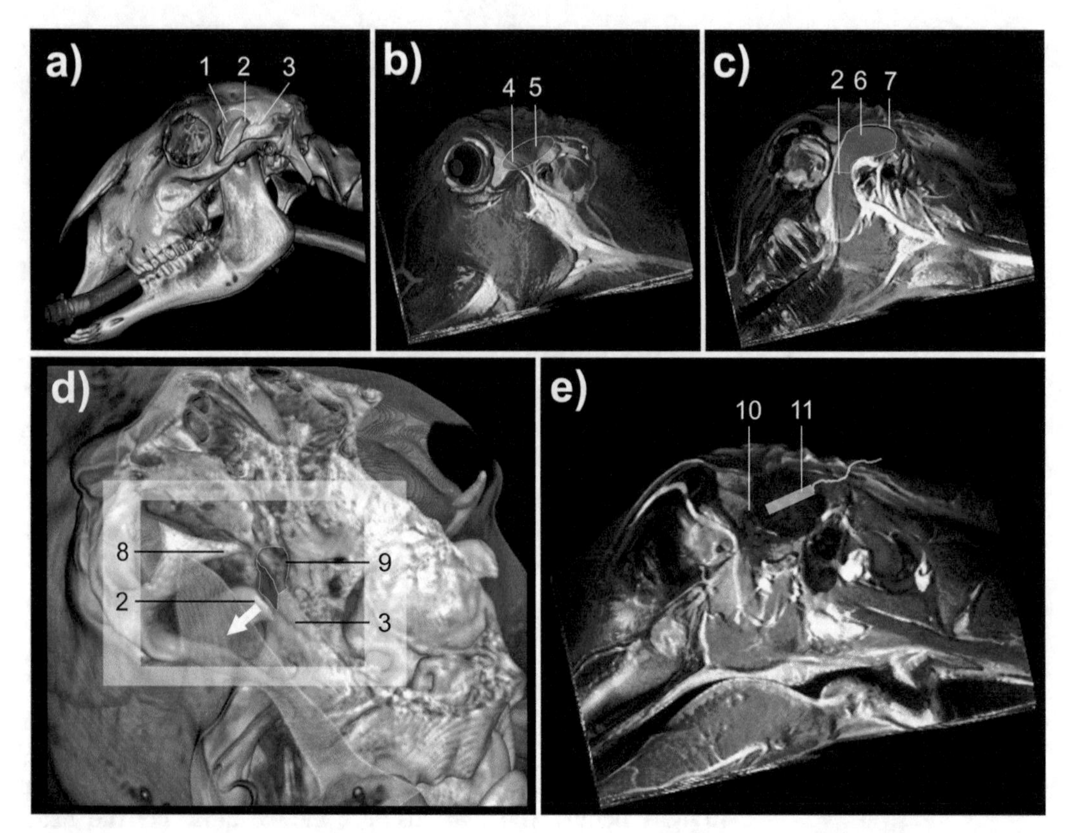

Fig. 4 Detailed scheme of surgical access to the ovine cranium and illustration of relevant steps during MCAO surgery. (**a**) Three-dimensional reconstruction of the skull from a CT data set: (*1*) area for trepanation, (*2*) the coronoid process, and (*3*) bone suture between temporal and parietal skull plate. (**b**) The approximate position of superficial temporal artery is indicated by (*4*). The artery and its accompanying vein are located in the surgical field and need to be cauterized after elliptic skin resection (*5*). (**c**) The coronoid process of the mandibula (*2*), being situated within the temporal muscle (*6*), needs to be elevated with the muscle after cutting the temporal muscle at the temporal line (*7*; *black line*). (**d**) Detailed visualization of the craniotomy: Due to the limited space directly behind the orbital rim (*8*), the coronoid process of the jaw (*2*) needs to be lifted with the temporal muscle (lifting direction is indicated by *white arrow*). The area of trepanation (*9*, indicated by *hemi-transparent overlay*) is limited in the depth by the suture between temporal and parietal skull plates (*3*). (**e**) After extending the trepanation and opening of the dura mater, the MCA can be found in the cranioventral field of trepanation (*10*). Place a brain cotton pad (*11*) on the brain surface to protect it during further manipulation

10. The temporal muscle is fixed to the bone by connective tissue. Remove this connection using a 2 mm Williger raspatory. Then further lift the temporal muscle. Repeat these steps until the surgical field is wide enough for further proceeding. This is the case when the bone suture between temporal and parietal skull plates is clearly visible.
11. Further expose the skull bone surface behind the orbital rim using the Williger retractor. The area of craniotomy is placed directly behind the orbital rim, ventral to the temporal line and dorsal to the bone suture between temporal and parietal skull

plates (*see* Fig. 4). Ensure elevation of the major part of the temporal muscle. This requires removal of connective tissue fixing the muscle also in the caudal part (*see* step 10 of this section).

12. Perform craniotomy with a 6.0 mm Barrel burr at 10,000 rpm in the specified region (*see* Figs. 1 and 4). Hold the burr at an angle of 30°–45° to the bone surface. Use the headlight and magnification glasses for better control and visualization. Avoid any opening the retroorbital space (*see* **Note 13**).
13. Extend the drill hole to all directions using Kerrsion rongeurs. If necessary, remove dura connections to the inner side of the skull bone using the 2 mm Williger raspatory. Make sure not to open the dura at this stage! In particular, extend frontal part of the transcranial approach. Remove all bone fragments (*see* **Note 14**).
14. Lift the dura with the Adson forceps. Carefully open the dura mater with a dorsoventral incision using the microscissors. Lifting is important to avoid accidental damage of underlying cortical vessels (*see* **Note 15**). Carefully widen the dura hole using the scissors. Collect the cerebrospinal fluid using eye swab or surgical pads.
15. Place the neurosorb® patties on the brain surface. Use the back of the blunt 2 mm Williger raspatory in your left hand to slowly apply very gentle pressure to the brain surface (always covered by neurosorb® patties). This may help to visualize the MCAO. DO NOT apply intense pressure and DO NOT move rapidly. Keep the Williger raspatory in place manually until step 18.
16. Collect cerebrospinal fluid using eye swab or surgical pads immediately before electrosurgical occluding the MCA. Use the nonadherent bayonet-shaped neurosurgical bipolar forceps and apply <50 W. *See* **Note 16** for MCAO.
17. You may occlude the proximal MCA branch for a large territorial stroke or distal MCA branches for smaller stroke lesions. For details, please refer to [14].
18. Cover the brain by repositioning of the dura mater.
19. It is possible to suture the dura (use 6-0 surgical threats) and/or close the bone defect with bone cement. However, this will result in dangerous intra cranial pressure (ICP) peaks due to the concomitant brain edema after larger strokes in the subacute phase following MCAO. Thus, just reposition the dura in case of a large territorial stroke. Step 20 is sufficient to prevent any damage to the area.
20. Relocate the temporal muscle, now covering the drill hole, and fix it to connective tissue that was left at the muscle insertion at the temporal line. Use 2-0 resorbable filament. A Kirschner suture with 2-0 filament should be performed for readapting of the subcutis, followed by a Reverdin suture to close the skin wound.

3.3.2 Postsurgical Care

1. Perform antibiotic and analgesic treatment by intramuscular injection of enrofloxacin and flunixin-meglumine before surgery and for at least 5 days (optimum: 7 days) postsurgical (*see* Table 3).
2. Intramuscular injections of buprenorphine (*see* Table 3) are performed for additional analgesic treatment for 48 h following MCAO. Repeat injection every 8 h (*see* **Note 17**).
3. Perform wound treatment with disinfection wound spray followed by covering with silver or aluminum wound spray.
4. Protect venous lines by covering with cotton, followed by a bandage. The bandage must then be covered with isolator tape to avoid damage induced by the subject or flock mates.

3.4 Imaging Procedures

3.4.1 General Procedure

1. Wrap the animal in folio drape to protect it from cooling down. The drape also prevents soiling of the scanner.
2. Place the subject in a prone position upon the scanner table with the head resting on the neck crest. The nose faces into the scanner bore (*see* Fig. 5).
3. Fix the head and the body of the animal to the table using adhesive tape to avoid any movement artifacts. Be sure not to hinder breathing movements.

3.4.2 MR Imaging

MRI is used to monitor the impact of MCAO, to visualize the lesion and to control its development by repeated measurement during a longer observation period. All means of modern MR imaging [20, 21] can be applied to the sheep model (*see* Table 4). While a 1.5 T is sufficient to monitor the lesion with basic parameters like T2 TSE, DWI, perfusion weighted imaging (PWI), or time of flight (TOF)-MRA, application of 3.0 T MRI provides further options including diffusion tensor imaging (DTI) for fiber track reconstruction [22].

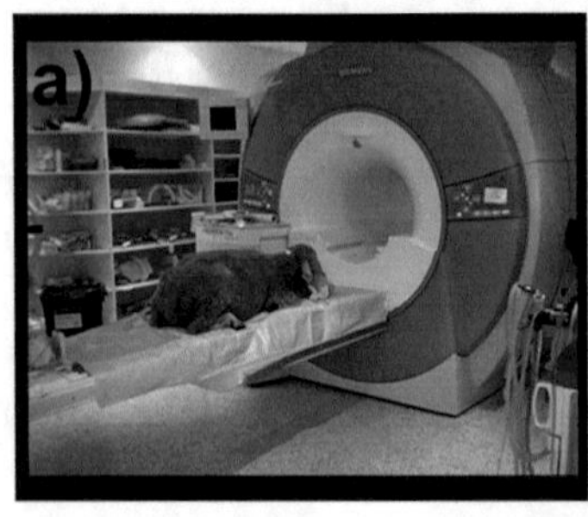

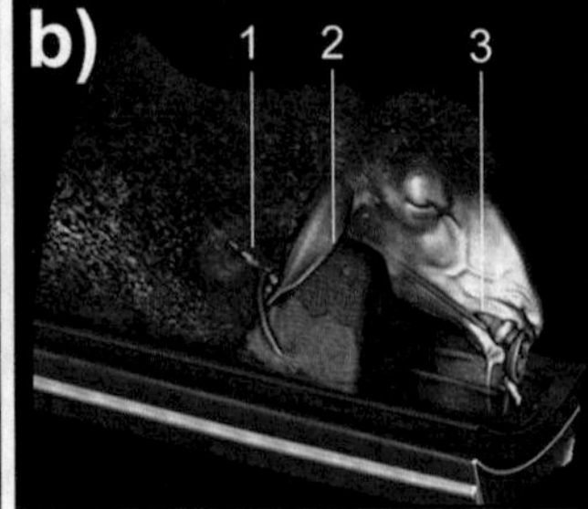

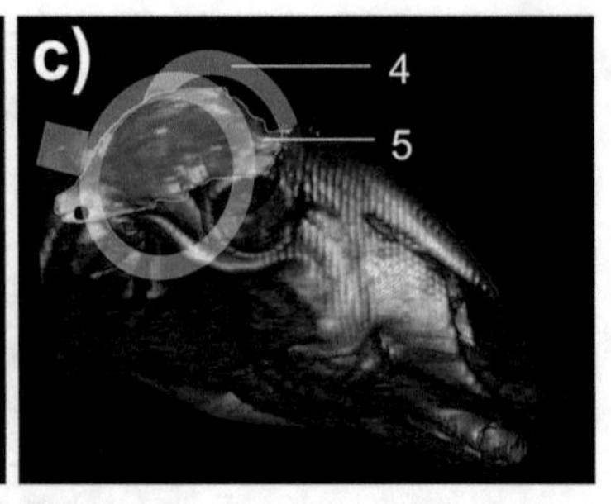

Fig. 5 Animal preparation and positioning for MRI session. (**a**) Animal on table of a clinical scanner in prone position with head facing into the scanner: The animal is not yet covered with folio drape. (**b**) Three-dimensional CT dataset reconstruction of animal before imaging: (*1*) Intravenous perfusion as described in **Note 8**. The use of a (*2*) neck-rest fixates the head. To prevent movement artifacts during imaging, the head should further be fixed by friction tape. Intubation (*3*) is recommended during the scanning procedure. (**c**) Illustration of SenseFlexM coil placement during MRI procedure: (*4*) Scheme of head coil SenseFlexM placement in a CT-MRI reconstruction corresponding to the position of the brain (*5*)

Table 4
Recommended imaging protocol for 3 T MRI Scanner

Sequence	Parameters
T2 TSE	Voxel size: 0.5 × 0.4 × 2.0 mm; slices: 50; TR: 6000; TE: 105; acquisition time: 8:26 min
T2*	Voxel size: 0.8 × 0.7 × 3.0 mm; slices: 35; TR: 700; TE: 20; acquisition time: 5:24 min
TOF MRA	Voxel size: 0.5 × 0.3 × 0.5 mm; slices: 40; TR: 24; TE: 4.43; acquisition time: 13:50 min
DTI	Voxel size: 1.9 × 1.9 × 1.9 mm; slices: 70; TR: 10600; TE: 100; acquisition time: 12:02 min; diffusion directions: 64
PWI	Voxel size: 1.6 × 1.6 × 5 mm; slices: 13; TR: 1450; TE: 45; measurements: 50; acquisition time: 1:18 min
T1 MPRAGE	Voxel size: 0.6 × 0.6 × 1.0 mm; slices: 160; TR: 1900; TE: 2.83; acquisition time: 7:58 min

Planning the sequences includes three short T2 HASTE sequences, performed in addition to the scout imaging
TSE turbo spin echo, *TOF MRA* time of flight magnetic resonance angiography, *DTI* diffusion tensor imaging, *PWI* perfusion weighted imaging, *MPRAGE* magnetization prepared rapid gradient echo, *TR* time of repetition, *TE* time of echo

Preparation for MRI Acquisition

1. Fix a MRI-positive marker laterally at the head to ensure adequate spatial orientation after imaging (*see* **Note 18**).
2. Put the coil around the head and fix it using adhesive tape (*see* Fig. 5 for positioning).
3. Focus the scanner to the sheep brain.

Imaging Process

1. Imaging starts with a scout image plus three additional T2 Half-Fourier Acquisition Single-shot Turbo Spin Echo (HASTE) sequences in three different planes, which facilitate exact planning of the scans. Always plan scans in orthogonal direction (corresponds to the axis of the brain and brain stem).
2. For basic monitoring, perform scanning sequences as given in Table 4 (*see* **Note 19**).
3. Proceed with MRA to identify the occlusion of the MCA (postsurgical control, compare to Fig. 3).

3.4.3 PET Imaging

General Remarks

1. Measure tissue attenuation using three rotating ^{68}Ge rod sources previous to tracer injection (transmission scan, scan time: 10 min).
2. Place a NeuroShield® in the neck region of the sheep for brain imaging to minimize scatter radiation.
3. Extend the venous line to the animal with additional tubing. Pay attention for the tubing to be placed outside of the field of view during tracer administration.
4. Use [^{15}O]H_2O for CBF measurement. Inject a dose of ~1000 MBq per sheep and scan. The injection system (*see* Sect. 2.4.3) realizes tracer administration automatically (over ~30 s).

Table 5
Exemplary blood sampling protocols for PET analysis

$[^{15}O]H_2O$						
Sample no.	1	2	3	4	5	6
Time p.i. (min)	0:00–2:00[a]	2:50	3:00	3:50	4:00	5:00
$[^{18}F]FDG$						
Sample no.	1	2	3	4	5	6
Time p.i. (min)	0:00–2:45[a]	3:00	4:00	5:00	7:00	10:00
Sample no.	7	8	9	10	11	12
Time p.i. (min)	15:00	20:00	30:00	40:00	50:00	60:00

no. number, *p.i.* post injection
[a]Time period of continuous blood sampling

5. Use $[^{18}F]FDG$ to measure cerebral metabolic rate for glucose (CMR_{Glu}). Inject a dose of ~370 MBq per sheep and scan. After i.v.-injection (over approximately 90 s), the perfusion line is washed with 0.9% sterile sodium chloride solution.
6. Link every PET scan with arterial blood sampling (*see* Table 5), which is needed for further kinetic modeling, especially if absolute CBF quantification is desired. If conducted manually, arterial blood samples should be taken in meaningful intervals with regard to individual tracer metabolism (*see* **Note 20**).

Basic Information on PET Data Processing

1. Perform dynamic emission scans according to the following parameters: axial field of view: 155 mm; number of parallel transverse slices: 63; slice thickness: 2.4 mm; image resolution: 7.1 mm (transverse), 6.7 mm (axial); matrix: 128 × 128; acquisition mode: 3D; acquisition time: 60:00 min for $[^{18}F]FDG$ and 5:00 min for $[^{15}O]H_2O$.
2. PET data obtained must initially undergo standard correction for radioactive decay, death time, scatter, and attenuation. Images are finally reconstructed by means of iterative Ordered Subset Expectation Maximisation (OSEM) algorithm (for instance 10 iterations, 16 subsets) (*see* **Note 21**).

3.4.4 Imaging Data Interpretation and Analysis

Relevant examples of frequently used imaging data analysis are given in this section.

MR-Based Analyses of Hemispherical Atrophy

A territorial infarct results in cerebral edema. Therefore, calculation of the lesion should always be performed corresponding to the remaining brain tissue [23]. Numerous software-based methods have been developed for automated or semiautomated calculation. An approach using the open source software ImageJ is feasible for many researchers, and can be used even in case professional (and expansive) MR image processing and analyzing software is not available (for alternatives *see* **Note 22**). The work flow is as follows:

1. Open all images of a certain MRI sequence (e.g., T2 TSE) using stack image.
2. Normalize gray-scale values by automatic histogram function.
3. Use threshold function to separate different regions step-by-step. (1) Left and right hemispheres including ventricles and infarction, (2) the infarct area, and (3) left and right ventricle. Use always a native stack for each structure (e.g., left ventricle). Save all resulting stacks separately in binary file-mode and name them for a randomized code (*see* **Note 23**).
4. Use the "wand-tracing-tools" to select the separated structure in the binary file and transfer it to the region of interest (ROI)-manager (*see* **Note 24**).
5. Measure ROI size and calculate the volume of a structure by the formula:

$$V_{\text{part},i} = \frac{A_{i-1} + A_i + \sqrt{A_{i-1} \times A_i}}{3} \times d_{i-1,i} \text{ and } V_{\text{total}} = \sum_{i=1}^{n} V_{\text{part},i}$$

(n—number of partial volume; $V_{\text{part},i}$—partial volume; V_{total}—total volume; Ai—Area of specific ROI measurement; di_{-1},i—distance between two slices)

6. Calculate hemispherical atrophy using the following formula:

Hemispherical atrophy $= (V_{\text{left hemisphere}} - V_{\text{infarction}} - V_{\text{left ventricle}})/(V_{\text{right hemisphere}} - V_{\text{right ventricle}})$

(V—*Volume*)

Postprocessing of DTI Sequences

A fast and practicable way for Fractional Anisotropy (FA) and Apparent Diffusion Coefficient (ADC) analysis with DTI studio (Version 3.0.2) is as follows (*see* Fig. 6):

1. Open DTI Studio and select "File" > "DTI Mapping."
2. Specify your data according to your scanner (i.e., Siemens Mosaic) > "Continue."
3. Check slice orientation, slice sequencing, slices to be processed (we use "all slices").
4. Specify the *b*-values according to your data (important for correct ADC calculation).
5. Select the folder with your DICOM data by choosing "Add a Fold."
6. Click "Get gradient from DICOM file header" followed by "OK."

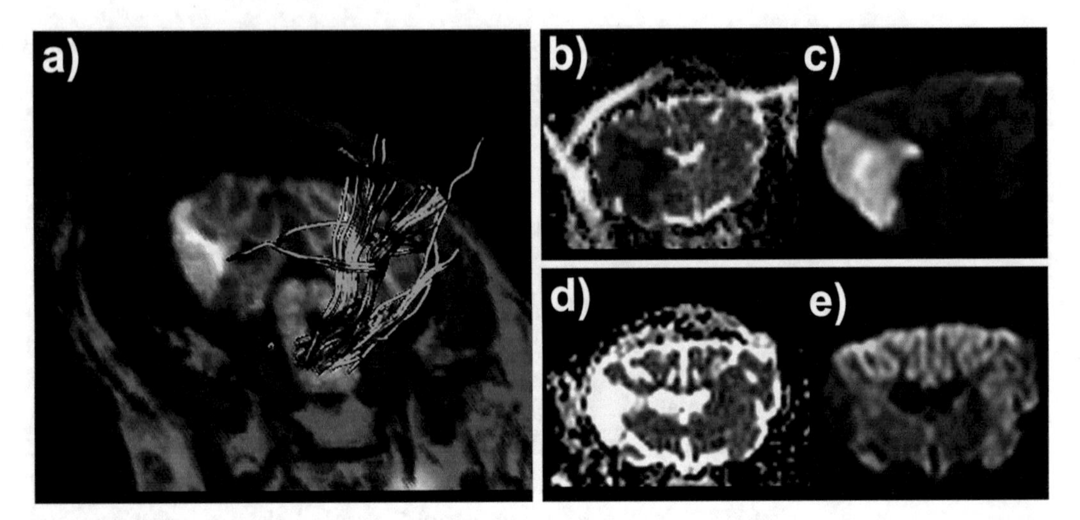

Fig. 6 MRI of the ovine brain following stroke. Images (**a**–**c**) were obtained 24 h after MCAO, whereas (**d** and **e**) represent scans conducted 6 weeks following experimental stroke. All images were obtained at 3 T MR. (**a**) Fiber tracking after acute MCAO in sheep: Fiber reconstruction based on Diffusion Tensor Imaging (DTI) fused with an anatomical 3D T1 and Diffusion Weighted Imaging (DWI). High signal intensities in DTI show a higher anisotropy of diffusion, indicating a higher density and homogeneity of fiber tracts. DTI signal loss in the region of a high DWI signal occurs due to post ischemic tissue destruction. (**b**) Apparent diffusion coefficient (ADC) map of DWI in acute stage of the ovine infarction: The typical dark signal indicating impaired diffusion can be seen at the site of the infarction. (**c**) DWI of acute MCAO: The circumscribed, increased signal intensity at the altered hemisphere indicates reduced diffusion. (**d**) ADC map of a DTI sequence in chronic stage of stroke: Higher signal intensity indicates a higher diffusivity. The tissue defect in the impaired hemisphere is filled with cerebrospinal fluid. Thus, the signal is equal to the signal within the ventricles. (**e**) DWI of a chronic infarct: Compare the signal characteristics with (**c**) to note the differences of the imaging signs at both stages

7. In the next window click on the “DTiMap”-tab and generate maps of FA (Tensor, Color Map, etc.) and ADC (ADC-Map) using the defaults.
8. Change to the tab “Image” and choose the map to be analyzed.
9. Change to the tab “ROI” and draw the ROI in the regions to be analyzed.

Perform tractography (*see* Fig. 6) using Siemens Syngo MR B15 (Vendor Specific) as follows:

1. Load DTI data and 3D MPRAGE into “Neuro 3D” application.
2. Click “fusion” followed by double click on “3D.”
3. Draw a ROI into the region where fiber tracking starts by simultaneously pressing “Control + left mouse button.”
4. Select the ROI and press the right mouse button. Choose “Start Tractography.”

PET-Based Analysis of Brain Perfusion

[^{15}O]H_2O PET is the gold standard method for determining CBF. In case of ischemic stroke, follow-up examinations allow for the determination of different infarct stages (*see* Fig. 7). It is also of use to investigate acute infarct evolution. Because of its short half-life time (approximately 2 min for ^{15}O), PET imaging with [^{15}O]H_2O can be performed in a serial scanning mode. Note that follow-up PET should not be started earlier than 20 min (10 half-life times of the tracer) after the previous tracer application. Data processing is performed as follows:

1. For a semiquanitative approach, sum up image frames with early cerebral tracer uptake up to a total time of 1 min (start with frame of first observable tracer activity in brain tissue). The interpretation of these data should be preferentially based on standardized uptake values (SUVs). Normalize images for injected activity (ID) and bodyweight as follows: SUV = ROI activity [Bq mL^{-1}] × bodyweight [g] × injected dose (ID) [Bq]$^{-1}$.
2. To obtain absolute CBF values (units: mL min^{-1} 100 g^{-1}), use dynamic PET data in conjunction with individually derived arterial input function (*see* **Note 25**) and quantify CBF by applying the method described by Alpert et al. [24] (*see* **Note 26**).
3. As most therapeutically interventions aim to preserve and/or rescue the ischemic penumbra, the visualization of that "tissue at risk" is of special interest. CBF quantification offers the opportunity to separate stroke-related tissue regions by applying established CBF thresholds [25] <8 mL 100 g^{-1} min^{-1} for infarction core, 8–22 mL 100 g^{-1} min^{-1} for the ischemic penumbra and >22 mL 100 g^{-1} min^{-1} for normal brain tissue (*see* Fig. 7, *see* **Note 27**).

PET-Based Assessment of Cerebral Glucose Consumption

Especially for long-term studies of brain vitality, indicated by glucose metabolization, it is useful to perform [^{18}F]FDG-PET imaging. Data processing is performed as follows:

1. Create a summed image from the late frames of the PET scan (frames should cover the last 30 min of acquisition time), which can be used for semiquantitative analysis approaches.
2. Perform the SUV quantification for subsequent qualitative and quantitative assessment of glucose consumption: This method is easy to implement in ovine FDG-PET image analysis without the need of arterial blood sampling.
3. For that purpose, convert the prepared single-frame image (the sum-image from late frames) to SUV data by normalizing data for bodyweight and injected activity (see semiquantitative [^{15}O]H_2O-PET analysis in Sect. "PET-Based Analysis of Brain Perfusion").

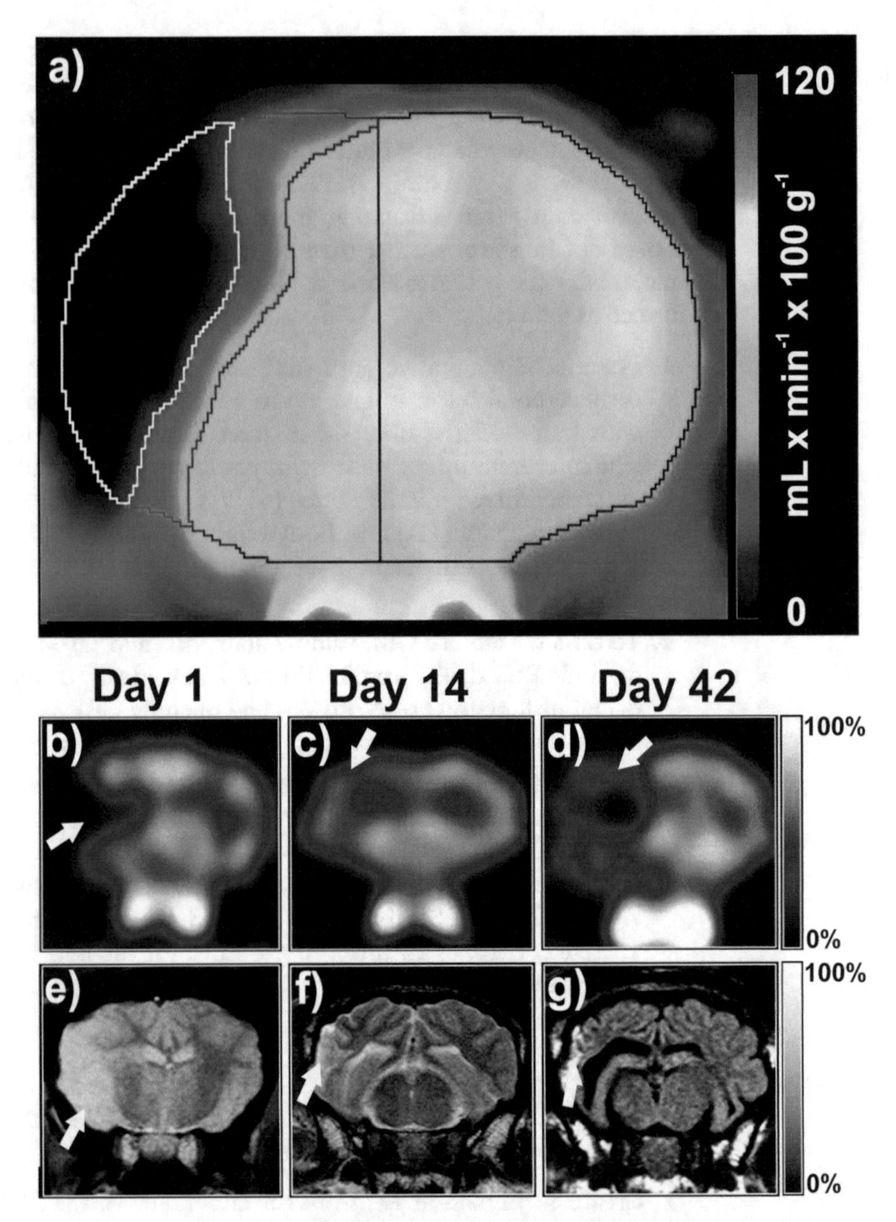

Fig. 7 [^{15}O]H_2O PET of the ovine brain after MCAO. (**a**) Parametric CBF map, derived from [^{15}O]H_2O PET 2 h after MCAO. Left hemispheric infarct resulted in different perfusion-based stroke compartments: infarct core (*white line*, CBF <8 mL min^{-1} 100 g^{-1}), ischemic penumbra (8–22 mL min^{-1} 100 g^{-1}), and normal brain tissue (>22 mL min^{-1} 100 g^{-1}). Outer delimination of brain tissue was done by means of individually superimposed MRI images (not shown). (**b–g**) show results from a long-term ovine stroke study including follow-up [^{15}O] H_2O-PET imaging. Representative coronal brain slices of [^{15}O]H_2O PET at day 1 (**b**), 14 (**c**), and 42 (**d**) correlated well with findings obtained with MR imaging using T2 TSE at day 1 (**e**) and 14 (**f**), and FLAIR on day 42 (**g**), respectively. The *white arrows* mark the area of perfusion deficit in (**b–d**) and corresponding MR findings in (**e–g**). The value bar of PET images represent 0 (*bottom*) to 120 (*top*) mL min^{-1} 100 g^{-1}

4. Outline the stroke-affected volume of interest (VOI) with decreased glucose consumption and low SUV units with PMOD software, using automated threshold function. Take the unaffected hemisphere as a reference region or mirror the selected VOI to the contralateral side. With a calculation of SUV ratios ($SUVR = SUV_{stroke\ VOI}/SUV_{reference\ VOI}$), regional deficits of CMR_{Glu} can be quantified (*see* **Note 28**).

3.5 Sacrifice and Post Mortem Analyses

1. Sacrifice subjects in deep anesthesia (*see* Sect. 3.2) by intravenous injection of pentobarbital (*see* Table 3). Control the absence of cardiac action, breathing, and reflexes over a period of at least 2 min.
2. Turn the animal on the back and decapitate at the atlanto-occipital junction using a sharp and robust knife. Dispose the trunk (or use it for any kind of other investigation) and proceed with head preparation.
3. Carefully expose both external carotid arteries, place steel cannulas in both carotids and fix them with stout thread.
4. Perfuse the head manually via cannulas with 3 L PBS using 100 or 500 mL syringes (*see* **Note 29**). In the case vitality staining is requested *see* **Note 30**.
5. Connect the roller pump system to the steel cannulas and perfuse the head with 20 L 4% PFA (rate: 15 mL min^{-1}).
6. Remove the skin and muscles from the cranial roof and the neck using a sharp knife.
7. Remove the upper part of the skull with an oscillating saw. Avoid damaging the dura mater and the brain.
8. Incise the dura mater and carefully expose the brain.
9. Perform immersion fixation of the entire head with 4% PFA for at least 48 h.
10. Carefully remove the brain from the cranial cavity and perform further immersion fixation in 4% PFA for 3 additional days.
11. Cut the brain consecutively into coronal slices of 4 mm thickness. Photograph and number the specimens together with a scale. Measure ventricle, tissue, and infarct of all photographed specimens (*see* **Note 31**).
12. Compartmentalize the specimens for further embedding procedure (*see* Fig. 8). Paraffin embedding after 18 h dehydration of the specimens is recommended for conventional histology, as well as for immunohistochemistry using antibodies for paraffin embedded material (e.g., rabbit-anti-GFAP, code: Z0344, Dako Cytomation AG, Germany, or mouse-anti-NSE, code: M 0873, Dako Cytomation AG) (*see* **Note 32**). Specimens are stored in 30% sucrose solutions until further processing. Cryoprotection is recommended when labeling with fluorescence-based techniques will be performed.

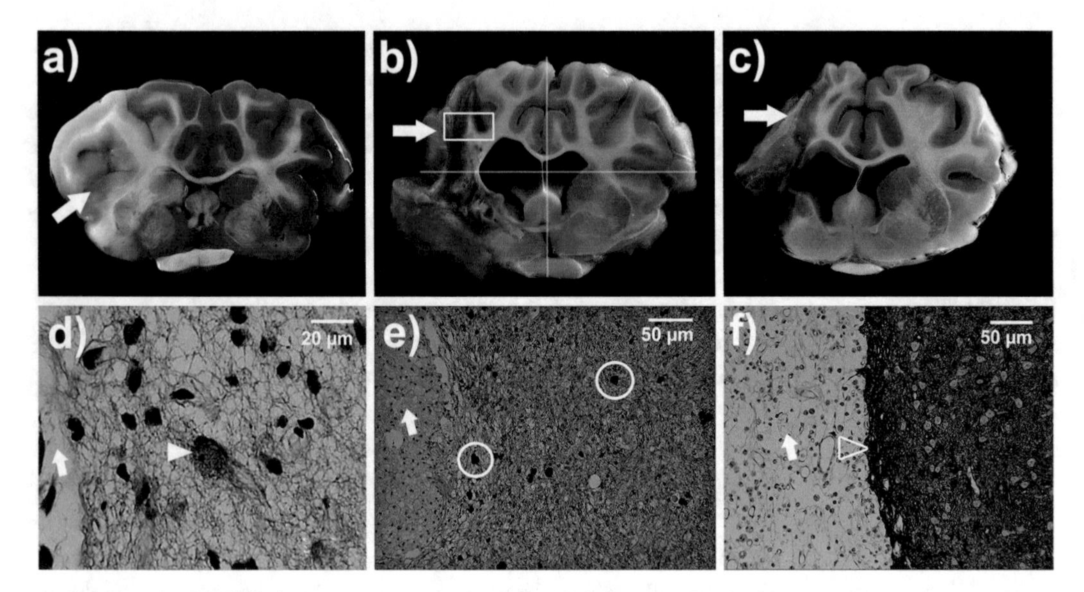

Fig. 8 Gross brain pathology and examples for histological evaluations. Pathologic-anatomical (**a–c**) and histological findings (**d–f**) following MCAO-caused infarction (*white arrows*) in sheep: (**a**) Ischemic brain tissue 6 h after MCAO: The infarct area in the coronal brain slice (at the level of the optic chiasm) is indicated by absent TTC staining. (**b**) A sharp demarcation of the infarct area can be seen macroscopically 6 weeks after MCAO. Atrophy in the impaired hemisphere is clearly associated with an enlargement of the lateral ventricle. For further analysis, a compartmentalization of specimen (*gray lines*) is recommended. (**c**) Macroscopic findings were confirmed 26 weeks after MCAO. Histological findings in the area next to the infarct (*box insert* in **b**) can be described as follows: (**d**) 6 week after MCAO, Nissl staining reveals alterations of neurons (*white arrow head*) and neuropil. (**e**) Labeling with an antibody against neuron specific enolase (monoclonal, mouse-anti-NSE) shows axonal alterations (*white circle*) in the white matter next to the impaired area (*white arrows* in **d–f**). (**f**) Astrogliosis (*white edged arrow head*) is indicated by increased glial fibrillary acid protein (polyclonal, rabbit-anti-GFAP) in the cortex next to the infarct

4 Notes

4.1 Introduction

1. Recently, a transient MCAO model using sheep was developed by an Australian group [26]. This model uses an aneurysm clip to induce a 2-h MCA blockage resulting in lesion extension and morphology being comparable to those seen in the described approach. The model was also reported to be highly reproducible and is also expected to be associated with a low incidence of intraoperative complications.
2. In addition to conventional surgical techniques, stereotaxic interventions become popular. The sheep model can also be adapted to hemorrhage induction by application of autologous blood using the Brainsight™ stereonavigation and stereotaxic system (Rogue Research Inc., Canada). The system can further be used to administer therapeutic compounds or cells stereotaxically [27].

4.2 Topography

3. Functional organization of the neocortex is described elsewhere [28–30] (Please also *see* Fig. 2).

4.3 Anesthesia

4. Considering the venous drainage system an interspecies comparison between sheep, dogs, and rats suggests that the cerebral venous angioarchitecture in large animals is better comparable with the human anatomy although substantial differences remain [31].
5. Detomidine (0.02–0.06 mg kg^{-1} i.v., Domosedan®, Orion Corporation, Finland) can be used instead of xylazine. Note the high receptor affinity of alpha-2-agonists for small ruminants.
6. Choose a 16G butterfly cannula (Braun Melsungen AG) to access the V. saphena lateralis distal to the tarsus in case an additional venous line is needed, especially during surgery.
7. Never apply carrier gas (such as nitrous oxide) for any reason as the gas will accumulate in the rumen.
8. Alternatively, intravenous infusion during MR imaging can be performed in case a MR-compatible respirator with isoflurane vapor is not available. For this, fill one 50 mL syringe (Original-Perfusor® syringe, Braun Melsungen AG) with propofol. Prepare another one with 3 mL 10% ketamine, 3 mL midazolam ad 30 mL 0.9% sterile sodium (*see* Table 3). Place the syringes in the perfusion pumps (e.g., Perfusomat® compact S, Braun Melsungen AG) and connect the demanded length of perfusion lines (Braun Melsungen AG) with the 3-way-stopcock (Discofix®, Braun Melsungen AG) upon the vein cannula. Apply the propofol at 0.2 mL h^{-1} kg^{-1} BW and the ketamine-midazolam mix at 0.26 mL h^{-1} kg^{-1} BW. If necessary, first increase infusion speed of the ketamine-midazolam mix stepwise, then that of the propofol. Ensure continuous monitoring of adequate breathing!

4.4 Surgery

9. The elliptic wound helps to avoid post surgical complications. Readapting the skin will result in a tight suture that prevents seroma formation.
10. The tissue which covers the region of the transcranial access is supplied by the Nervus auriculopalpebralis (originates from the facial nerve). The nerve can be cut.
11. Dissect carefully! The artery and veins supplying the temporal muscle should not be damaged during the procedure.
12. Dissect carefully! The coronoid process of the jaw (*see* Fig. 4) must not be exposed from the covering muscle.
13. Avoid incision of the retroorbital space, as a massive bleeding can result from the venous sinuses situated in the retroorbital space.

14. The thickness of the dorsocaudal part of the Os parietale is only 2–3 mm while the ventrofrontal part is about 5–10 mm (Fig. 4). First, carefully ablate the bone stepwise in the area of craniotomy by circling the burr without applying much pressure. Use a 4 mm Rosen burr to widen the area. Finally, only a very thin bone lamina is remaining. Now open the cavity in the middle of the ablation by using forceps or by very gentle drilling.

In the case bone bleeding occurs: Stop the bleeding using bone wax before opening the dura.

15. Due to the relative small space for the surgical access, Adson forceps are sometimes not feasible for dura lifting (not enough space to pick the dura). In such cases, the dura can be lifted alternatively by a dura hook. The ovine dura has multiple "layers," so the hook can be placed safely in an upper layer for lifting the entire dura without uncontrolled incision and/or alterations of the underlying brain tissue and vessels (use sharp and angled Fisch dura retractor, length 185 mm, catalog-no: FD376R, Braun Melsungen AG for that purpose, see http://www.surgical-instruments.info/en/products.html). Place the hook at the dura in an angle of <10° in relation to the dural surface. After penetrating of an upper dura layer with the retractor tip, lift the hook about 5 mm and do not move it anymore! Cut the dura in the described manner. Make sure to lift the flap all the time while cutting the dura.
16. In a few cases of own experiments, a noticeable variation of MCA architecture (doubling at the origin of the CW, preferentially in female subjects) necessitated occlusion of both MCA branches to induce a stroke.
17. After occlusion of the MCA main branch, a larger stroke and more severe brain edema may occur in the first days following stroke. An open burr hole reduces the intracranial pressure, therefore, lethal herniation is a very rare complication. In the case of a reduced state of activity and depressed motor functions (usually to be seen between day 2 and 5 post MCAO), provide stimulation of ruminal/gastrical digestion with proprionic acid (oral), butafosfan (intramuscular), menbuton (intramuscular), and Amynin® (intravenous infusion) (*see* Table 3).

4.5 MR Imaging

18. The use of small capsules of glycerol nitrate is recommended for this purpose.
19. Due to the size of the sheep brain, different sequences should be used to describe the infarct status or other brain alterations (e.g., DWI, PWI, T2 TSE) [32].

4.6 PET Imaging

20. Alternatively, an automated sampling device can be used. Latter equipment (blood sampler) is highly commended for

short acquisition times, as in the case of [^{15}O]H_2O-PET (scan time: 5 min).

In contrast to small animal models, the ovine blood volume is similar to that of humans (approximately 60–70 mL kg^{-1} bodyweight), allowing blood sampling (*see* Table 5) without affecting circulation.

21. Other methods, like filtered backprojection algorithms result in lower spatial resolution and should therefore be replaced by iterative reconstruction algorithms.

4.7 Imaging Data Interpretation and Analysis

22. Alternatively, freeware (OsiriX, Pixmeo, Geneva, Switzerland) can be used on another computer. The scans have to be exported from the scanner either via burning them to a CD-ROM or sending them to a picture archiving system, depending on local configurations [33].
23. Using the OpenSource software ImageJ offers several plugins for MRI analyses (e.g., Quickvol 2) [34]. Alternatively, OsiriX can be used.
24. ImageJ names the created ROI automatically. The ROI name identifies region of interest in a specific slice. Manually changes of the ROI name disturb the defined allocation!
25. Various methods for CBF quantification not requiring arterial blood sampling have been published. To avoid the need for arterial input function data, most alternative methods determine several parameters for kinetic modeling, such as the partition coefficient (V_d) [35]. However, due to the existence of heterogenic infarct compartments (V_d differs relevantly in infarct core, penumbra and tissue of benign oligemia), these alternative approaches may not give reliable results in acute stroke imaging.
26. Use the PMOD tool "PKIN" for that purpose. The software corrects individual arterial input functions for delay and dispersion and creates parametric image data for CBF (voxel values = mL 100 g^{-1} min^{-1}) and distribution volume (V_d; [mL g^{-1}]). These three-dimensional parametric images can be processed further (VOI based, voxel based, or other approaches).
27. Use automated threshold function of PMOD for an operator-independent VOI determination (*see* Fig. 7).
28. As FDG is intensively taken up into the brain, local deficits show a high target-to-background image contrast. Considering the fact of increased glucose consumption is a typical indicator of inflammatory processes as well as the knowledge about post stroke luxury perfusion and inflammation, it is recommended to interpret FDG-PET images from acute infarct stages with special care. For long-term stroke studies, FDG-PET offers reliable data on neuronal integrity, i.e., brain tissue viability.

4.8 Post Mortem Specimen Processing

29. Perfusion should be done simultaneously via both cannulas.
30. Immediate vitality staining using triphenyltetrazolium chloride (TTC) of acute cerebral infarction can be performed alternatively by incubation of 4 mm brain specimens in 1% TTC (diluted in PBS) at 37 °C for 1.5 h (*see* Fig. 8). This requires removal of the brain immediately after perfusion. Do not apply an additional fixation period. Remove the brain very carefully!
31. Gross pathology: Acute infarctions (up to 10 h following MCAO) can hardly be discriminated from the surrounding tissue without any staining procedures. Therefore, TTC staining can be used to visualize ischemic altered gray and white matter (*see* Fig. 8).
32. Histological findings 6 weeks after MCAO are comparable to those of human [36] and nonhuman primate species [37]. Common histological staining procedures as well as immunohistochemistry and fluorescent techniques using a portfolio of antibodies against neurons (e.g., NSE), astrocytes (e.g., GFAP), microglia (e.g., CD11b, IBA1), vessels (e.g., Glut-1), and collagen can be performed (*see* Fig. 8).

Acknowledgments

The authors want to thank Dr. Karl-Titus Hoffmann, professor of neuroradiology, and Dr. Osama Sabri, professor of nuclear medicine, for the allowance to use the scanners in their departments at Leipzig University. The authors are further grateful to Dr. Uwe Gille, Dr. Johannes Seeger, and Dr. Heinz-Adolf Schoon, professors at the Faculty for Veterinary Medicine at Leipzig University.

References

1. O'Donnell MJ, Xavier D, Liu L et al (2010) Risk factors for ischaemic and intracerebral haemorrhagic stroke in 22 countries (the INTERSTROKE study): a case-control study. Lancet 376:112–123
2. Hacke W, Kaste M, Bluhmki E et al (2008) Thrombolysis with alteplase 3 to 4.5 hours after acute ischemic stroke. N Engl J Med 359:1317–1329
3. Wahlgren N, Ahmed N, Davalos A et al (2008) Thrombolysis with alteplase 3–4.5 h after acute ischaemic stroke (SITS-ISTR): an observational study. Lancet 372:1303–1309
4. Hachinski V, Donnan GA, Gorelick PB et al (2010) Stroke: working toward a prioritized world agenda. Cerebrovasc Dis 30:127–147
5. Traystman RJ (2003) Animal models of focal and global cerebral ischemia. ILAR J 44: 85–95
6. STAIR-group (1999) Recommendations for standards regarding preclinical neuroprotective and restorative drug development. Stroke 30:2752–2758
7. Savitz SI, Chopp M, Deans R et al (2011) Stem cell therapy as an emerging paradigm for stroke (STEPS) II. Stroke 42:825–829
8. Amiridze N, Gullapalli R, Hoffman G et al (2009) Experimental model of brainstem stroke in rabbits via endovascular occlusion of the basilar artery. J Stroke Cerebrovasc Dis 18:281–287
9. Kang BT, Lee JH, Jung DI et al (2007) Canine model of ischemic stroke with permanent middle cerebral artery occlusion: clinical and histopathological findings. J Vet Sci 8:369–376
10. Garcia JH, Kalimo H, Kamijyo Y et al (1977) Cellular events during partial cerebral ischemia. I. Electron microscopy of feline cerebral cortex

after middle-cerebral-artery occlusion. Virchows Arch B Cell Pathol 25:191–206
11. Imai H, Konno K, Nakamura M et al (2006) A new model of focal cerebral ischemia in the miniature pig. J Neurosurg 104:123–132
12. Marshall JW, Ridley RM (2003) Assessment of cognitive and motor deficits in a marmoset model of stroke. ILAR J 44:153–160
13. Shuaib A, Lees KR, Lyden P et al (2007) NXY-059 for the treatment of acute ischemic stroke. N Engl J Med 357:562–571
14. Boltze J, Forschler A, Nitzsche B et al (2008) Permanent middle cerebral artery occlusion in sheep: a novel large animal model of focal cerebral ischemia. J Cereb Blood Flow Metab 28:1951–1964
15. Behrens H, Ganter M, Hiepe T (2009) [Textbook of ovine diseases]. Parey, Stuttgart, Germany, Chapter in German
16. Nitzsche B, Frey S, Collins L et al (2015) A stereotaxic, population-averaged T1w ovine brain atlas including cerebral morphology and tissue volumes. Front Neuroanat 9:69. doi:10.3389/fnana.2015.00069
17. Salomon FV, Geyer H, Gille U (2008) [Textbook of anatomy for veterinary medicine]. Enke Verlag, Leipzig, Germany, Chapter in German
18. Forschler A, Boltze J, Waldmin D et al (2007) [MRI of experimental focal cerebral ischemia in sheep]. Rofo 179:516–524, Article in German
19. Ashwini CA, Shubha A, Jayanthi KS (2008) Comparative anatomy of the circle of Willis in man, cow, sheep, goat, and pig. Neuroanatomy 2008:54–65
20. Muir KW, Santosh C (2005) Imaging of acute stroke and transient ischaemic attack. J Neurol Neurosurg Psychiatry 76:19–28
21. Wechsler LR (2011) Imaging evaluation of acute ischemic stroke. Stroke 42:12–15
22. Nucifora PG, Verma R, Lee SK et al (2007) Diffusion-tensor MR imaging and tractography: exploring brain microstructure and connectivity. Radiology 245:367–384
23. Swanson RA, Morton MT, Tsao-Wu G et al (1990) A semiautomated method for measuring brain infarct volume. J Cereb Blood Flow Metab 10:290–293
24. Alpert NM, Eriksson L, Chang JY et al (1984) Strategy for the measurement of regional cerebral blood flow using short-lived tracers and emission tomography. J Cereb Blood Flow Metab 4:28–34
25. Baron JC (2001) Perfusion thresholds in human cerebral ischemia: historical perspective and therapeutic implications. Cerebrovasc Dis 1:2–8
26. Dreyer A, Stroh A, Nitzsche B et al (2012) Frameless stereotaxy in sheep—neurosurgical and imaging techniques for translational stroke research. INTECH Open Access Publisher, London
27. Wells AJ, Vink R, Blumbergs PC, Brophy BP, Helps SC, Knox SJ, Turner RJ (2012) A surgical model of permanent and transient middle cerebral artery stroke in the sheep. PLoS One 7(7):e42157
28. Cooley KR, Vanderwolf CH (2004) The sheep brain—a photographic series. A.J. Kirby Co, London, ON
29. Mitchell JF (1958) The characteristics of some points of cardiovascular and respiratory representation in the cerebral cortex of sheep. J Physiol 144:17–8P
30. Gierthmuehlen M, Wang X, Gkogkidis A et al (2014) Mapping of sheep sensory cortex with a novel micro-electrocorticography grid. J Comp Neurol 522(16):3590–3608
31. Hoffmann A, Stoffel MH, Nitzsche B et al (2014) The ovine cerebral venous system: comparative anatomy, visualization, and implications for translational research. PLoS One 9(4):e92990
32. Latchaw RE (2004) Cerebral perfusion imaging in acute stroke. J Vasc Interv Radiol 15:29–46
33. Rosset A, Spadola L, Ratib O (2004) OsiriX: an open-source software for navigating in multidimensional DICOM images. J Digit Imaging 17:205–216
34. Schmidt KF, Ziu M, Schmidt NO et al (2004) Volume reconstruction techniques improve the correlation between histological and in vivo tumor volume measurements in mouse models of human gliomas. J Neurooncol 68:207–215
35. Watabe H, Itoh M, Cunningham V et al (1996) Noninvasive quantification of rCBF using positron emission tomography. J Cereb Blood Flow Metab 16:311–319
36. Mena H, Cadavid D, Rushing EJ (2004) Human cerebral infarct: a proposed histopathologic classification based on 137 cases. Acta Neuropathol 108:524–530
37. Garcia JH, Kamijyo Y (1974) Cerebral infarction. Evolution of histopathological changes after occlusion of a middle cerebral artery in primates. J Neuropathol Exp Neurol 33:408–421

Chapter 16

A Nonhuman Primate Model of Delayed Cerebral Vasospasm After Aneurismal Subarachnoid Hemorrhage

Ryszard M. Pluta, John Bacher, Boris Skopets, and Victoria Hoffmann

Abstract

Animal modeling of human disease has a long history but remains controversial especially when using a sensitive species. Despite these controversies, in a stroke research, a nonprimate model has been recognized as a most successful and useful for developing new treatments. Among stroke models, a nonhuman primate model of delayed cerebral vasospasm after subarachnoid blood clot placement has a very unique position as it revealed important pathomechanisms and led to development of several crucial clinical trials. In 1989, we adopted this model to study pathophysiology and develop a treatment against delayed cerebral vasospasm after intracranial aneurysm rupture (aSAH). In this chapter, we presented detailed descriptions of the animal treatment according to the National Institutes of Health guidance, techniques of anesthesia, cerebral arteriography, suboccipital puncture for cerebrospinal fluid collection, a surgical clot placement along the middle cerebral artery as well as postoperative care, euthanasia, and autopsy. Moreover, to accommodate the recent clinical findings, strongly suggesting a limited role of delayed cerebral vasospasm on the outcome of aSAH, we proposed a modification of the model, which addressed some mechanisms of ultra and early damage to the brain evoked by an intracranial aneurysm rupture.

Key words Animal modeling, Human disease, Stroke research, Nonprimate model

1 Introduction

1.1 Aneurismal Subarachnoid Hemorrhage (aSAH) and Delayed Cerebral Vasospasm: Challenges and Solutions

While securely located in the skull, brain is additionally protected by three layers of connective tissue of different thicknesses known as a dura matter, arachnoid, and pia; the latter being the closest to the brain surface. Under normal conditions, there is no space between the bone and dura matter and only a hairline thick space between the dura and arachnoid membrane. The space between the arachnoid and pia known as subarachnoid space consists of several cisterns, which in man contains about 140 ml of cerebrospinal spinal fluid (CSF) produced by the intraventricular choroid plexus at about 0.5 ml/min. Detailed surgical anatomy of these cisterns has been described in a seminal paper by Professor Gazhi Yasargil [1]. The CSF plays several physiological and pathophysiological

Miroslaw Janowski (ed.), *Experimental Neurosurgery in Animal Models*, Neuromethods, vol. 116,
DOI 10.1007/978-1-4939-3730-1_16,

roles by providing mechanical support for the brain, an exchange pathway for brain metabolites, as well as providing a space for cerebral arteries and veins running on the surface of brain. Because of its presence, pressure, and pulsatile flow, the CSF influences cerebral blood flow, facilitates metabolite exchange between blood and brain, and when obtained via lumbar puncture or ventricular drainage it serves as useful diagnostic tool of diseases affecting the brain. Furthermore, it provides a route for drug administration but can also facilitate spreading of infection and some tumors [2–23]. Arteries supplying blood to brain enter the skull and pierce the dura matter to access the subarchnoid space and the cisterns; vertebral arteries enter the cisterna magna and internal carotids enter the basal cisterns. In those cisterns, both vertebral arteries give off extracranial, intracranial, and spinal branches before coalescing into the basal artery. Internal carotids after sending the posterior communicating artery along the edge of tentorium to the posterior cerebral artery bifurcate and become the anterior and middle cerebral arteries. All these cerebral vessels run in the subarachnoid space and cisterns on the base of brain. In some people, an intracranial aneurysm develops at the bifurcation of these conductive vessels. Most often, the first symptom of such an aneurysm is the worse ever experienced headache when the aneurysm ruptures in about 2–10 of 100,000 people [24, 25]. Such a rupture produces aneurismal subarachnoid hemorrhage (aSAH). At the beginning of aSAH, violently flowing arterial blood, which is highly oxygenated and under high pressure, rapidly fills the subarachnoid cisterns. Rapidly clotting blood displaces cerebrospinal fluid and increases intracranial pressure until it almost equals the systemic blood pressure, which effectively stops the blood outflow from the aneurysm but at the same time ceases blood flow in the artery harboring the aneurysm and adjacent or little bit more distant vessels evoking so-called stop flow phenomenon [26–28]. In about 50% of cases platelet thrombus plugs the opening in the aneurysm allowing for clot formation and the patient survives the initial aSAH [25]. These lucky people when admitted to emergency room are diagnosed by CT, CT arteriography, or digital cerebral arteriography as having intracranial aneurismal bleeding and their aneurysms can be successfully treated with endovascular coiling or surgical clipping [29–36]. Unfortunately, even those patients are not out of the woods yet because the presence of blood clot and a release of vasogenic metabolites acting via mechanical and chemical pathways evoke constriction of the cerebral arteries, which causes a delayed cerebral vasospasm [37–46]. For a well over a half of century, this enigmatic phenomenon has been widely accepted as a central culprit responsible for a poor outcome after aSAH due to delayed ischemic neurological deficits. But this view on the role of delayed cerebral vasospasm has been recently changing as new venues of aSAH-related research are pursued [40, 47, 48]. This renewed interest in mechanisms contributing to development of delayed cerebral vasospasm, separately from ischemic neurological deficits, has increased the interest in aSAH models.

Many experimental models of aSAH have been developed during the second part of the last century in pursuit of elucidating pathomechanism(s) of vasospasm. The review of these models is beyond the scope of this chapter but interested readers may find excellent summaries, descriptions, and analyses of aSAH models in several recent reviews [43, 49] and original papers. Among these models, a nonhuman primate model with a direct surgical placement of blood clot around the middle cerebral artery, as developed by Espinoza in Bryce Weir laboratory, has proven to be the most consistent in mimicking development of vasospasm after aSAH [43, 49–52]. This model led to current development of two of the most promising clinical trials focused on prevention of vasospasm: CONSCIOUS-1/3 (Clazosentan to Overcome Neurological iSChemia and Infarction OccUring after Subarachoid hemorrhage Study) [29, 53] and sodium nitrite for prevention of vasospasm currently at Phase IIB (www.clinicaltrials.gov). The Clazosentan study showing a clinical efficacy confirmed the earlier—but brushed away—observations [40, 47, 48, 54] that vasospasm, despite being a tremendous challenge and a possible contributor to a poor outcome, may not be as crucial as it has been thought for years. New culprits, including early ischemia and vasospasm, inflammation, gene, and protein changes as well as combination of cortical spreading depressions and cortical spreading ischemias [47, 48, 55–57] have been proposed to be responsible for a poor outcome after successful elimination of an aneurysm surgically or using endovascular techniques. This new perception of aSAH and its consequences has opened new perspectives for research and new venues for aSAH modeling.

2 Nonhuman Primate Subarachnoid Hemorrhage Model

2.1 Animal Selection and Preparation

In 1984, F. Espinoza in the Bryce Weir's laboratory in Alberta, Canada published his seminal description of a surgical subarachnoid hemorrhage model [50] that quickly become recognized as a best fitting standard experimental model to study a delayed cerebral vasospasm after aSAH [49]. This model incorporated combined advantages of (1) providing superior control of SAH conditions, (2) allowing for repeated cerebral arteriographies, (3) producing reliable and consistent the middle cerebral artery spasm in above 90% of animals without (4) evoking neurological deficits. Despite it costs, it become widely accepted as the model of choice to study vasospasm pathophysiology and has been used to validate new proposed treatments in many preclinical studies. As mentioned above, positive results of studies using a nonhuman primate model of aSAH led to development of promising current clinical trials with Clazosentan [53], nitroprusside [58], and sodium nitrite [59]. Nevertheless, as mentioned above, the results of CONSCIOUS-1/3 studies dramatically exposed the weakness of this model that for a long time was treated as its strength. Clazosentan successfully, as predicted from primate studies, prevented development of vasospasm but, unfortunately, it did not improve the overall outcome in patients after aSAH [60]. This result that was further investigated and analyzed in the CONSCIOUS-2 and -3 studies [61] opened the discussion about the value of disease modeling especially using sensitive species [47, 48]. Nevertheless, despite some still existing controversy, it has become clear that the nonhuman primate model successfully delivered what it was design for. The investigators were able to establish pathomechanisms of delayed cerebral vasospasm and subsequently addressed them successfully [60]. The problem was, and remains valid, that some early warning reports that vasospasm might not be the only, or even the most important, contributor to poor outcome after technically perfect treatment of a ruptured intracranial aneurysm, were overlooked [62]. So now, the research is refocusing on explaining this unexpected phenomenon and the nonhuman primate model of aSAH should continue to play a leading role in testing hypotheses and treatments.

2.2 Animals

In the Espinoza's pioneered model, the nonhuman primates, Cynomologus monkeys (*Macaca fasccularis*) were most often used because they are smaller, easier to handle, and in the 1980s were cheaper than rhesus and other nonhuman primates. These monkeys vascular neuroanatomy corresponds to that of human except that they have, a characteristic for primates, the single pericallosal artery. For our experiments, initially we used wild caught animals but later when breeding colonies were established in the United States we used purpose bred animals. We have also used, if adequate, less expensive recycled monkeys obtained from different sources.

2.3 Quarantine, Husbandry, and Treatments of SAH Monkeys

Adult monkeys (6–17 years old) of both sexes were used and their weights ranged from 2.5 to 7.5 kg. Animals come from four domestic breeding/quarantine facilities. Before arrival to the research facility on main NIH campus in Bethesda, MD, animals underwent at least 13-weeks quarantine. During quarantine, the animals were examined at least twice by a facility veterinarian, TB-tested five times at 2 week intervals, received two courses of a broad spectrum parasiticide, and their blood was tested for hematology and chemistry as well as by serology and/or PCR for a comprehensive viral panel (i.e., serology for measles, Herpes B, SRV-1, 2, 3, and 5, SIV, STLV-1; and PCR for SRV-1, 2, and 5). Animals that did not have protective measles antibodies were vaccinated.

Additional tests and treatment were conducted if necessary depending on animal health problems at discretion of facility veterinarian. Within 2 weeks of arrival to the quarantine facility each animal had a behavior examination, which became a part of the individual animal record. During quarantine, caretakers wore a respirator, shoe covers, Tyvek coveralls, eye protection, cap or bonnet, and latex gloves when entering the quarantine rooms or working with monkeys. Protective clothing was immediately disposed of upon leaving the room.

After completion of quarantine and upon arrival to the research facility on main NIH campus, animals received an entrance physical examination by a facility veterinarian and started on a three times per year TB-testing program.

During and after quarantine, animals were singly housed in appropriate size cages and their health checks were performed at least twice daily. Any problems were brought to the immediate attention of the supervisor and the veterinarian. Animals were fed 8788 Harlan/Teklad primate diet twice daily in quantities developed by a NIH Laboratory Animal Nutritionist according to animal body weights. For example, animals weighing 3–6 kg received 5–8 biscuits and animal weighing 6–10 kg received 8–11 biscuits of 8788 diet. Animals had constant access to water, which was supplied through automatic watering system and were given daily fruit supplements. Cages including pans were rinsed at least once daily. Cages were sanitized at least every 14 days by passing through the cage wash. Animal room floors were mopped at least once daily with appropriate Tuberculocidal disinfectant/detergent.

As a part of environmental enrichment program, each animal was offered at least two toys: one inside and one outside the cage. The toys were rotated every 2 weeks at the times of cage changes. Additionally, approximately once weekly, animals were shown cartoons or nature movies on a TV/DVD screen mounted on a cart.

2.4 Preparation and Anesthesia

Cynomologus monkeys weighing from 2.5 to 7.5 kg were used as an animal model to study the delayed cerebral vasospasm after aSAH. Monkeys were held off food overnight before surgery. The morning of surgery the animals were transported to the surgical facility. The transport cage was placed inside a Plas-Labs Intensive Care System (Plas-Labs, Inc. Lansing, MI 48906) and the oxygen was turned on and set at 8 l/min and the temperature set at 85 °F to stabilize the animals temperature prior to surgery. Next, the monkey was given Ketamine (10 mg/kg IM) and atropine (0.04 mg/kg IM) as a preanesthetic. Once sedated, the animal was moved to the preparation room and the groin(s) and head areas were clipped in preparation for surgery. During preparation, the animal's body temperature was maintained by a heat lamp suspended over the preparation table as well as an electric heating pad placed on the tabletop. A 22 gauge, 1-in. angiocatheter was inserted into the cephalic or saphenous vein for hydration with normal saline (10–15 ml/kg/h) and drug delivery. Ketofen (2 mg/kg IM) and Cefazolin (50 mg/kg IV) were given. Propofol (10 mg/ml) to effect was given at 0.2 ml increments until proper level of anesthesia was reached for intubation. Depending on the size of the monkey a 4 or 5 mm endotracheal tube was inserted to maintain the airway. Animals were then transported to the surgery room and placed in supine for cerebral arteriogram or left lateral recumbence for the subarachnoid hemorrhage clot placement.

For cerebral angiograms, the head was placed in a homemade head holder made of foam, which had a radiopaque ruler on top so accurate measurements of the middle cerebral artery could be made (Fig. 1). The cerebral arteriograms were done using an OEC mobile digital imaging system (Series 9600) set on the subtraction mode at eight frames per second (Fig. 2). Arteriographic runs were played back and one or two images were selected for measurement of the middle cerebral artery.

The surgical sites were scrubbed with Alcare (Steris Corporation, St. Louis, MO 63110) or chlorhexidine surgical scrub and draped for aseptic surgery. During surgery, anesthesia was maintained with a Narkomed 2B anesthesia machine (Fig. 3) using Isoflurane (0.5–2 %) and oxygen at a total flow rate of 2 l/min. A 40-in. Universal F anesthetic hose, which connected the

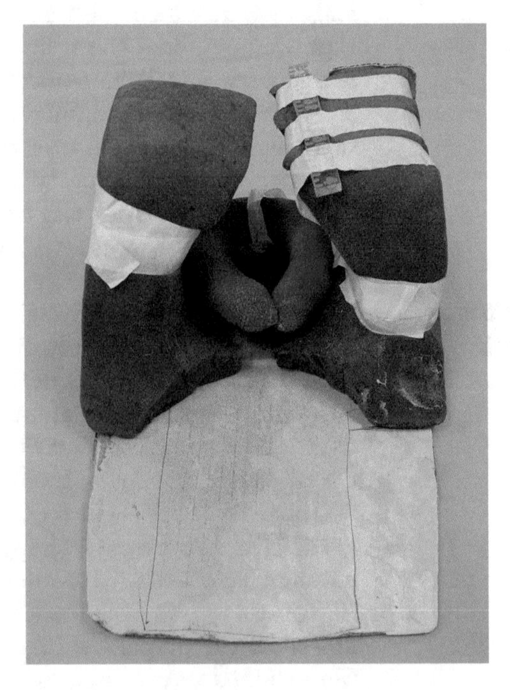

Fig. 1 Head positioner with radiographic ruler used for cerebral angiograms

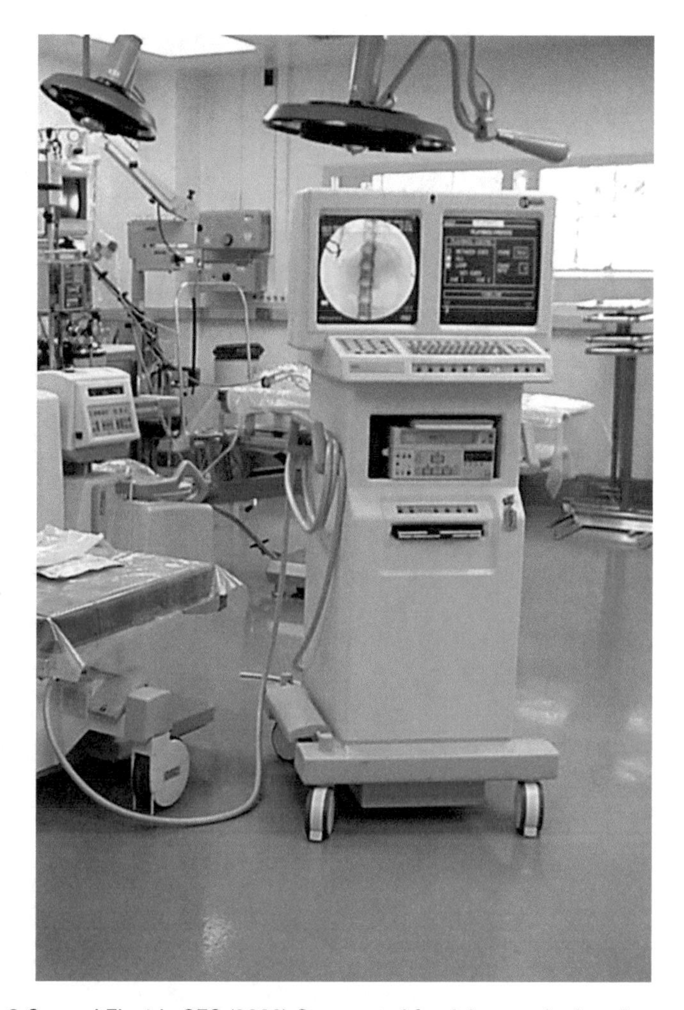

Fig. 2 General Electric OEC (9600) C-arm used for doing cerebral angiograms

anesthesia machine to the animal, was used to conserve heat loss and maintain proper humidity of gasses delivered to the monkey. The use of these hoses prevents the need for the addition of a heat and moisture exchanger and prevents the problems associated with the added dead space and increased flow resistance. Body temperature was maintained during surgery by use of a piped-in warm water heating blanket (Hemotherm by Cincinnati Sub-Zero Cincinnati, OH 45241) set at 44 °C, Hotline Fluid Warmer to maintain IV fluids at 37–42 °C (Smith Medical, Rockland, MA 02370), and portable heat lamps as needed. During surgery, ECG, SpO_2/Pleth, end tidal carbon dioxide ($ETCO_2$), respiratory rate, minute volume, airway peak pressure, tidal volume, I:E ratio, oxygen percentage, temperature, and direct or indirect blood pressures as appropriate were monitored.

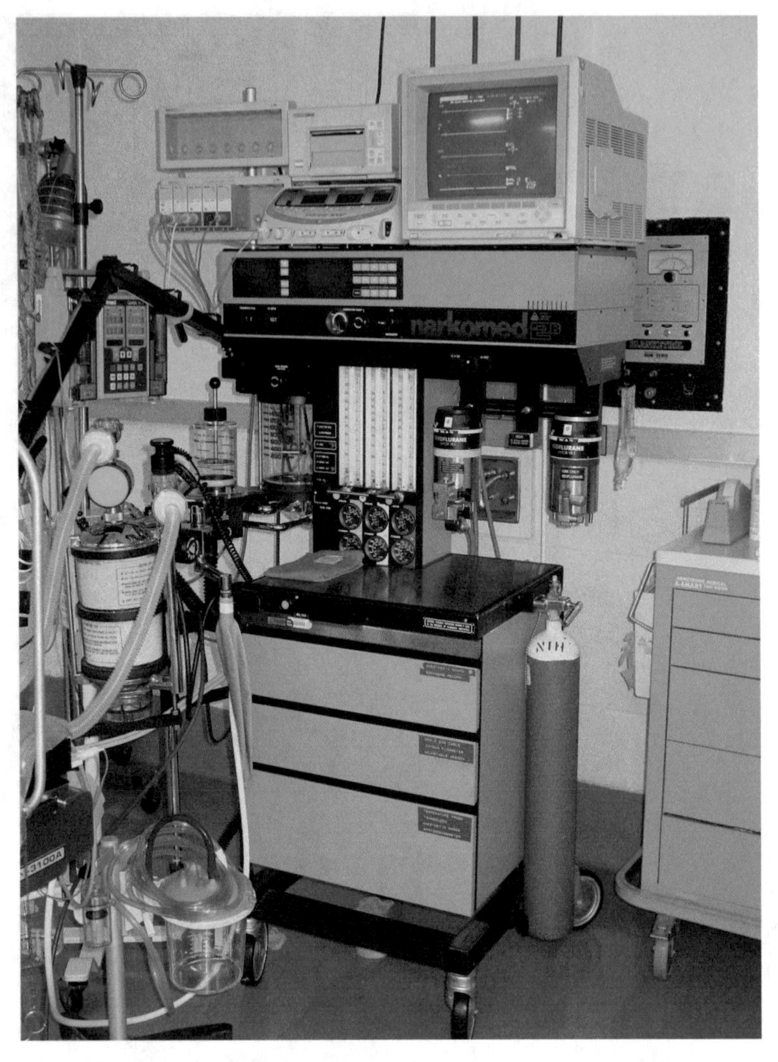

Fig. 3 Narkomed 2B anesthesia machine and Hewlet Packard monitors used for surgical procedures

3 Cerebral Arteriography

Each animal after quarantine but before being accepted for a protocol had a baseline arteriography and, if necessary, a baseline transcranial Doppler study.

3.1 Equipment

Ketamine-Xylazine-atropine induction, clippers, alcohol wipes, 22 gauge IV angiocath, saline, propofol, endotracheal tube size 4–5 mm, clippers, antibiotic, analgesic, rebreathing anesthesia machine with isoflurane anesthetic, scrubbing solution, cut down tray, sterile drapes, #10 or #15 scalpel blade, iris scissors, fine dissecting scissors, vascular forceps, bipolar coagulation, vascular retractor, straight and L-shaped vascular forceps, 3, 5, and 12 cc syringes, 18–25 gauge needles, 3 or 5 French catheter preshaped to the "lazy S" with guide wires, basin, local anesthetics, gealfoam, papaverine, heparin, half-by half inch cottonoids (patties), 5 ml of iodine contrast, surgical loupes, head holder with radio-opaque ruler, digital arteriography C-arm, 3-0 silk ties, 4-0 Vicryl sutures for skin closure, bandage material for pressure dressing, and Buprenorphine.

4 Technique of Cerebral Arteriography

Animal is placed in a supine position on the operating room table with the lower extremities slightly bent, rotated externally, and secured with small sand bags and surgical ties. Usual skin preparation is performed and surgical field is covered with four small sterile drapes and then with a larger pediatric laparotomy drape. The French 3–5 angiographic catheter is flushed with heparinized saline and put with the guide wire in a sterile basin with saline. Sterile saline (10 ml) with 100 units of heparin is prepared in a small cup for periodic flushing of the catheter. Five milliliter of contrast (iopamidol injection 61 %, Bracco Diagnostics) is drawn up into a 6 ml syringe. A 2-in. long skin incision is performed with a #10 blade about half an inch above the knee in the crease between a vastus medialis and an abductor muscle. Bupivicaine (1–2 cc of 0.25 %) is injected subcutaneously around the incision site for additional analgesia. Dissection is continued between the muscles to expose the neurovascular bundle and the femoral artery is isolated with small dissecting scissors. Two 3-O silk ties are placed under the artery about 1 in. (2 cm) apart. If the artery gets into spasm, a half-by-half inch patty soaked in papaverine is placed on the vessel for 1–2 min. Then, a small V-shaped incision is made in the artery; blood flow is controlled by tension on both ties and the intra-arterial catheter with the wire exposed by 1–2 mm is advanced into the artery. To avoid accidental perforation and use a curvature on the tip of the catheter, the wire is moved back about 1 in. as soon as the catheter is safely in the artery. After advancing the catheter to the level of the heart under radiological visualization, the tip of catheter is rotated up and catheter advances to the right brachiocephalic trunk and then to the common carotid artery. At this point, the head of the monkey is rotated to the right and the tip of the catheter is guided to the internal carotid artery (Fig. 4). The head is moved back to the straight position and placed in the head holder. 0.2 ml of contrast in injected to confirm correct position of the catheter tip in the internal carotid artery. The catheter is flushed with 1 ml of heparinized saline followed by 1 ml of contrast agent, which is injected during the eight frames per second subtraction run for arteriography (Fig. 5a, b). After obtaining the arteriogram,

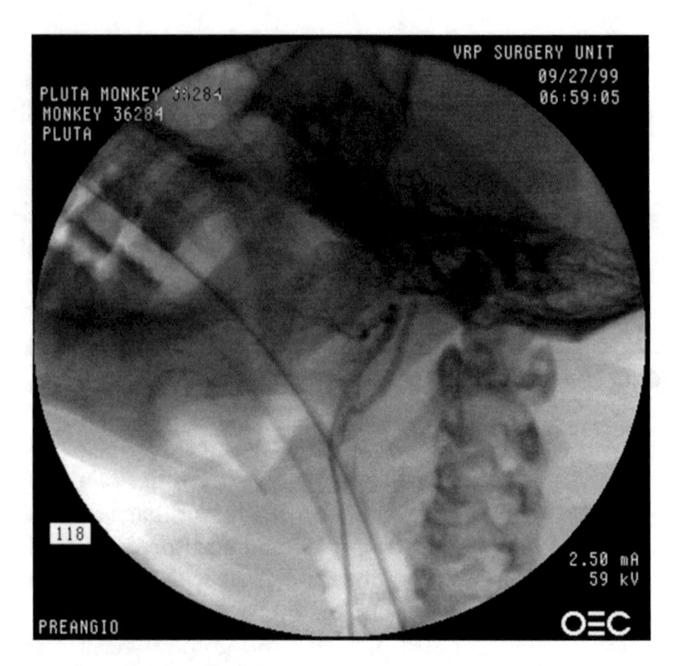

Fig. 4 The angiocatheter is advanced to he common carotid and the head of monkey is rotated to the right to facilitate the navigation of catheter into the right internal carotid artery

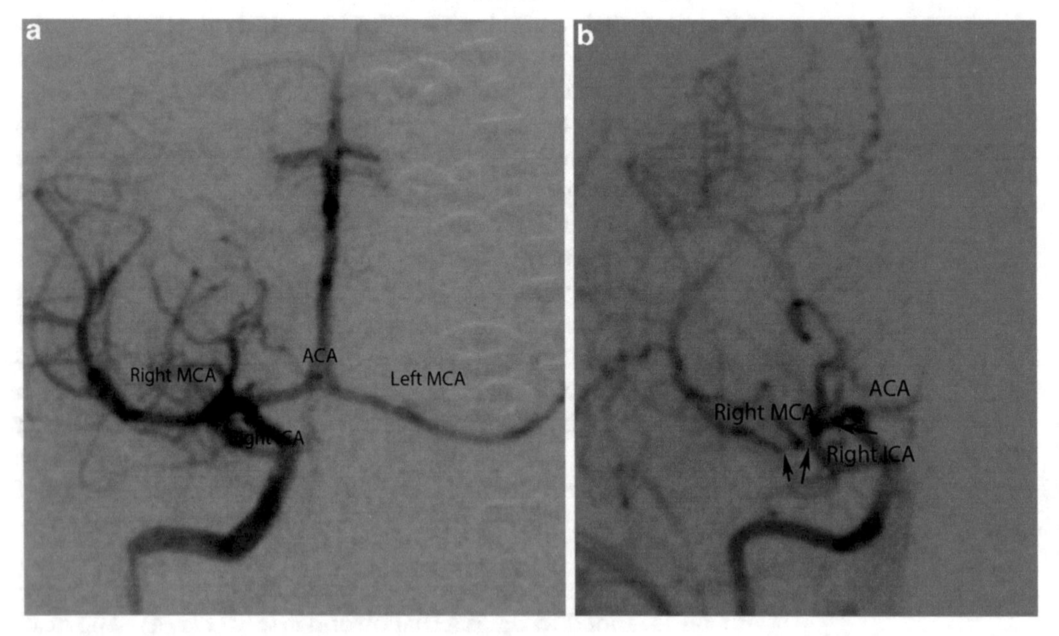

Fig. 5 (**a**) Right internal carotid arteriogram. Sometimes all four cerebral arteries of circle of Willis can be visualized with 1 ml of contrast as on this baseline arteriograms. (**b**) This animal developed moderate vasospasm of the right middle cerebral artery (*arrows*) on the 7th day after a clot placement

the catheter is quickly removed and the opening in the artery is covered with a tiny piece of gelfoam and kept under pressure through a small half-by-half inch cottonoid patty for about 5 min. After the patty is removed, the skin incision is closed with an intradermal a 4-0 vicryl suture. The wound is covered with antibiotic cream and a pressure dressing is applied. Gas anesthesia is discontinued.

The monkey is allowed to gain a gag reflex and then the endotracheal tube, rectal temperature probe, ECG leads, SpO_2, and IV line are removed and the monkey is returned to the transport cage that remains in the Plas-Labs Intensive Care Unit. The cage temperature is maintained at 85 °F unless the monkey is hypothermic. Oxygen is delivered to the cage at the rate of 8 l/min. Buprenorphine (0.01 mg/kg IM) is given postoperatively for additional analgesia. The animal is monitored until recovery is complete and then returns to its home cage.

Usually, arteriography is repeated on the 7th day after a clot placement (SAH) and subsequently, if necessary on days 14, 21, 28. For subsequent arteriograms, the same side femoral artery is exposed but 1-in. higher than the initial incision. Usually, after the incision of the artery with iris scissors and removal of the white clot, the patency of the vessel is restored allowing for catheter insertion and navigation. If this is too difficult and survival of the animal is anticipated then the left femoral artery is exposed and catheterized.

If this is a terminal arteriogram, the dissection of the right femoral artery could be continued higher along the femoral artery to the iliac arteries or if this is impossible and the left femoral artery is also thrombosed, a direct access to the common carotid artery can be used. We have also used the brachial artery in a couple of instances to gain access to the internal carotid.

For some experiments when repeated arteriographies are necessary, contrast reinjections are performed within minutes or hours after the baseline. Between injections the catheter has to be moved back to the common carotid to avoid thrombosis and while not used it is slowly infused with heparinized saline. Each subsequent contrast injection is with no more than 0.75 ml of contrast and the animals are properly hydrated with continuous monitoring of blood pressure and heart rate.

5 Arteriographic Measurements: Vasospasm Assessment

All cerebral arteriograms were performed as described above. The two dimensional area of the proximal 14 mm of the right middle cerebral artery on the anterioposterior projection was recorded for all arteriograms using an image analysis system (NIH Image, version 1.62 or NIH Image J; National Institutes of Health, Bethesda,

MD). First, the ruler on the arteriogram was measured three times and the mean value of measurements was assessed and calibrated. Then, the length of the right middle cerebral artery from the internal carotid bifurcation to the most lateral exposed M2 part was measured on the baseline arteriogram. Next, the area of the exposed to clot middle cerebral artery was outlined and measured. Then, all the steps were repeated with the post SAH arteriogram making sure that the starting point of the length measurement was exactly the same as on the baseline arteriogram. The degree of vasospasm for each animal at the time of assessment of the vascular responses was determined by comparing to the initial preoperative baseline arteriogram. The presence of significant vasospasm was defined as a 25% or greater reduction in the proximal 14 mm of the right middle cerebral artery area as measured on the anterioposterior projection of cerebral arteriogram. Arteriographic results are reported as the average of measurements performed by three independent, blinded observers before the experimental code is broken.

6 Cerebrospinal Fluid Collection via Suboccipital Puncture

6.1 Equipment

General anesthesia or heavy sedation, clippers, scrubbing solutions, alcohol, drape, 1 1/2 in. long 22 gauge short bevel needle or 2 in. 25 gauge needle, 3 ml syringe, 5 ml red top blood collection tube.

6.2 Technique

Animal under general anesthesia (with or without intubation) is shaved and prepped in usual way and then positioned on the right side with the head bent to the chest. A 1 1/2 in. 22 gauge needle is attached to the 3 ml syringe. The needle is introduced in the neck muscles a short distance below the palpable edge of the occipital protuberance, in the midline and slowly advanced toward the cistern magna while a slight negative pressure is kept in the syringe. If resistance is encountered, the needle is turned more caudally and advanced 2–3 mm until the CSF flaws in the syringe. One to 3 ml of CSF is collected. The needle is then removed and skin covered by antibiotic cream. If blood enters the needle, the fluid is immediately centrifuged and only supernatant is removed and used for further investigation.

7 A Clot Placement, Subarachnoid Hemorrhage

7.1 Anesthesia

During the subarachnoid hemorrhage procedure, the animal is placed on the pediatric ventilator with the tidal volume adjusted to approximately 15 ml/kg and respiratory rate set to 10–12 breaths/min. Once the peak pressure has stabilized, the bellows on the ventilator is adjusted to reach a peak pressure of 13 cm H_2O and later adjusted to maintain an $ETCO_2$ of about 40. Prior to exposing the brain the $ETCO_2$ is lowered to about 22 mmHg by increasing the respirations to 17–22 and/or increasing the tidal volume by raising the bellows to decrease blood flow and to limit cranial swelling. This level of $ETCO_2$ is maintained until closure when the $ETCO_2$ is allowed to go back to normal. When necessary dexamethason (1–2 mg/kg IV or IM) or furosemide (1–2 mg/kg IV) is used to decrease cranial swelling during surgery or during the recovery phase.

7.2 Equipment

Ketamine-Xylazine-atropine induction, clippers, alcohol wipes, 22 gauge IV angiocath, NaCl fluids for IV administration, propofol, endotracheal tube size 4–5 mm, antibiotic, analgesics, antibiotic, rebreathing anesthesia machine with isoflurane anesthetic, tape to secure head on foam cushion, scrubbing solution, general instrument tray, micro instruments, microforceps, microscissors curved and straight, arachnoid knife and #15 scalpel blade, sterile drapes, rubber bands, monopolar and bipolar coagulations, suction tubing, cottonoids, gelfoam, Hall drill with cutting and diamond burrs (1–3 mm), 1, 3, and 6 cc syringes, 25 gauge angiocath for CSF collection, local anesthetics, gelfoam, 1/4 × 1/4 in. cottonoids (patties), surgical loupes and operating microscope, bone wax, sutures for closure (monofilament 7/8-0 for dural closing; vicryl 3-0 for muscle closing, 4-0 for intradermal skin closure), topical antibiotic ointment for the incision site, and Buprenorphine.

8 SAH: Surgical Technique

Intubated monkey is placed on the heating pad on the operating room table in a left lateral oblique position. Extremities are secured to the sides of the operating table exposing the left groin area. The head is positioned in a homemade foam head holder with several layers of 4 × 4 gauze put in the cut out area of the foam to tip of the head to a 10–15° angle. The position of the head is secured with masking tape running through the base of the nostrils (**Note 1**). The head and both groins, which are shaved immediately following induction of anesthesia, are prepped for aseptic surgery. Bupivicaine (11/2 to 2 cc of 0.25%) is injected subcutaneously around incision site for additional analgesia. After placement of sterile drapes, a C-shaped skin incision is made in the right supra-orbital area from the zygomatic arc to close to the midline just above the orbital ridge. The skin flap is turned up over the eye and kept in position by 3-0 vicryl suture. Next, a shallow T-shaped incision with a Bovie electocautery is made through the fascia from the zygomatic arch to 1–1 1/2 in. above the "key-hole" area at the base of the orbital rim and then the muscle incision is carried out from the orbital rim perpendicularly to the first one above an area of the Sylvian fissure. The frontal part of the temporalis muscles is separated from the bone and after covering with wet gauze secured with the 2-0 vicryl suture. Then, a deeper incision through the superficial temporalis muscle toward the zygomatic arc is performed and superficial muscle flap is separated and secured with the suture/rubber band to the drapes. The incision is carried down to the temporal bone and after resection of the part of the deep temporalis muscle and coagulation of the deep branch of temporal artery the rest of the muscle is secured with the superficial muscle (**Note 2**). A 2 in. long and 1 in. wide bone flap with its base at the bottom on the orbital ridge is marked with the cutting drill bit and then drilled to the internal bone plate under constant irrigation (**Note 3**). Any bone bleeding is topped with bone wax. A diamond drill is used to remove the internal plate. At this moment, 1–2 mg of furosemide is administrated intravenously (**Note 4**) and hyperventilation is started to lower and then maintain the $ETCO_2$ at 22 mmHg. With Penfield number 2 or 4, depending on the thickness of the bone, the bone flap is carefully elevated and removed. The sphenoid wing bone is drilled out under magnification of the operating microscope (Fig. 6). Next, under microscope magnification, the dura is incised about ½ in. above the orbital ridge with a #15 blade avoiding the opening of the arachnoid. The incision is carried up to the orbital ridge above the frontal lobe and then down above the Sylvian fissure toward tip of the temporal lobe under the zygomatic arch (Fig. 7). The orbital edge of the dura is turned up and if necessary kept against the bone with wet cottonoid. After identification of the upper, frontal lip of the Sylvian fissure, the arachnoid knife (a 27-gauge slightly bend needle) is used to open the arachnoid, which then is further opened sharply with microscissors or divided with microforceps. The CSF is gently collected using 1-in. long 25-gauge IV catheter on a 3 ml syringe. The frontal lobe is gently elevated using a Penfield retractor #4 and single micro-patties are individually placed for retraction of the brain partially achieved by furosemide and CSF removal. The same maneuver is repeated with the temporal lobe; usually four to five micro-patties are placed under each lobe. Occasionally, small perforating veins crossing from the temporal lobe need to be coagulated with bipolar coagulation and divided. Then the arachnoid over the Sylvian fissure is removed exposing the middle cerebral artery (Fig. 8), its branches, and the internal carotid artery down below the anterior clinoid. Attention is then shifted to prepare the blood clot for placement around the middle cerebral artery. A skin incision is made to expose the femoral artery in the left groin similar to what was done for the cerebral arteriography. After exposure of femoral artery, a 25-gauge IV angiocath is introduced into the artery and 5 ml of blood is removed and set aside to clot. The artery and the wound are secured as described for arteriography. At this moment, hyperventilation is stopped and $ETCO_2$ gradually increases back to normal of 38–40 mmHg. Under magnification, the micro-patties are removed, CSF is collected, and the first small

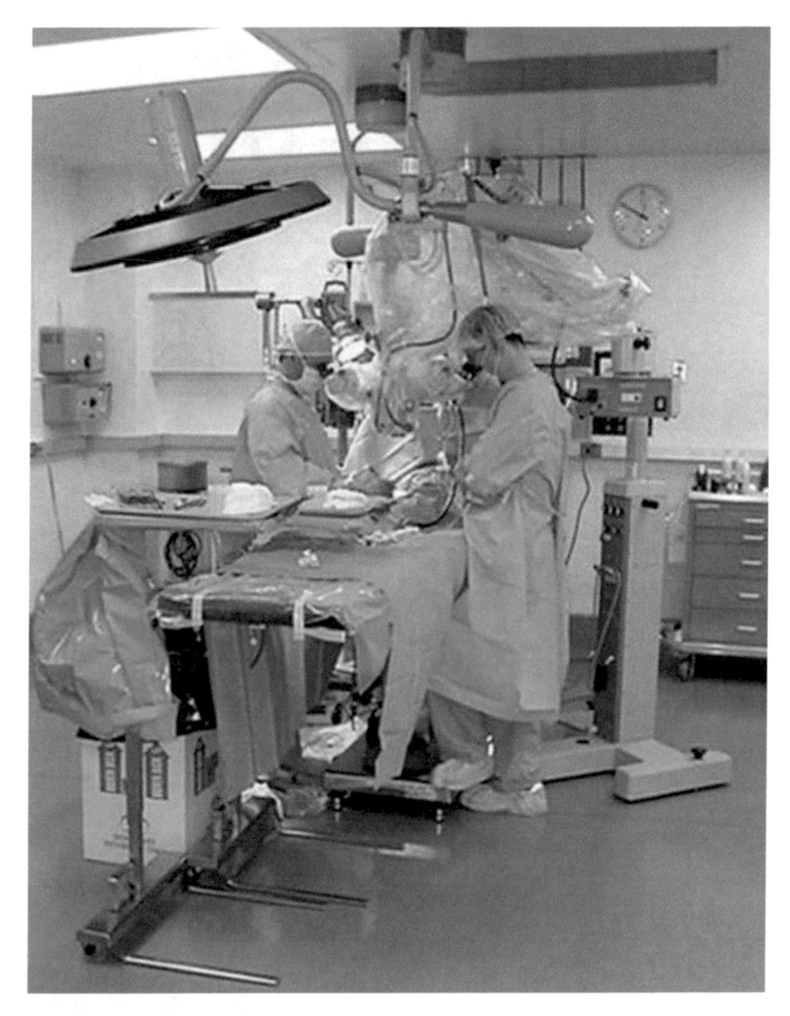

Fig. 6 The operating microscope is in position for dissection of the Sylvian fissure to expose the right middle cerebral artery

pieces of a clot are "locked" in the front and back of the internal carotid artery below the bifurcation (**Note 5**). Then, bigger pieces of the clot are inserted under the frontal and temporal lobes to keep the Sylvian fissure open, and the last piece of clot is placed on the middle cerebral artery and if a significant subdural space persists it is filled up with the rest of clot (Fig. 9). Usually a clot from 5 ml of blood is enough to cover adequately the middle cerebral artery and it has been enough to produce vasospasm in 95% of animals. The next step is closing the dura with 7-0 or 8-0 monofilament continuous suture (**Note 6**). After removal of the microscope, the temporalis muscle is approximated with 3-0 vicryl suture and the skin is sutured (**Note 7**) with 4-0 intradermal sutures (**Note 8**). The wound is covered with a topical antibiotic cream, anesthesia is turned off, and the monkey extubated after regaining a gag reflex. An initial neurological assessment is done before the animal is sent back to the heated cage.

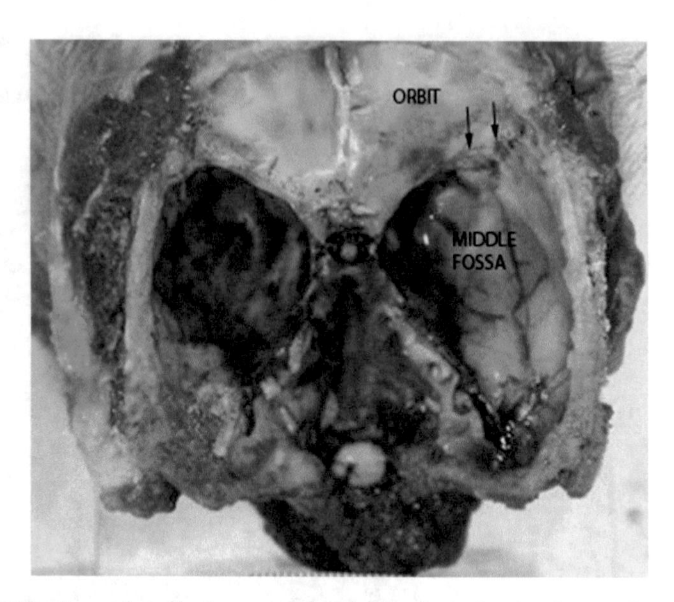

Fig. 7 The base of skull after removal. *Arrows* point to the dural suture line. Note hemosiderin deposits in the dura of anterior and temporal fossae

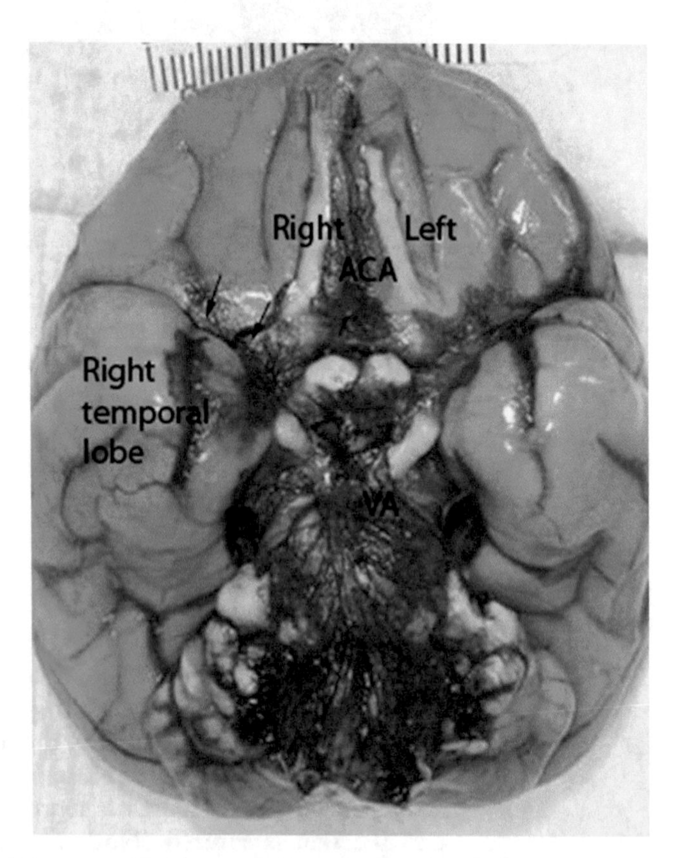

Fig. 8 The base of the brain, *arrows* indicate the area of removed arachnoid over the right middle cerebral artery

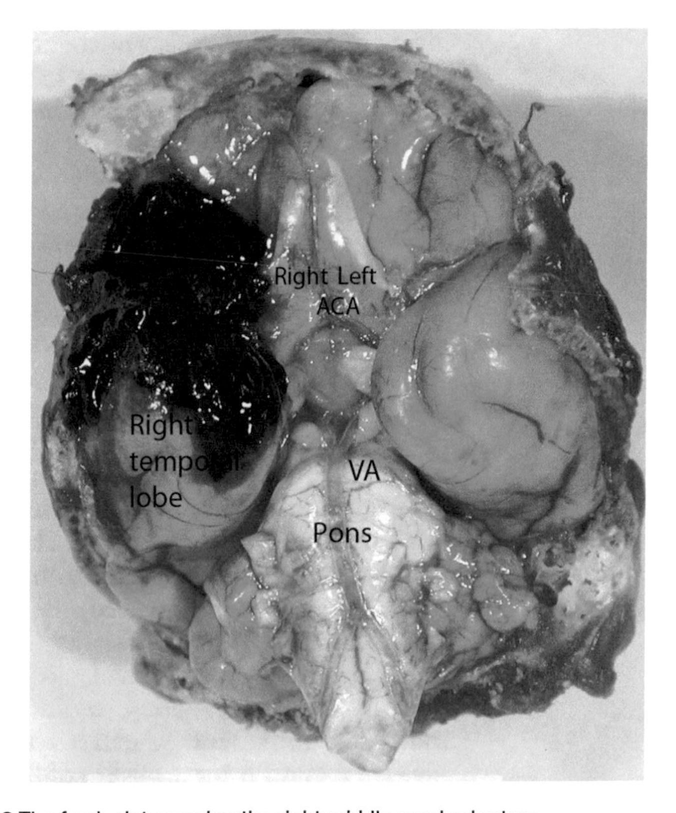

Fig. 9 The fresh clot covering the right middle cerebral artery

9 Modification of Model to Examine aSAH-Evoked Delayed Ischemic Deficits

As mentioned above, the results of the recent study and earlier observations suggest that vasospasm may not be as crucial in development of delayed ischemic deficits as has been thought for a long time. To provide an insight in the aSAH-related mechanisms responsible for delayed ischemic deficits and/or delayed cerebral vasospasm, we propose a slight modification of the model, in which after opening the dura as described before with the #15 blade, the culvinear incision of dura is performed using a microscissors 1 in. above the frontal bone rim extending from the frontal lobe to the edge of the temporal pole. The crucial part of this surgery is that the incision has to be carried out avoiding the opening the arachnoid. Then, a small nick in the arachnoid is performed with the arachnoid knife just above the edge of the Sylvian fissure. The tip of 25-gauge intravenous catheter attached to 6 cc syringe with normal saline is introduced through the arachnoid opening and 1–2 ml of warm saline is slowly infused under lower magnification toward the frontal lobe. This maneuver usually produces an elevation of arachnoid over the posterior aspect of the inferior and/or middle frontal gyrus. After another small nick of the elevated arachnoid, using an arachnoid knife or microscissors and microforceps in an atraumatic way, at least a 1 × 1 cm area of arachnoid over the frontal gyri is removed without any bleeding. If any bleeding starts, one or two micro-patties are inserted under the arachnoid and left for couple minutes. Then the removal of arachnoid is continued in this area or if more convenient in another close-by area. After removal of arachnoid over a small aspect of the lateral frontal lobe, the attention is directed toward opening the Sylvian fissure and exposure of the middle cerebral artery. At first, 1 ml of saline is injected through the previous opening in arachnoid above the Sylvian fissure which allows for separation of arachnoid from the pia usually more on the side of the frontal than temporal lobe. Then, using a microscissors/microforceps and/or arachnoid knife the arachnoid is opened along the distal first and then proximal MCA, the ICA bifurcation, and the ICA below the tentorial edge; if easily achieved, the Lilliequist membrane is also sharply opened. The arachnoid is removed from above the vessels. Each time, an additional part of arachnoid, if safely accessible, is removed exposing additionally about 0.5–1 cm of a posterior frontal and sometimes temporal lobe parallel to the Sylvian fissure. After exposure of the internal carotid artery and its bifurcation, further dissection and removal of arachnoid is focused on the orbital part of frontal lobe and under the highest magnification the arachnoid over the anterior cerebral artery and pericallosal artery is removed exposing additional part of orbital gyrus and olfactory nerve. The area of the brain depleted of arachnoid is described carefully in the operative notes to assure an easy recognition of the relevant areas during autopsy and histopathological studies. This is especially important for the control group because in the SAH group at the time of autopsy, the denuded area, which has been exposed to blood has a significantly more yellowish tint from hemosiderin than the area that remained covered by arachnoid.

Upon completion of surgical procedure the monkey is allowed to regain a gag reflex and then the endotracheal tube, rectal temperature probe, ECG leads, SpO_2, and IV line is removed and animal returned to the transport cage that is still in the Plas-Labs Intensive Care Unit. The cage temperature is maintained at 85 °F unless the monkey is hypothermic at which time the cage temperature is adjusted between 85 and 100 °F (**Note 9**). Oxygen is delivered to the cage at the rate of 8 l/min. Buprenorphine (0.01 mg/kg IM) is given postoperatively for additional analgesia. The animals are monitored until recovery is complete and then returned to their home cage (**Note 10**).

10 Posttreatment Handling

Upon return from the surgery the animals are started on a treatment course consisting of an analgesic Buprenorphine at 0.03 mg/kg twice a day for total of 3 days and antibiotic Cefazolin at 25 mg/kg IM twice a day for 7 days. Animals are closely monitored for alertness, activity level, integrity, swelling or infection of surgical incisions, hydration, appetite, urination, defecation, and presence of neurological deficits. If animal appears in pain, Ketoprofen is administered at 2 mg/kg IM as needed. In cases of possible infection of incision site, Gentamicin is added to the treatment plan at 4 mg/kg IM twice a day for 7 days. If animal appears dehydrated, 100–200 ml of Lactated Ringers Solution is given subcutaneously. If animal appears hypoactive, it is placed in the temperature and relative humidity controlled oxygen cage with oxygen flow up to 10 l/min. Additional diagnostic workup (e.g. CBC, blood chemistry, X-rays) is ordered as needed.

11 Experiment Termination and Autopsy

The procedure for performing necropsies on macaques and other nonhuman primates is similar to that for other mammals. However, additional safety considerations are necessary. Monkeys can harbor many zoonotic agents such has various Mycobacterium species. The most important pathogen is *Cercopithicine herpesvirus 1* (Herpes B). This virus can cause oral and genital ulceration in macaque species or be carried asymptomatically. Animals can harbor the virus and remain serum negative, so, there is no way to be sure that an animal in not infected. Herpes B is highly fatal in humans with 70–80% fatality rate reported in untreated human cases. Fatalities are much lower in cases treated immediately with antiviral drugs.

Because of the zoonotic risk associated with primate necropsies appropriate personal protection (PPE) and biosafety equipment is necessary. PPE includes a full-face shield, surgical mask, a laboratory coat made of nonabsorbent material, surgical scrubs, shoe covers, and disposable gloves. Steel gloves are also available for the nondominant hand. For removing tissue from the central nervous system (CNS), a Biosafety cabinet is recommended. Necropsies can be also performed on down draft tables; so, that pathogens are less likely to be spread by air currents.

Necropsies are performed with the cadaver in supine position. First an external exam is performed noting general body condition, hydration status, surface lesions, eyes, and oral cavity. A ventral midline incision is made in the skin from the mandibular symphysis to the pubic bone and the skin is reflected from underlying body wall. A ventral midline incision is made through the abdominal muscles. An incision is made through the diaphragm following the contour of the rib cage. Bone cutting shears are used to remove the rib cage. After the thoracic and abdominal cavities are opened, a systematic examination is performed including lymph nodes, thoracic viscera, and abdominal viscera. Any abnormal tissues are noted. Tissue samples then can be collected as needed for study protocol as well for diagnostic purposes in the case of unexpected lesions. Tissues of interest can be also sliced at 1 cm intervals for the presence of deep parenchymal lesions.

To remove the brain, the head is severed from neck. The skin and muscles overlying the cranial vault are removed. A vibrating (or Stryker) saw is used to cut through the skull starting with a transverse cut through the frontal bone, extending ventrally, and caudally through the temporal bone and then the occipital bone to the foramen magnum. The top of the skull in pried off and the brain gently dissected away from the cranial floor. The remaining skull can be hemisectioned if necessary to examine the nasal cavity and remove the pterygopalatine ganglia. Because the brain is soft and easily macerated, sectioning of the brain is recommended until after tissue fixation. After fixation, the brain can be sectioned transversely beginning at the frontal cortex and slices from the various levels examined for abnormalities.

12 Notes

1. As with a surgical approach to the intracranial aneurysm, the head position is crucial to facilitate the approach and access to the basal cisterns with no or only minimal brain retraction. But the problem is we do not have a Mayfield frame to facilitate and stabilize the head. So, make sure that the monkey head is properly positioned and secured with the tape.
2. The incision of temporalis fascia and removal of the deep temporalis muscle should be performed very carefully for a watertight closing later on.
3. When you start drilling the bone, start hyperventilation to give the brain time for relaxation. While removing the bone flap, make sure that you carefully secure all the bleeds from the muscle and the bone. It is crucial to avoid any trauma to the brain.
4. Before the dura incision, give the dose of furosemide. While opening the dura, it is crucial to keep the arachnoid intact and the surgical field bloodless. Any bleeding from the muscle, bone, or dura should be stopped before dura opening. Coagulation of the middle meningeal artery should be performed directly before its cutting; avoid evoking the shrinkage

of dural edges. This part is crucial for a water-tight closure of dura.

5. While placing the clot around the middle cerebral artery, remember about putting the tiny pieces of clot around the internal carotid artery below the optic nerve and below the posterior communicating artery. This maneuver will protect the rest of clot from being "washed" away from the surface of cerebral arteries. As the second step, lay a significant amount of clot under the frontal and temporal lobes to keep the Sylvian fissure open, and then cover the middle cerebral artery with the rest of clot.
6. The dural closure had to be water tight. If you need, use the fascia/muscle graft or tissue glue to secure it.
7. The anatomical closure of the muscle and fascia is important for several reasons; it accelerates wound healing and animal recovery, it allows animal for faster restarting drinking and eating and it protects against collection of fluid under the skin alleviating the risk of infection.
8. ALWAYS use the intradermal sutures on the skin.
9. During preparation, surgery, and after surgery remember that monkey has only a thin layer of fat; it may become hypothermic very easily; so, monitor closely the body and environment temperature to avoid it.
10. During the first 24 h, do not leave any easy to choke on things or food in the cage. Make sure that animal is regularly checked and given a proper analgesia. If the animal does not drink/eat, try a softer food (banana) and orange juice before starting the subcutaneous hydration and tube feeding.

Acknowledgment

This research was supported in part by the Intramural Research Program of the NIH, NINDS.

References

1. Yasargil M, Kasdaglis K, Jain K, Weber H (1976) Anatomical observations of the subarachnoid cisterns of the brain during surgery. J Neurosurg 44:298–302
2. Bergstrand G, Larsson S, Bergstrom M, Eriksson L, Edner G (1983) Cerebrospinal fluid circulation: evaluation by single-photon and positron emission tomography. AJNR Am J Neuroradiol 4:557–559
3. Brandt L, Ljunggren B, Anderson K et al (1981) Vasoconstrictive effects of human post-hemorrhagic cerebro-spinal fluid on cat pial arterioles *in situ*. J Neurosurg 54:351–356
4. Clark R, Carcillo J, Kochanek P, Obrist W, Jackson E, Mi Z, Wisniewski S, Bell M, Marion D (1997) Cerebrospinal fluid adenosine concentration and uncoupling of cerebral blood flow and oxidative metabolism after severe head injury in humans. Neurosurgery 41:1284–1293
5. Fuijmori A, Yanagisawa M, Saito A, Goto K, Masaki T, Mima T, Takakura K, Shigeno T

(1990) Endothelin in plasma and cerebrospinal fluid of patients with subarachnoid haemorrhage. Lancet 336:633

6. Hardemark HG, Almqvist O, Johansson T, Pahlman S, Persson L (1989) S-100 protein in cerebrospinal fluid after aneurysmal subarachnoid haemorrhage: relation to functional outcome, late CT and SPECT changes, and signs of higher cortical dysfunction. Acta Neurochir (Wien) 99:135–144
7. Jung CS, Oldfield EH, Harvey-White J, Espey MG, Zimmermann M, Seifert V, Pluta RM (2007) Association of an endogenous inhibitor of nitric oxide synthase with cerebral vasospasm in patients with aneurysmal subarachnoid hemorrhage. J Neurosurg 107:945–950
8. Kim P, Yaksh T, Romero S, Sundt T (1987) Production of uric acid in cerebrospinal fluid after subarachnoid hemorrhage in dog: investigation of possible role of xanthine oxidase in chronic vasospasm. Neurosurgery 21:39–44
9. Kırış T, Erden T, Sahinbas M, Omay B, Esen F (2005) CSF drainage for prevention and reversal of cerebral vasospasm after surgical treatment of intracranial aneurysms. In: Macdonald RL (ed) Cerebral vasospasm. Thieme, New York, pp 255–258
10. Latour LL, Warach S (2002) Cerebral spinal fluid contamination of the measurement of the apparent diffusion coefficient of water in acute stroke. Magn Reson Med 48:478–486
11. Moriyama E, Matsumoto Y, Meguro T, Kawada S, Mandai S, Gohda Y, Sakurai M (1995) Combined cisternal drainage and intrathecal urokinase injection therapy for prevention of vasospasm in patients with aneurysmal subarachnoid hemorrhage. Neurol Med Chir (Tokyo) 35:732–736
12. Naff N, Williams M, Rigamonti D, Keyl P, Hanley D (2001) Blood clot resolution in human cerebrospinal fluid: evidence for first-order kinetics. Neurosurgery 49:614–621
13. Penn RD (2003) Intrathecal medication delivery. Neurosurg Clin N Am 14:381–387
14. Pluta R, Boock R, Afshar J, Clouse K, Bacic M, Ehrenreich H, Oldfield E (1997) The source and cause of endothelin 1 release to CSF after SAH. J Neurosurg 87:287–293
15. Pluta RM (2006) Dysfunction of nitric oxide synthases as a cause and therapeutic target in delayed cerebral vasospasm after SAH. Neurol Res 28:730–737
16. Proescholdt MG, Hutto B, Brady L, Herkenham M (2000) Studies of cerebrospinal fluid flow and penetration into brain following lateral ventricle and cisterna magna injections of the tracer [14c] inulin. Neuroscience 95:577–592
17. Sato S, Ishihara N, Yuniki K (1990) The changes in endothelin concentration in CSF following subarachnoid hemorrhage. In: Sano K et al (eds) Cerebral vasospasm. University of Tokyo Press, Tokyo, pp 269–271
18. Shore PM, Berger RP, Varma S, Janesko KL, Wisniewski SR, Clark RS, Adelson PD, Thomas NJ, Lai YC, Bayir H, Kochanek PM (2007) Cerebrospinal fluid biomarkers versus glasgow coma scale and glasgow outcome scale in pediatric traumatic brain injury: the role of young age and inflicted injury. J Neurotrauma 24:75–86
19. Suzuki H, Muramatsu M, Tanaka K, Fujiwara H, Kojima T, Taki W (2006) Cerebrospinal fluid ferritn in chronic hydrocephalus after aneurysmal subarachnoid hemorrhage. J Neurol 253:1170–1176
20. Suzuki Y, Osuka K, Noda A, Tanazawa T, Takayasu M, Shibuya M, Yoshida J (1997) Nitric oxide metabolites in the cisternal cerebral spinal fluid in patients with subarachnoid hemorrhage. Neurosurgery 41:807–812
21. Voldby B, Petersen OF, Buhl M, Jakobsen P, Ostergaard R (1984) Reversal of cerebral arterial spasm by intrathecal administration of a calcium antagonist (nimodipine). Acta Neurochir (Wien) 70:243–254
22. Woszczyk A, Deinsberger W, Boker A-K (2003) Nitric oxide metabolities in cisternal CSF correlate with cerebral vasospasm with a subarachnoid hemorrhage. Acta Neurochir 145:257–264
23. Xu L, Wang B, Kaur K, Kho MF, Cooke JP, Giffard RG (2007) NOx and ADMA changes with focal ischemia, amelioration with the chaperonin groel. Neurosci Lett 418:201–204
24. Kassell NF, Torner JC (1984) The international cooperative study in timing of aneurysm surgery—an update. Stroke 15:566–570
25. Stapf C, Mohr J (2004) Aneurysms and subarachnoid hemorrhage—epidemiology. In: Le Roux PD, Winn HW, Newell DW (eds) Management of cerebral aneurysms. Elsevier Inc, Philadelphia, PA, pp 183–187
26. Asano T, Sano K (1977) Pathogenic role of no reflow phenomenon in experimental subarachnoid hemorrhage in dogs. J Neurosurg 46:454–466
27. Nornes H (1973) The role of intracranial pressure in the arrest of hemorrhage in patients with ruptured intracranial aneurysms. J Neurosurg 39:226–234
28. Trojanowski T (1984) Early effects of experimental arterial subarachnoid haemorrhage on the cerebral circulation. Part II: Regional cerebral blood flow and cerebral microcirculation after experimental subarachnoid haemorrhage. Acta Neurochir (Wien) 72:241–259

29. Macdonald RL, Higashida RT, Keller E, Mayer SA, Molyneux A, Raabe A, Vajkoczy P, Wanke I, Frey A, Marr A, Roux S, Kassell NF (2010) Preventing vasospasm improves outcome after aneurysmal subarachnoid hemorrhage: rationale and design of CONSCIOUS-2 and CONSCIOUS-3 trials. Neurocrit Care 13:416–424
30. Al-Tamimi YZ, Ahmad M, May SE, Bholah MH, Callear J, Goddard A, Quinn AC, Ross SA (2010) A comparison of the outcome of aneurysmal subarachnoid haemorrhage before and after the introduction of an endovascular service. J Clin Neurosci 17:1391–1394
31. Alnaami I, Saqqur M, Chow M (2009) A novel treatment of distal cerebral vasospasm. A case report. Interv Neuroradiol 15:417–420
32. Ryttlefors M, Enblad P, Kerr RS, Molyneux AJ (2008) International subarachnoid aneurysm trial of neurosurgical clipping versus endovascular coiling: subgroup analysis of 278 elderly patients. Stroke 39:2720–2726
33. Kaku Y, Watarai H, Kokuzawa J, Tanaka T, Andoh T (2007) Cerebral aneurysms: conventional microsurgical technique and endovascular method. Surg Technol Int 16:228–235
34. Lanzino G, Fraser K, Kanaan Y, Wagenbach A (2006) Treatment of ruptured intracranial aneurysms since the international subarachnoid aneurysm trial: practice utilizing clip ligation and coil embolization as individual or complementary therapies. J Neurosurg 104:344–349
35. Wijdicks EF, Kallmes DF, Manno EM, Fulgham JR, Piepgras DG (2005) Subarachnoid hemorrhage: neurointensive care and aneurysm repair. Mayo Clin Proc 80:550–559
36. Flett LM, Chandler CS, Giddings D, Gholkar A (2005) Aneurysmal subarachnoid hemorrhage: management strategies and clinical outcomes in a regional neuroscience center. AJNR Am J Neuroradiol 26:367–372
37. Asano T, Matsui T (1993) Cerebral vasospasm: a disorder of the local mechanotransduction system in cerebral arteries. Crit Rev Naurosurg 3:284–294
38. Dorsch N (1994) A review of cerebral vasospasm in aneurysmal subarachnoid hemorrhage. III. Mechanisms of action of calcium antagonists. J Clin Neurosci 1:151–160
39. Fox J (1983) Vasospasm. II. Clinical consideration. In: Fox J (ed) Intracranial aneurysms. Springer, New York, pp 250–271
40. Hansen-Schwartz J, Vajkoczy P, Macdonald RL, Pluta RM, Zhang JH (2007) Cerebral vasospasm: looking beyond vasoconstriction. Trends Pharmacol Sci 28:252–256
41. Jahromi B, Macdonald R (2004) Vasospasm: diagnosis and medical treatment. In: Leroux P, Winn W, Newell D (eds) Management of cerebral aneurysms. Elsevier Inc., Philadelphia, PA
42. Kassell N, Sasaki T, Colohan A, Nazar G (1985) Cerebral vasospasm following aneurysmal subarachnoid hemorrhage. Stroke 16:562–572
43. Macdonald R, Weir B (2001) Cerebral vasospasm. Academic, San Diego, pp 449–450
44. Mayberg M (1998) Cerebral vasospasm. Neurosurg Clin N Am 9:615–627
45. Ohta T (2004) Cerebral vasospasm revisited: SAH syndrome. In: Macdonald RL (ed) Cerebral vasospasm: advances in research and treatment; 8th international symposium of cerebral vasospasm (Chicago, IL, 2003). Thieme, New York, pp 106–111
46. Weir B, Grace M, Hansen J et al (1978) Time course of vasospasm in man. J Neurosurg 48:173–178
47. Macdonald RL, Pluta RM, Zhang JH (2007) Cerebral vasospasm after subarachnoid hemorrhage: the emerging revolution. Nat Clin Pract Neurol 3:256–263
48. Pluta RM, Hansen-Schwartz J, Dreier J, Vajkoczy P, Macdonald RL, Nishizawa S, Kasuya H, Wellman G, Keller E, Zauner A, Dorsch N, Clark J, Ono S, Kiris T, Leroux P, Zhang JH (2009) Cerebral vasospasm following subarachnoid hemorrhage: time for a new world of thought. Neurol Res 31:151–158
49. Megyesi JF, Vollrath B, Cook DA, Findlay JM (2000) In vivo animal models of cerebral vasospasm: a review. Neurosurgery 46:448–460, discussion 460–441
50. Espinosa F, Weir B, Overton T, Castor W, Grace M, Boisvert O (1984) A randomized placebo-controlled double-blind trial of nimodipine after SAH in monkeys. Part I: Clinical and radiological findings. J Neurosurg 60:1167–1175
51. Macdonald RL, Weir BKA, Runzer TD, Grace MGA, Findlay JM, Saito K, Cook DA, Mielke BW, Kanamaru K (1991) Etiology of cerebral vasospasm in primates. J Neurosurg 75:415–424
52. Pluta R (2005) Delayed cerebral vasospasm and nitric oxide: review, new hypothesis, and proposed treatment. Pharmacol Ther 105:23–56
53. Vajkoczy P, Meyer B, Weidauer S, Raabe A, Thome C, Ringel F, Breu V, Schmiedek P (2005) Clazosentan (axv-034343), a selective endothelin a receptor antagonist, in the prevention of cerebral vasospasm following severe aneurysmal subarachnoid hemorrhage: results

of a randomized, double-blind, placebo-controlled, multicenter phase IIa study. J Neurosurg 103:9–17

54. Treggiari M, Walder B, Suter P et al (2001) Systematic review of the prevention of delayed ischemic neurological deficits with hypertension, hypervolemia, and hemodilution therapy following subarachnoid hemorrhage. J Neurosurg 98:978–984
55. Dreier JP, Woitzik J, Fabricius M, Bhatia R, Major S, Drenckhahn C, Lehmann TN, Sarrafzadeh A, Willumsen L, Hartings JA, Sakowitz OW, Seemann JH, Thieme A, Lauritzen M, Strong AJ (2006) Delayed ischaemic neurological deficits after subarachnoid haemorrhage are associated with clusters of spreading depolarizations. Brain 129:3224–3237
56. Leng LZ, Fink ME, Iadecola C (2010) Spreading depolarization: a possible new culprit in the delayed cerebral ischemia of subarachnoid hemorrhage. Arch Neurol 68:31–36
57. Sehba F, Bederson J (2006) Mechanisms of acute brain injury after subarachnoid hemorrhage. Neurol Res 28:381–398
58. Vajkoczy P, Hubner U, Horn P, Bauhuf C, Thome C, Schilling L, Schmiedek P, Quintel M, Thomas JE (2000) Intrathecal sodium nitroprusside improves cerebral blood flow and oxygenation in refractory cerebral vasospasm and ischemia in humans. Stroke 31:1195–1197
59. Pluta RM, Dejam A, Grimes G, Gladwin MT, Oldfield EH (2005) Nitrite infusions prevent cerebral artery vasospasm in a primate model of subarachnoid aneurismal hemorrhage. JAMA 293:1477–1484
60. Macdonald RL (2008) Clazosentan: an endothelin receptor antagonist for treatment of vasospasm after subarachnoid hemorrhage. Expert Opin Investig Drugs 17:1761–1767
61. Macdonald RL, Kassell NF, Mayer S, Ruefenacht D, Schmiedek P, Weidauer S, Frey A, Roux S, Pasqualin A (2008) Clazosentan to overcome neurological ischemia and infarction occurring after subarachnoid hemorrhage (CONSCIOUS-1): randomized, double-blind, placebo-controlled phase 2 dose-finding trial. Stroke 39:3015–3021
62. Schatlo B, Dreier JP, Glasker S, Fathi AR, Moncrief T, Oldfield EH, Vortmeyer AO, Pluta RM (2010) Report of selective cortical infarcts in the primate clot model of vasospasm after subarachnoid hemorrhage. Neurosurgery 67:721–728, discussion 728–729

INDEX

Miroslaw Janowski (ed.), *Experimental Neurosurgery in Animal Models*, Neuromethods, vol. 116, DOI 10.1007/978-1-4939-3730-1, © Springer Science+Business Media New York 2016

N

O

P

Q

R

S

T

V

W

X

Z

Printed in the United States
By Bookmasters